Lecture Notes in Mechanical Engineering

Lecture Notes in Mechanical Engineering (LNME) publishes the latest developments in Mechanical Engineering—quickly, informally and with high quality. Original research or contributions reported in proceedings and post-proceedings represents the core of LNME. Volumes published in LNME embrace all aspects, subfields and new challenges of mechanical engineering.

To submit a proposal or request further information, please contact the Springer Editor of your location:

Europe, USA, Africa: Leontina Di Cecco at Leontina.dicecco@springer.com
China: Ella Zhang at ella.zhang@cn.springernature.com
India, Rest of Asia, Australia, New Zealand: Swati Meherishi at swati.meherishi@springer.com
Topics in the series include:

- Engineering Design
- Machinery and Machine Elements
- Mechanical Structures and Stress Analysis
- Automotive Engineering
- Engine Technology
- Aerospace Technology and Astronautics
- Nanotechnology and Microengineering
- Control, Robotics, Mechatronics
- MEMS
- Theoretical and Applied Mechanics
- Dynamical Systems, Control
- Fluid Mechanics
- Engineering Thermodynamics, Heat and Mass Transfer
- Manufacturing Engineering and Smart Manufacturing
- Precision Engineering, Instrumentation, Measurement
- Materials Engineering
- Tribology and Surface Technology

Indexed by SCOPUS, EI Compendex, and INSPEC.

All books published in the series are evaluated by Web of Science for the Conference Proceedings Citation Index (CPCI).

To submit a proposal for a monograph, please check our Springer Tracts in Mechanical Engineering at https://link.springer.com/bookseries/11693.

Libor Sitek · Joško Valentinčič · Franz H. Trieb ·
Sergej Hloch
Editors

Advances in Water Jetting II

Selected Papers from the International Conference on Water Jet 2023 - Research, Development, Applications, October 17–19, 2023, Kutna Hora, Czech Republic

Springer

Editors
Libor Sitek
Fluid Jets
Institute of Geonics
Ostrava-Poruba, Czech Republic

Franz H. Trieb
Kapfenberg, Austria

Joško Valentinčič
Faculty of Mechanical Engineering
University of Ljubljana
Ljubljana, Slovenia

Sergej Hloch
Faculty of Manufacturing Technologies
Technical University of Košice
Prešov, Slovakia

ISSN 2195-4356 ISSN 2195-4364 (electronic)
Lecture Notes in Mechanical Engineering
ISBN 978-3-031-72777-1 ISBN 978-3-031-72778-8 (eBook)
https://doi.org/10.1007/978-3-031-72778-8

This Springer imprint is published by the registered company Springer Nature Switzerland AG
The registered company address is: Gewerbestrasse 11, 6330 Cham, Switzerland

Preface

Dear Readers,

We are glad to offer you here the new edition of the proceedings of Advances in Water Jetting. Published again in the Lecture Notes in Mechanical Engineering book series, this time the book is entitled Advances in Water Jetting **II**. It includes selected papers from the 8th International Conference Water Jet 2023 - Research, Development, Applications held from 17 to 19 October 2023 in Kutná Hora, Czech Republic. The aim of the conference was – again - to present the latest achievements and provide an opportunity to discuss and exchange knowledge, information and experiences in the fields of water jet research and development, and a long list of applications such as cutting, machining, cleaning, surface treatment, milling, removal of coatings and layers, micro cutting and micro machining, jet grouting, repair of structures and buildings, hydrodemolition. Special applications of water jets, abrasive materials, jet observation and parameters measurement, erosion wear, high pressure technology, environmental aspects, and the improvement of process efficiency are also discussed. A special emphasis is given on sustainable practices, and of course on new trends in water jetting. The international conference Water Jet is organized every two years by the Institute of Geonics of the Czech Academy of Sciences, which has been conducting research in waterjet technology since 1985 and is one of the few institutes in the world that has been systematically and continuously dealing with this topic for such a long time. The 8th edition of the conference, saw 28 interesting and inspiring presentations covering either fundamentals or applied waterjet research, as well as related fluid mechanics problems. Many contributions focused on knowledge and technology transfer in this or related fields. The conference program also included commercial contributions from high-pressure equipment manufacturers and national and international sponsors, whose financial support enabled the conference to include a traditionally rich social program and technical excursions.

For these proceedings, only contributions that met the traditionally high quality of Springer Nature publications have been selected. Therefore, only original research and new challenges in the field of water jetting have been included in the proceedings; lower quality or commercially oriented contributions were excluded after a careful review process. Reviewers were selected from members of the International Advisory Board of the Water Jet Conference, other prominent and erudite experts in the field, and the editors of the proceedings. The reviewers came from countries in Europe, North America and Asia (a list of reviewers can be found in the Organisation section). To avoid potential conflicts of interest, each paper was reviewed by at least two independent reviewers, and in rare cases (especially in the case of contradictory reviews) by three reviewers. Careful attention was paid to the quality of the reviews, so that each reviewer could provide a maximum of five independent reviews. The reviewers had to evaluate each paper by answering questions about the originality of the paper, its quality, the significance of the results presented, ethical issues, chapter presentation, etc. Each article was checked for plagiarism using iThenticate, achieving a similarity index of less than 25%. After a thorough peer review process (single-blind review), the editors accepted 19 papers for this LNME book (acceptance rate of 68%).

This publication would not have been realized without the dedicated work of the editors and the organizing committee of the conference. I would like to express my sincere thanks to them, as well as to the participants of the conference, the authors of the papers and all the reviewers. I would also like to thank the management of the Institute of Geonics for their tireless support and patience with the sometimes-unusual demands of the organizing committee. I would also like to thank the Mayor of the beautiful historic town of Kutná Hora for his patronage of the conference. Many thanks to the professionals at Springer Nature Publishing, who were always helpful throughout the preparation of the publication and contributed significantly to the result i.e. the book you are now reading. On behalf of the entire team of editors and authors, I wish you an enjoyable and stimulating experience. We would be delighted if you find something in our publication that not only interests you, but also inspires your future work. Together we can contribute to the advancement of waterjet technology and even spread awareness of this technology in other areas of human endeavour.

June 2024

Libor Sitek
Joško Valentinčič
Franz H. Trieb
Sergej Hloch

Organization

Editors

Libor Sitek	Institute of Geonics of the Czech Academy of Sciences, Ostrava, Czechia
Joško Valentinčič	University of Ljubljana, Ljubljana, Slovenia
Franz H. Trieb	Kapfenberg, Austria
Sergej Hloch	Faculty of Manufacturing Technologies TUKE with a seat in Prešov, Prešov, Slovakia

International Advisory Board

Chair

Josef Foldyna	Institute of Geonics of the CAS, Czechia

Members

Lenka Bodnárová	Brno University of Technology, Czechia
Axel Henning	OMAX, Hypertherm Associates, USA
Sergej Hloch	Faculty of Manufacturing Technologies TUKE with a seat in Prešov, Slovakia
Zdenko Krajný	Slovak University of Technology in Bratislava, Slovakia
Marco Linde	ANT Applied New Technologies AG, Germany
Ralph Lindken	Bochum University of Applied Sciences, Germany
Giuseppe Modoni	University of Cassino and Southern Lazio, Italy
Andreas W. Momber	RWTH Aachen University, Germany
Marcello Papini	Toronto Metropolitan University, Canada
Guoyi Peng	Nihon University, College of Engineering, Japan
Jana Petrů	VSB – Technical University Ostrava, Czechia
Frank Pude	INSPIRE AG, Switzerland
Manuel Schüler	HydroMill GmbH, Germany
Franz H. Trieb	Kapfenberg, Austria
Joško Valentinčič	University of Ljubljana, Slovenia

Organizing Board

Chair

Libor Sitek	Institute of Geonics of the CAS, Czechia

Members

Lucie Gurková	Institute of Geonics of the CAS, Czechia
Daria Nováková	Institute of Geonics of the CAS, Czechia
Dagmar Sysalová	Institute of Geonics of the CAS, Czechia
Michal Zeleňák	Institute of Geonics of the CAS, Czechia
Jiří Starý	Institute of Geonics of the CAS, Czechia

External Reviewers

Lenka Bodnárová	Brno University of Technology, Czechia
Axel Henning	OMAX, Hypertherm Associates, USA
Sergej Hloch	Faculty of Manufacturing Technologies TUKE with a seat in Prešov, Slovakia
Zdenko Krajný	Slovak University of Technology in Bratislava, Slovakia
Giuseppe Modoni	University of Cassino and Southern Lazio, Italy
Andreas W. Momber	RWTH Aachen University, Germany
Marcello Papini	Toronto Metropolitan University, Canada
Guoyi Peng	Nihon University, College of Engineering, Japan
Jana Petrů	VSB – Technical University Ostrava, Czechia
Andrzej Perec	Jacob of Paradies University, Poland
Frank Pude	INSPIRE AG, Switzerland
Manuel Schüler	HydroMill GmbH, Germany
Franz H. Trieb	Kapfenberg, Austria
Joško Valentinčič	University of Ljubljana, Slovenia

Contents

Observation and Theory to Improve the Efficiency of Jet Grouting

Giuseppe Modoni[1(✉)], Lidia Wanik[2], Joanna Bzówka[3], Erminio Salvatore[1], and Michela Arciero[1]

[1] University of Cassino and Southern Lazio, 03043 Cassino, Italy
modoni@unicas.it
[2] Akademia Śląska, 40555 Katowice, Poland
[3] Silesian University of Technology, 44100 Gliwice, Poland

Abstract. This work describes the scientific process that transformed the pioneering empirical solution of jet grouting into an engineering technique. In its basic version, jet grouting involves drilling small holes in the ground and emitting jets of cement grout at very high speed while extracting and rotating the drill rods, to cut the soil, mix it with the binder and form cemented columns. The dimensions and mechanical properties of the columns depend on the ability of the jet to disintegrate the original soil fabric, mix particles and mortar homogeneously, and carry out the cement reaction. In this process the hydrodynamics of submerged turbulent jets play a fundamental role. Observation and theory intertwine to analyze, conceptualize, and simulate basic mechanisms. The result is a closed-form relationship, simple but complete, capable of capturing the role of the fundamental factors. Aim of this process is to reduce the subjectivity and uncertainties inherent in the applications and pave the way for the improvement of the technique. Finally, a practical application is reported to show the potential of the scientific approach on improving the technique.

Keywords: Submerged Jet · Turbulence · Velocity · Numerical Modelling · Theory

1 Historical Development of Jet Grouting

The jet grouting is one of the most popular ground improvement techniques, being used to solve geotechnical problems in a large variety of civil constructions. The different versions nowadays available are the result of a long ongoing history of technological developments started in the late 1960's, then continued with the contribution of technicians, practitioners, and scientists. Initially, a group of Japanese specialists [1] got inspired by the cutting with high-speed jets of rock and rocklike materials [2] to develop a ground improvement technique based on a similar principle. Their idea was to inject a fluid binder (grout) within boreholes previously drilled in the ground to destroy the soil fabric, separate particles, and mix with the binder to finally produce, upon hardening, bodies of cemented material. Simultaneous rotation of the monitor during withdrawal

L. Sitek et al. (Eds.): Advances in Water Jetting II, LNME, pp. 1–18, 2025.
https://doi.org/10.1007/978-3-031-72778-8_1

enables to create columns of circular section. The system was firstly patented in Japan, then in other countries [1] (Fig. 1a) to create "underground walls comprising a plurality of columns in the earth and soil formation". In the subsequent decades, the technique expanded worldwide, being applied to a larger number of applications, and becoming arguably the most common ground improvement technique.

Historically, the first patented version (Chemical Churning Pile - CCP – [3] - was based on the injection of chemical binders, but soon these products revealed to be not economically convenient and were replaced by the more cost-effective water-cement grout. An evolution of the technique called Jumbo Special Pile (JSP) was patented some year later. It was aimed at improving the dimensions of the jet grouted elements [4] by shrouding the grout jet with an annular jet of compressed air. The function of this annulus consists in reducing the transfer of hydrodynamic energy from the inner jet to the surrounding fluid, extending in this way the erosive power of the grout over a larger distance.

Thereafter, a third more complex system, referred to as the Jet Grout method, was conceived by another group of specialists [5, 6], where erosion and binding were accomplished by two separate jets. With this system, soil is firstly eroded by a high-speed jet of water, then binding of the remolded material is accomplished with a later jet of cement grout ejected from a lower nozzle.

CCP and Jet Grout methods attracted the attention of European companies, particularly during the international competition held for the stabilization of the Pisa Tower, in the early 1970's. Here the use of CCP (proposed by Konoike Co. Ltd.) was one of the five solutions considered worthy of mention. Commercial agreements between some Italian companies and the Nissan Freeze Company, owner of the patent of the CCP method, triggered the penetration of the technology into Europe.

A further development of the CCP technique (Column Jet Grout - CJG- also known as the Kajima method from the name of the Japanese company that developed it) shrouded the water jet with a ring jet of compressed to empower the erosive action of the upper nozzle [7].

Nowadays, all techniques and their evolution are categorized into three main groups, defined respectively single, double and triple-fluid systems, this name obviously depending on the number of injected fluids (grout, air, water).

In its former application, jet grouting was mostly used to form arrays of isolated columns reinforcing the foundation of large structures. However, considering that columns can be overlapped to form continuous structural elements of various shapes, (e.g. massive blocks, walls of variable thickness, arched vaults) the technique was also applied to support excavation, assist tunneling, create water barriers and underpin structures. After the 1980's, jet grouting gained popularity in Italy (e.g. [8]), Germany (e.g. [9]) and U.K. (e.g. [10]). In the 1990's the technique became popular in the North (U.S. and Canada) and South America (Brazil), being progressively used for excavation support, ground-water barrier, bottom-sealing of pollutants, protection of bridges against scour, stabilization of slopes and underpinning of existing foundations in commercial and industrial settings. Nowadays jet grouting is used all over the world including middle east and far east countries.

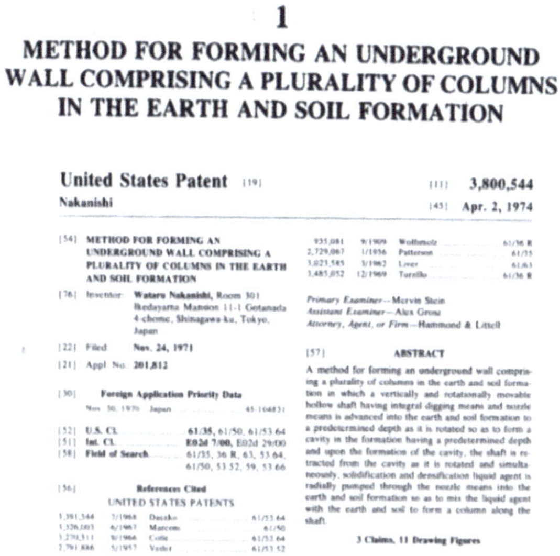

1

METHOD FOR FORMING AN UNDERGROUND WALL COMPRISING A PLURALITY OF COLUMNS IN THE EARTH AND SOIL FORMATION

United States Patent [19] — [11] **3,800,544**

Nakanishi — [45] **Apr. 2, 1974**

[54] METHOD FOR FORMING AN UNDERGROUND WALL COMPRISING A PLURALITY OF COLUMNS IN THE EARTH AND SOIL FORMATION

[76] Inventor: **Wataru Nakanishi,** Room 301 Ikedayama Mansion 11-1 Gotanada 4-chome, Shinagawa-ku, Tokyo, Japan

[22] Filed: **Nov. 24, 1971**

[21] Appl. No.: **201,812**

[30] **Foreign Application Priority Data**

Nov. 30, 1970 Japan 45-104831

[52] **U.S. Cl.** **61/35,** 61/50, 61/53.64

[51] **Int. Cl.** **E02d 7/00,** E02d 29/00

[58] **Field of Search** ..... 61/35, 36 R, 63, 53.64, 61/50, 53.52, 59, 53.66

[56] **References Cited**

UNITED STATES PATENTS

3,391,544	7/1968	Daczko	61/53.64
3,326,003	6/1967	Marcom	61/50
3,270,511	9/1966	Cottie	61/53.64
2,791,886	5/1957	Veder	61/53.52
935,081	9/1909	Wolfsholz	61/36 R
2,729,067	1/1956	Patterson	61/35
3,023,585	3/1962	Liver	61/63
3,485,052	12/1969	Turzillo	61/36 R

Primary Examiner—Mervin Stein
Assistant Examiner—Alex Grosz
Attorney, Agent, or Firm—Hammond & Littell

[57] **ABSTRACT**

A method for forming an underground wall comprising a plurality of columns in the earth and soil formation in which a vertically and rotationally movable hollow shaft having integral digging means and nozzle means is advanced into the earth and soil formation to a predetermined depth as it is rotated so as to form a cavity in the formation having a predetermined depth and upon the formation of the cavity, the shaft is retracted from the cavity as it is rotated and simultaneously, solidification and densification liquid agent is radially pumped through the nozzle means into the earth and soil formation so as to mix the liquid agent with the earth and soil to form a column along the shaft.

3 Claims, 11 Drawing Figures

(a)

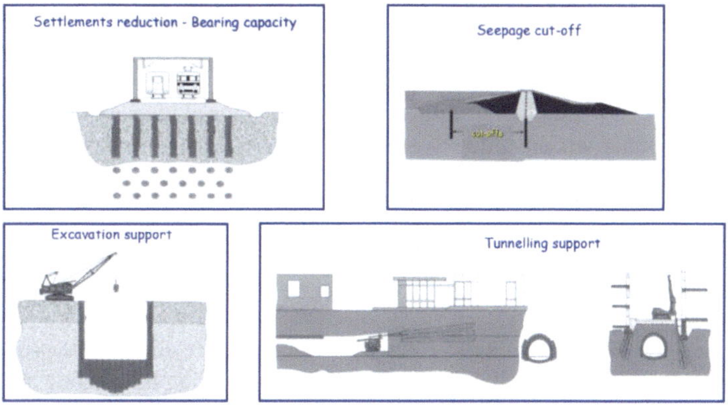

(b)

Fig. 1. US patent (1974, courtesy of Mr. Yasuharu Nakanishi) (a) and typical engineering application of jet grouting (b).

The present work describes the scientific process run to date, made of observation and theory, to transform the pioneer empirical solution into an engineering technique where mechanisms are analyzed, conceptualized, and simulated to produce predictive models that reduce subjectiveness and uncertainties in the application of the technique and pave the way for its improvement.

2 Hydrodynamics of the Submerged Jet

The jet grouting technique consists in drilling into the ground small holes, of almost any length and direction, and ejecting very high-speed jets of fluids while retracting and rotating the drilling bars. Jets are expelled with velocities variable between 200 and 250 m/s from one or more nozzles having diameter ranging between 2 and 8 mm positioned in the end of the drilling string (monitor). The jet erodes the soil all around, mix it with the binder and form cemented columns. Normally, the rotation of the monitor during injection produces columns of circular section, but recent developments enable to tune rotation along the different directions and form elliptical or thin-wall cross section columns.

Generally, ground improvement results are dictated by the combination of technological factors, composition, and state properties of the soil (density, overburden stress, water content). For jet grouting, effects consist in the dimension and mechanical properties of columns and depend on the capacity of the jet to disintegrate the original soil fabric, mix homogeneously particles and grout, accomplish cement reaction. Hence a paramount factor is the influence radius of the jet, i.e. the maximum distance at which the jet propagating in the space outside the nozzle can erode the soil. More precisely, the column's radius is a compromise between soil resistance to erosion and capacity of the fluid threads to maintain high velocities at larger distances coupled with the time length of their action, the latter directly proportional to the monitor withdrawal speed. Observation of jet grouting hydrodynamics has been mostly accomplished indirectly, performing treatments on site with variable parameters and measuring the diameter of discovered columns (e.g. [11]). Experimental observation of the jet characteristics in the laboratory are scarce and usually at a reduced scale (e.g. [12]), basically due to a limitation of the experimental facilities, unable to generate and measure turbulent phenomena at the energies typical of jet grouting. Additionally, the evidence of jets reported in literature refer to water jets only, due to obvious problems arising in the management of binding fluids. Therefore, results need to be extrapolated to other fluids (e.g. cement grout) by means of theoretical considerations.

2.1 Observation of Submerged Jet Diffusion

Normally, the turbulent motion of high velocity jets is characterized with the profiles of the mean velocity component parallel to the jet axis, i.e. neglecting the chaotic agitation velocity components. When a jet penetrates a still fluid mass, a free shear layer is formed on the jet boundary surface originating at the nozzle lip and extending downstream. In very high-speed jets, this boundary layer is turbulent from its origin and exhibits a rapidly growing instability. Ring vortexes are advected downstream, interact, merge and eventually break down. The jet evolution can be seen in Fig. 2.a [12]. Within a short length from the nozzle exit (few diameters), the flow keeps a rather constant cross section, and the velocity v is close to the nozzle outlet one (v_0). In this region the shear boundary layer does not affect the inner core and the radial velocity components are negligible. At larger distances, the flow becomes unstable, whirling threads and vortexes start to appear at the boundaries and progressively propagate toward the inner core. The turbulent mechanisms occurring at the external boundary in this region (Fig. 2b) can be summarized with: a mass exchange between moving and still fluid that increases the jet flow rate due to the entrainment of the surrounding fluid in the jet; a loss of energy caused by the vortexes and by the interface viscous shear stresses. These mechanisms force the jet to widen as a fan with progressively larger cross sections and, considering that energy cannot increase, a progressive reduction of velocity should necessarily occur.

An interesting and comprehensive study is provided by de Vleeshauwer and Maertens [13] who measured with a Pitot pipe the hydrodynamic pressure decay along the longitudinal axes of several submerged water jets expelled from a nozzle of 2.2 mm diameter with injection pressures p_o and outlet velocities typical of jet grouting ($p = 15$–40 MPa - $v_o = 200$–300 m/s). The axial velocity profiles can be computed as follows:

$$v_x(x, r = 0)/v_o = \sqrt{P_x(x, r = 0)/P_o} \quad (1)$$

where $v_x(x,r)$ and $P_x(x,r)$ are respectively the velocity and equivalent pressure computed at any position identified with the longitudinal distance from the nozzle (x) and the radial distance from the jet axis (r), v_o is outlet velocity form the nozzle, P_o the equivalent pressure. The plot versus scaled distance x/d_o (see Fig. 3) show that, despite a drop of about 80% at a distance $x/d_o = 100$, velocities are still high at larger distance as initial values v_o range from 200 to 300 m/s.

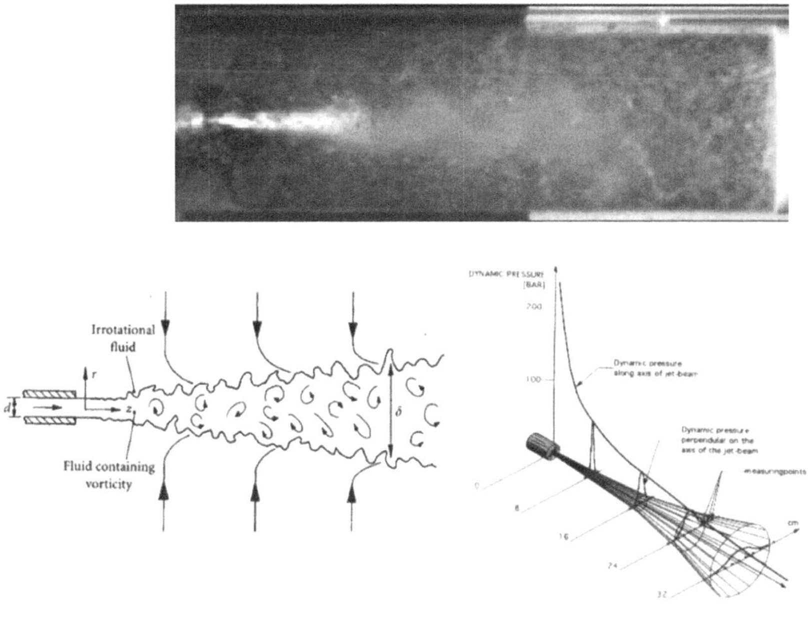

(a)
(b)

Fig. 2. Picture (a) and sketch (b) of a high speed submerged jet (from Berghschneider, 2002).

The longitudinal profiles are compared with those derived by Di Natale and Greco [14] with Particle Image Velocimetry technique on lower velocity jets (0.35–1.5 m/s) expelled from a larger nozzle ($d_o = 4$ mm). The most evident effect is the dependency of the profile shape on the inlet velocity at the nozzle. This effect, also seen on the data from de Vleeschauwer and Maertens, can be explained considering the phenomena involved in the jet propagation, inertia and shear stress exchange at the jet boundaries. Inertia can be quantified by the momentum M of the jet:

$$M \propto \rho \cdot \frac{\pi \cdot {d_o}^2}{4} \cdot {v_o}^2 \quad (2)$$

with ρ representing the fluid density, d_o the nozzle diameter. Shear stress exchange is proportional to the gradient between the velocities of the jet and of the surrounding fluid and can thus be quantified by:

$$S \propto \mu \cdot \pi \cdot d_o \cdot v_o \tag{3}$$

where μ represents the fluid viscosity. The ratio between these two quantities is proportional the Reynolds number at the nozzle (Re), which can hence be adopted to quantify the effects of turbulence:

$$Re = \frac{\rho \cdot v_o \cdot d_o}{\mu} \tag{4}$$

This figure is the most enlightening evidence of the role of inertia in jet grouting and explains the tendency of jet grouting manufacturers to adopt higher energy, basically increasing the power of pumps and the nozzle diameter. In fact, for low outlet velocities (the experiments of Di Natale and Greco infer $Re = 2 \div 4*10^3$) the tangential stresses at the jet contour prevail giving a faster decay of velocity, while for high outlet velocities and diameter (the experiments from de Vleeschauwer and Maertens infer $Re = 3.7 \div 6.0*10^5$) the inertial effects become predominant.

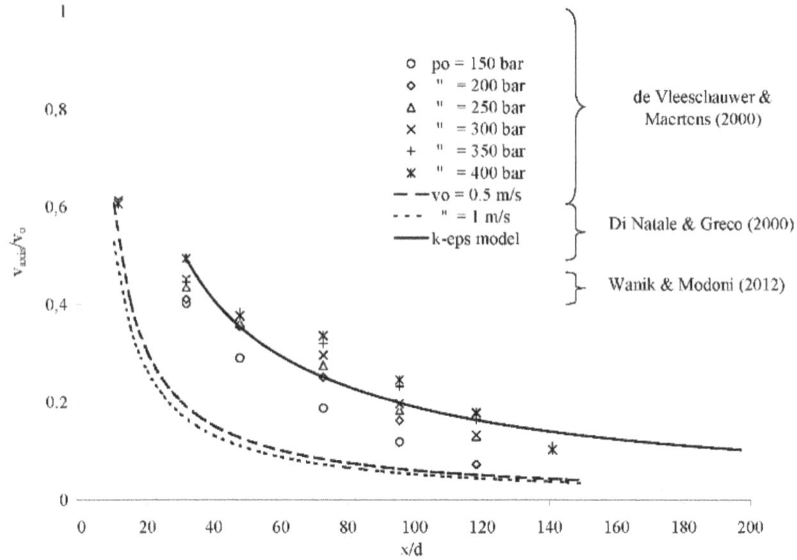

Fig. 3. Velocity profiles of a submerged jet at variable distances from the nozzle [15].

In addition to the above, the jet gains efficiency from a reduction of the mass and shear stress exchange at the jet contours. This is the principle inspiring the double fluid system, where an outer annular jet of air shrouds the inner jet of grout. The plots of Fig. 3 show an example of experiments recently performed in the laboratories of the Institute of Geonics of the Czech Academy of Science at Ostrava (Czechia), where a point sensor equipped with piezoelectric pressure transducer is adopted to measure the dynamic pressure generated by a submerged jet of water at variable standoff and radial distances from the nozzle. The figures show the longitudinal profiles of the mean velocity on the jet axis and the cross-sectional distribution of the mean velocity measured at standoff distances of 50, 100, 200 and 300 mm. Computation of velocity from the measured pressure was accomplished with the following equation:

$$v_x = \sqrt{2 \cdot \frac{P}{\rho}} \tag{5}$$

In both experiments water was ejected with an outlet velocity of 200 m/s from a nozzle of 2.2 mm but, in one case (Fig. 4a) water is the only ejected fluid, in the second case (Fig. 4b) an outer coaxial jet of air is ejected with a pressure of 0.5 MPa from an annular nozzle of 19 mm inner diameter and 2 mm thickness. The role of air shroud in increasing the cutting power of the jet is evident, mostly at longer distances. Looking for example at the longitudinal profile and footprints at 300 mm standoff distance, it is observed that axis velocity is double for the air assisted jet (75 versus 35 m/s).

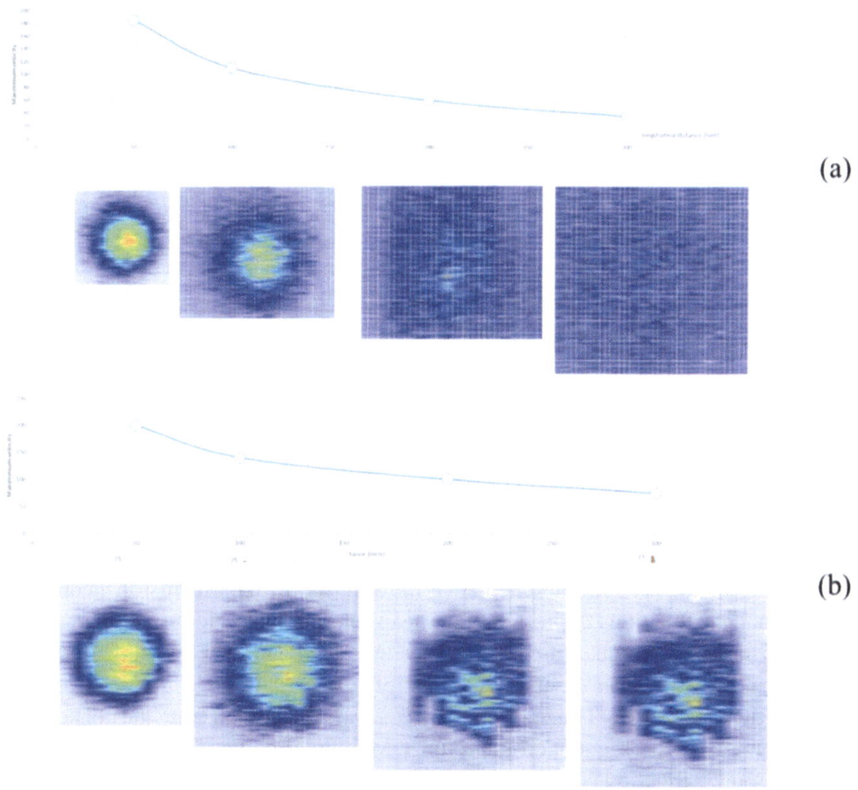

Fig. 4. Velocity profile of single (a) and double fluid (b) jets (courtesy of CRM and Geonics).

A last factor affecting the hydrodynamic efficiency of submerged jets is the nozzle shape. Experimental evidence of jet diffusion in air (Fig. 5, extracted from [16]) prove that a gentle reduction of diameter coupled with a straight portion in the last part of the nozzle infers more coherence to the ejected fluid threads. This effect facilitates the focusing of velocity in the central portion of the cross section as shown by the slower decay with distance of the axis velocity. However, it must not be neglected that the velocity profiles of Fig. 5 are scaled versus the outlet velocity v_o at the nozzle, the latter dictated by the energy losses throughout the flow path. With this regard, the role of local restriction must be carefully considered as they might induce additional losses to the system.

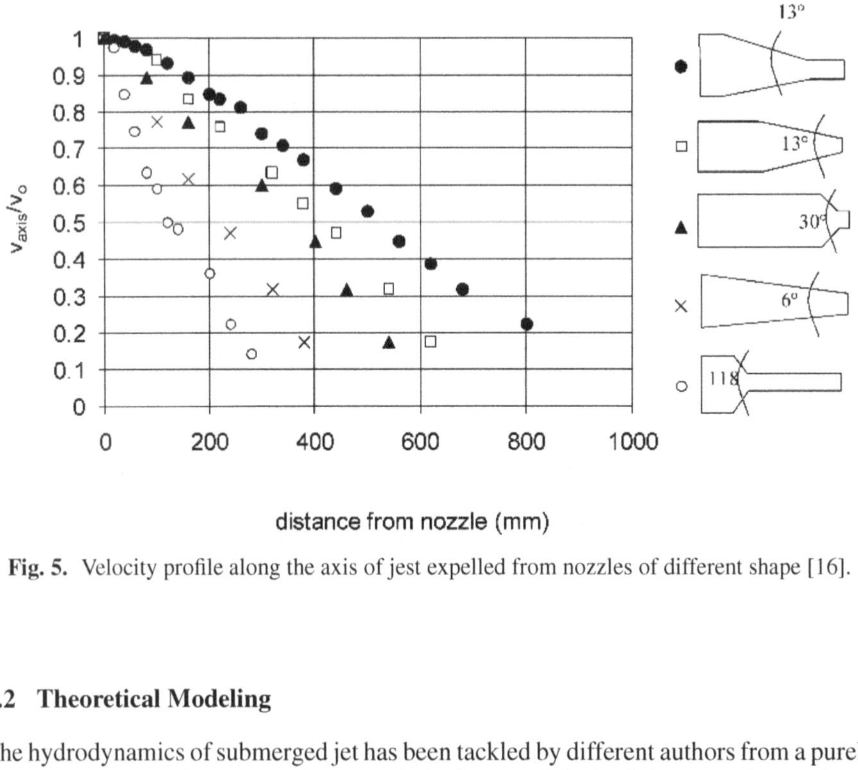

Fig. 5. Velocity profile along the axis of jest expelled from nozzles of different shape [16].

2.2 Theoretical Modeling

The hydrodynamics of submerged jet has been tackled by different authors from a purely theoretical perspective, e.g. [17, 18] or inferring empirical relations on the results of experiments [19, 20]. A common statement is the description of the diffusion mechanisms via mathematical functions expressing the mean longitudinal velocity component as a function of the position (quantified by x and r in Fig. 6) and of the assigned boundary conditions (nozzle diameter d_o and inlet velocity v_o).

The theory proposed by Hinze [30] derives the $v(x, r)$ function from the conservation of momentum in the subsequent cross sections. Manipulating this formulation, [21] and [22] expressed the function $v(x, r)$ with two relations describing respectively the velocity decay along the jet axis $v(x, r = 0)$ (Eq. 6) and the bell shaped distribution of velocity in cross sections positioned at variable distances x (Eq. 7). This analytical description

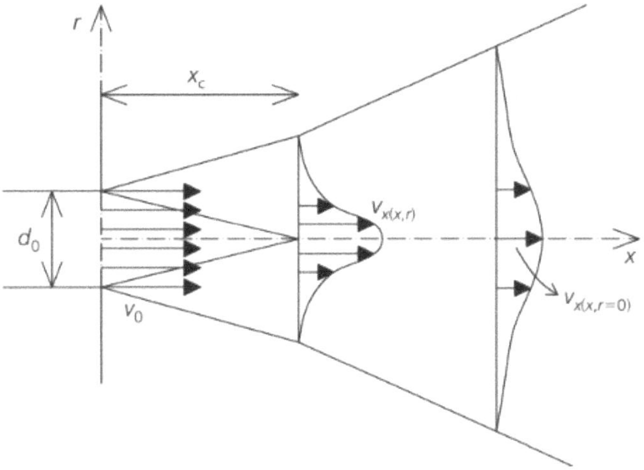

Fig. 6. Mechanical characterization of the jet diffusion (from Hinze 1948 [30]).

is valid for the region where the jet is fully developed, i.e. after the distance x_c from the nozzle where velocity along the jet axis starts its decay (see Fig. 6).

$$\frac{v(x, r = 0)}{v_o} = \Lambda \frac{d_o}{x} \qquad (\text{for } x > x_c) \tag{6}$$

$$\frac{v(x, r)}{v(x, r = 0)} = \frac{1}{\left[1 + 1.33 \cdot \left(\Lambda \frac{r}{x}\right)^2\right]^2} \qquad (\text{for } x > x_c) \tag{7}$$

An advantage of this formulation is simplicity, as both profiles depend on a unique dimensionless parameter Λ, which from Eq. (5) is the ratio x_c/d_o, i.e. the normalized distance where $v(x, r = 0) = v_o$.

Equations (6) and (7) have been adopted by Modoni et al. (2014) to infer the results of a series of numerical simulation of submerged jets in general conditions. After calibrating the rheological model of the fluid (water) with the experimental observation previously reported in Fig. 3 [13], the authors performed a parametric numerical study varying nozzle diameter ($d_0 = 2.2 \div 6$ mm), inlet velocity ($v_o = 200 \div 300$ m/s) and fluid composition (water and grout with cement/water ratio $\Omega = 1$). Figure 7 shows the sample interpolation of longitudinal and transverse profiles at different distances from the nozzle ($x = 0.25$, 0.50 and 0.75 m) for water (Fig. 7a) and cement grout (Fig. 7b) ejected with the same inlet velocity ($v_o = 300$ m/s) from nozzles of the same dimension ($d_o = 5$ mm).

Interestingly from this parametrical study, a strong relation is inferred between the values of Λ fitting the outcomes of each numerical simulation and the correspondent input parameters grouped into the Reynolds number at the nozzle (defined in Eq. 4).

$$\Lambda = 0.0216 \cdot Re^{0.5} \tag{8}$$

This relation enables to generalize the theoretical model of the jet diffusion, capturing the dependency on the properties of nozzle (d_o), outlet velocity (v_o) and fluid composition

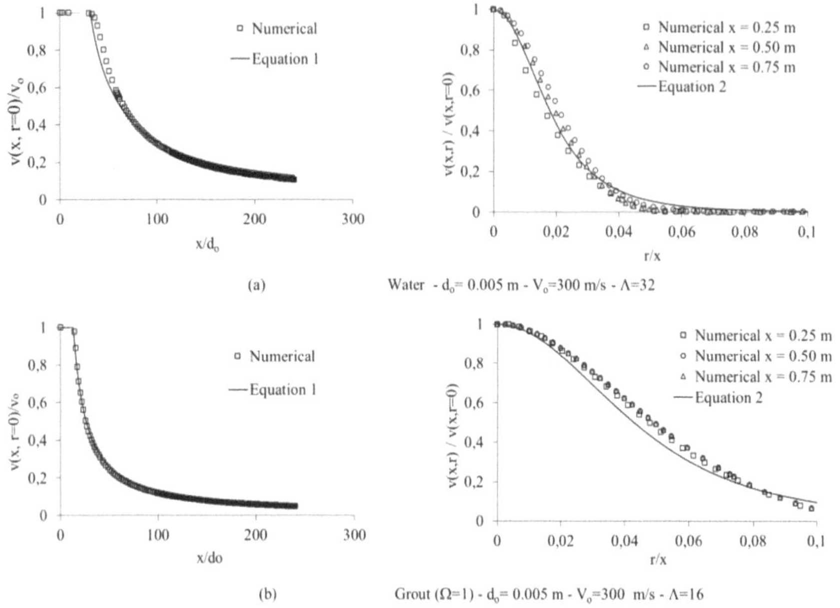

Fig. 7. Examples of numerical simulation and fitting with Eqs. (5) and (6).

(ρ and μ). It is worth noting that, according to Eqs. (5) and (6), a slower velocity decay along the longitudinal axis (higher Λ) corresponds to a narrower cross-sectional profile, i.e. a concentration of the velocity profile near the central axis. Λ can thus be seen as an efficiency parameter of the submerged jet. Its dependency on the Reynolds number at the nozzle (Eq. 7) confirms the importance of the jet inertia for jet grouting (Fig. 8).

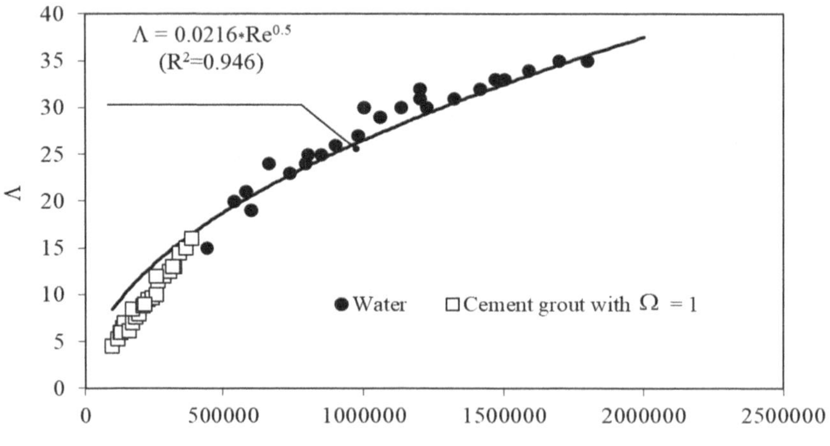

Fig. 8. Fitting of Λ versus Reynold's number at the nozzle with Eq. (8).

3 Prediction of the Column's Diameter

Examining the available field evidence, the effects of jet grouting on the different soils can be classified considering three jet–soil interaction mechanisms (Fig. 9).

- In very coarse materials like clean gravel, grout seepage is the predominant mechanism and increasing jet pressure or flow rate may not produce meaningful advantage as permeation is fed by pressure, not by velocity. The material strength can reach very high values although they are largely variable throughout the treated soil mass.
- In non-plastic materials like sand, alone or with gravel or silt, particles are small enough to be dragged out by the jet, thus erosion is the predominant mechanism. Strength of material is very high and relatively homogenous throughout the treated soil mass.
- In plastic soils like silt and clay, the jet is unable to separate particles and cutting produces clods that are typically found in the fabric of the treated material that possesses relatively low strength.
- These mechanisms have been postulated by [22] in their model developed to predict the diameter of single fluid jet grouting columns. Willing to make the use of Modoni et al. model more practical and extend this study to double and triple fluid jet grouting, [23] estimate the diameter of columns with a closed form relation where the role of all injection parameters is summarized by the specific energy at the nozzles E'_n, i.e. the kinetic energy of the jet per unit length of treatment:

$$E'_n = \frac{\pi}{8} \cdot \frac{M \cdot \rho \cdot {d_o}^2 \cdot {v_o}^3}{v_s} \tag{9}$$

where M is the number of nozzles simultaneously adopted and v_s the withdrawal speed of the monitor. In the formula, the soil resistance to erosion is quantified with the results of typically performed geotechnical in situ tests. In particular, the results of standard penetration tests (N_{SPT}) ([24]) have been assumed to characterize the strength of coarse-grained soils while the results of cone penetration tests CPT (q_c) ([25]) have been assumed to characterize the strength of fine-grained soils.

Finally, the following formulas are written to compute the diameter of columns in the different soils:

$$D_a = D_{ref} \cdot \left(\frac{\alpha \cdot \Lambda \cdot E'_n}{7.5 \cdot 10}\right)^{\beta} \cdot \left(\frac{N_{SPT}}{10}\right)^{\delta} \tag{10a}$$

for coarse grained soils, with E'_n in MJ/m.

$$D_a = D_{ref} \cdot \left(\frac{\alpha \cdot \Lambda \cdot E'_n}{7.5 \cdot 10}\right)^{\beta} \cdot \left(\frac{q_c}{1.5}\right)^{\delta} \tag{10b}$$

for fine grained soils, E'_n in MJ/m and q_c in MPa.

The reference diameter D_{ref} depends on soil properties and, along with the two exponents β and δ (Table 1), values being calibrated on a large dataset of experimental data.

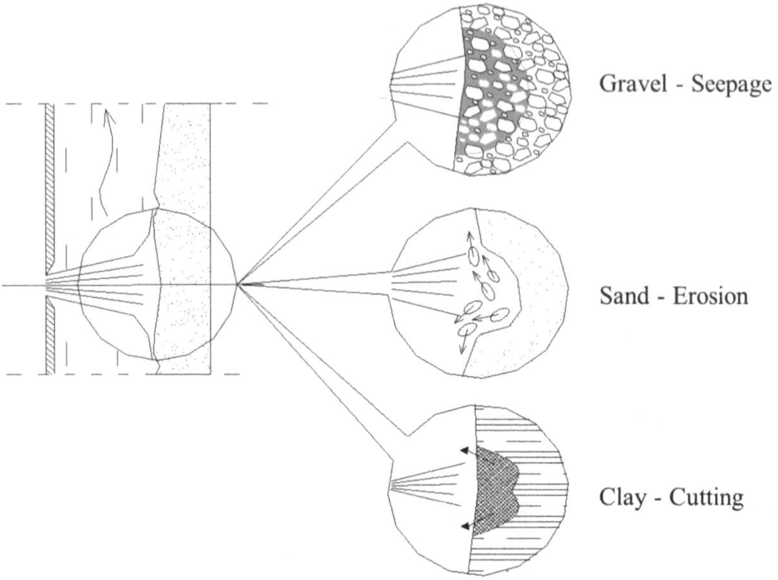

Fig. 9. Interaction mechanisms between jet and different soil types.

In the original formulation the parameter Λ^* was computed for grouts of different composition as proposed by Modoni et al. [21]:

$$\Lambda = \Lambda_w \cdot \sqrt{\frac{\mu_w}{\mu_g} \cdot \frac{\rho_g}{\rho_w}} \quad (11)$$

where $\Lambda_w = 16$ is the parameter found by Modoni et al. fitting the data of de Vleeshauwer and Maertens [13] for water jets, μ is the dynamic viscosity coefficient, and the subscripts g and w respectively indicate grout or water (for water, $\mu_w = 0.001$ Ns/m^2 and $\rho_w = 9.81$ kN/m^3). The viscosity μ_g of a cement grout has been studied for instance by Raffle and Greenwood (reported by [26]) as a function of the cement-water weight ratio ω (Fig. 10).

The density of a grout ρ_g can be calculated again as a function of ω with the following equation derived from mass balance:

$$\rho_g = \frac{\rho_c \cdot (1+\omega)}{\omega + \frac{\rho_c}{\rho_w}} \quad (12)$$

Following the presented developments on the study of submerged jets, Λ can be computed as a function of the Reynold's number with Eq. (8) (Fig. 8).

The comparison shown in Fig. 11 reveals that the diameter of columns can be predicted adopting Eq. (9) with an accuracy of ±20% which is quite fair considering the uncertainties inherent with the considered database.

A slightly better fitting is obtained by Ochmanski et al. [27] inferring the relation among the considered variables with an artificial neural network instead of Eq. (10a, 10b). However, despite a lower accuracy, the use of a closed form solution has the main advantage of being very practical.

Table 1. Values of the parameters to be adopted in Eq. (9), calibrated on the experimental data collected in the field trials.

Soil type		ASTM D2487 classification	D_{ref} (m)	β	δ	α single fluid	α double /triple fluid
Coarse grained	without fine	Gravels and sands with less than 5% fines GW-GP-SW-SP	1.00	0.2	−0.25	1	6
	with fine	Gravels and sands with more than 5% fines GM-GC-SM-SC	0.80				
Fine grained		Silts, clay and organic soils CL-ML-OL-CH-MH-OH-Pt	0.50				

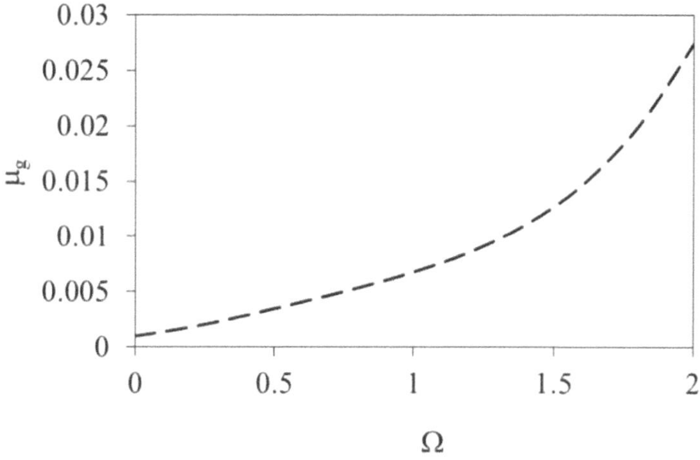

Fig. 10. Viscosity of grout as a function of the cement-water ratio ω (Raffle and Greenwood (1961, reported by [26]).

4 Improvement of the Jet Grouting Efficiency

The predictive formulation of Eq. (9), coupled with the general expression of Λ given in Eq. (7) enables to infer a dependency of the diameter of jet grouting columns on the most relevant technological factors. Among them, a study recently performed by Spagnoli et al. [28] has proposed to modify the rheological properties of cement grout with chemical additives. In particular, the authors adopted a special superplasticizer based on polycarboxylates normally used in the concrete industry. In some preliminary tests the viscosity of the grout-additive mixture has been quantified for variable composition of the grout. In these experiments grouts with largely variable cement/water ratios (Ω

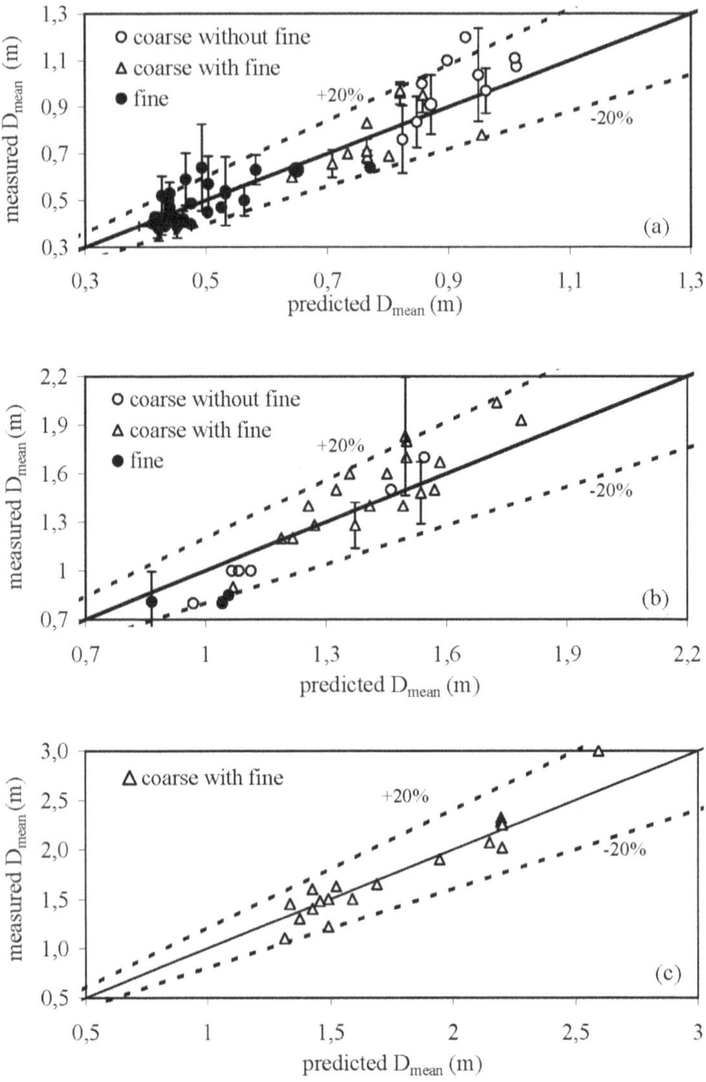

Fig. 11. Prediction of jet grouting columns mean diameter with Eq. (9) versus measurement (from [23]).

equal to 0.71, 0.83, 1.0, 1.25 and 1.67) have been considered in a pure state (without additive) or adding fractions of the superplasticizer equal to 0.3 and 0.5%, these values computed versus the dry cement weight.

The grout density can be computed using Eq. (12). The viscosity has been measured with a viscometer (Thermo Haake VT500) formed of two coaxial steel bodies, an outer fixed cylindrical vessel and an inner rotating drum connected to a SV DIN 53019 sensor. The tests performed at variable rotation velocity of the drum consist in measuring the torque reaction provided by the fluid filling the annular space between cylinder and drum. Finally, rotation speed is thus converted into shear strain rate and torque into shear stress. It is worth noting that for cement grout, the ratio between shear stress and shear strain rate is not constant, as an initial stress (yield) is needed to trigger shear strains. However, in the present study the fluid has been assumed as Newtonian, computing an equivalent viscosity coefficient μ as the ratio between the above quantities measured at high rotational velocity of the viscometer. At such levels the ratio becomes less dependent on strain rate, i.e. the role of initial yield becomes negligible.

The results plotted in Fig. 12 show that viscosity increases significantly with the cement/water ratio. Being this value proportional to the strength of the cemented soil, while viscosity inversely affect the columns diameter, it comes immediate out that a compromise must be sought. Interestingly, this balance can be positively modified by the additive, as it reduces the grout viscosity and makes higher cement/water rations more appealing. Viscosity reduction is economically advantageous from several perspectives, as it reduces the energy losses throughout the system and thus the feeding power of pumps, and the consumption of equipment, which represent a meaningful voice of cost for jet grouting.

The following equation has been adopted to interpolate the experimental dots and express the dependency of viscosity on the water/cement ratio and on the chemical product:

$$\mu = 9.548 \cdot (\omega)^{2.801} \cdot e^{(-1.15 \cdot SP)} \tag{13}$$

The above formula enables, together with Eqs. (8) and (10), to estimate the increase of column's diameter given using the superplasticizer. From this analysis (Fig. 13) an increase of the column diameter by 12.5% is estimated with the addition of 1% fraction of superplasticizer. A field trial has recently been performed by the authors of the present paper creating two columns with double fluid jet grouting in loose sand.

The columns have been formed adopting the same set of parameters (nozzle = 1xØ5 mm, grout flow rate = 340 l/min, monitor lifting speed = 5 mm/s, air flow rate = 7000 l/min), one using pure grout ($\omega = 1$, without additive), the other adding 1% fraction by dry cement weight of the above considered superplasticizer. The direct measurement of diameter upon discovery of columns (Fig. 14) shows increases of diameter for about 25% produced by the superplasticizer, i.e. double than estimated.

5 Conclusions

Engineering is the art of reducing complex physics into simple schemes. Pursuing the trade-off balance between a complete and accurate description of complex mechanisms and its transformation into simplified yet reliable, sound and flexible models is the playground of engineering science (as simple as possible, but not simpler). Geotechnical engineering, in particular ground improvement, deals with a natural, non-homogenous and often poorly investigated material, the soil, this situation often leaving space to the imagination of practitioners. While in a pioneering phase, an empirical approach has triggered the birth of different methods and often led to a considerable success, in many other cases thumb rules have demonstrated their limitation, sometimes with dramatic

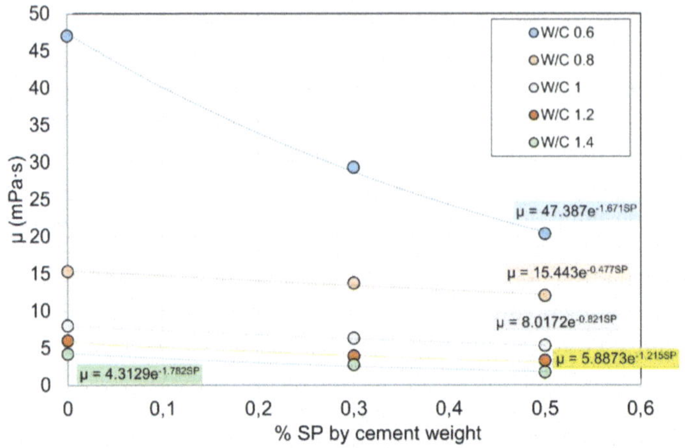

Fig. 12. Results of viscosity tests on grout having variable composition [28].

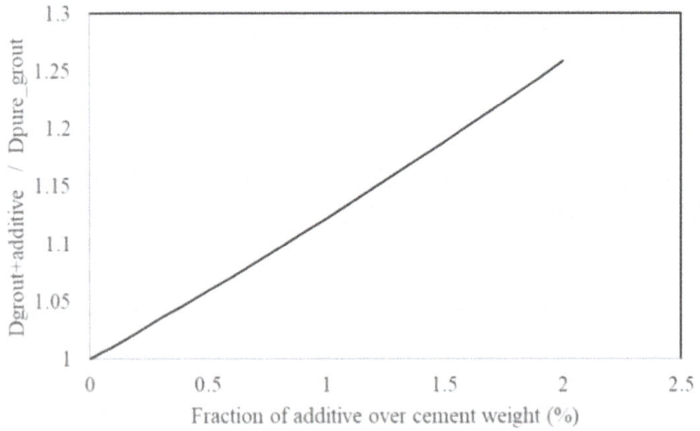

Fig. 13. Estimated increase of column diameter given by the additive.

consequences. On the other hand, a scientific approach, although less immediate, laborious and time consuming, has the advantage of circumscribing the epistemic uncertainty with the rigor and generality of its solutions. This article has applied the above concepts to one of the most popular ground improvement techniques, the jet grouting. From its early application in the 1970s to the more advanced solutions, a paramount factor is the hydrodynamics of submerged jets. Laboratory observation have shown the turbulent character of the submerged jets diffusion, highlighting the fundamental role of inertia compared with viscosity. Their analysis forms the scientific basis for the development of theoretical relations. In details, the transition from observation to theory has been accomplished generalizing few available experiments, obtained in peculiar conditions, with the results of a numerical model. A simple mathematical relation is thus inferred to simulate the mean velocity profile of submerged jets. A main advantage of the proposed

Fig. 14. Jet grouting columns formed in a field trial without (left) and with (right) 1% fraction of superplasticizer.

analytical formula (Eqs. 5, 6 and 7), is that it captures the role of the main technological factors (nozzle diameter, fluid density and viscosity, jet velocity) on the jet profile. This relation is a key factor to predict with a fair approximation the diameter of columns, which is one of the most difficult steps for the execution design of jet grouting treatments. The validity of the proposed approach has been demonstrated with an application where the predicted benefit given by the reduction of the grout viscosity with additives has been confirmed by a campaign of in situ tests. As is typical of the scientific approach, the process is still ongoing and more effort is needed to reduce the epistemic and aleatory uncertainties inherent with the technology. Concerning the first issue, a better knowledge of the role of nozzle shape may, as an example, lead to develop more efficient solutions that improve the cutting efficiency of the jet. Concerning the aleatory uncertainty, which is still relevant on the prediction of the column diameter, a strategy based on experimental observation (field trial) and interpretation with statistical analyses is proposed by some authors (e.g. [29]).

References

1. Nakanishi, W.: Method for forming an underground wall comprising a plurality of columns in the earth and soil formation. US Patent 3,800,544 (1974)
2. Farmer, L.W., Attewell, P.B.: Rock penetration by high velocity water jet. Int. J. Rock Mech. Min. Sci. **2**, 135–153 (1965)
3. Miki, G.: Chemical stabilization of sandy soils by grouting in Japan. In: Proceedings of the 8th ICSMFE, pp. 395–405, Moscow (1973)
4. Xanthakos, P., Abramson, L.W., Bruce, D.A.: Ground Control and Improvement. John Wiley & Sons inc. (1994)
5. Yahiro, T., Yoshida, H.: Induction grouting method utilizing high speed water jet. Proceedings of the 8th ICSMFE, pp. 402–404, Moscow (1973)

6. Yahiro, T., Yoshida, H., Nishi K.: Soil improvement utilizing a high speed waterjet and air jet. In: Proceedings of the 6th International Symposium on Water Jet Technology, pp. 397–428, Surrey (1974)
7. Yahiro, T., Yoshida, H., Nishi, K.: The development and application of Japanese grouting system. Water Power Dam Constr. **27**, 55–69 (1975)
8. Garassino, A.L.: Uso delle colonne in terra stabilizzata con jet grouting come elemento provvisionale. In: Proceedings of the XV National Geotechnical Conference, pp. 155–161, Associazione Geotecnica Italiana, Spoleto (1983)
9. Bell, A.L.: Control of column geometry. Jet grouting in sand. GKN Keller Technical Report (1983)
10. Coomber, D.B., Wright, D.W.: Jet grouting at Felixstowe docks. Ground Eng. **17**(5), 19–24(1984)
11. Croce, P., Flora, A.: Analysis of single fluid jet-grouting. Geotechnique **50**(6), 739–748 (2000)
12. Bergschneider, B.: Zur Reichtweite beim Düsenstrahlverfahren im sand. PhD thesis. Bergische Universität Wuppertal Fachbereich Bauingenieurwesen [in German] (2002)
13. de Vleeshauwer, H., Maertens, G.: Jet-grouting: state of the art in Belgium. In: Proceedings of the Conference 'Grouting – Soil improvement - Geosystem Including Reinforcement', pp. 145–156, Finnish Geotechnical Society, Helsinki (2000)
14. Di Natale, M., Greco, R.: Misure di velocità in un getto sommerso ad asse verticale mediante la tecnica PIV. In: Proceedings of the Symposium promoted by Associazione Italiana di Anemometria Laser (AIVELA), 11 pages, Ancona (2000)
15. Modoni, G., Wanik, L., Giovinco, G., Bzòwka, J., Leopardi, A.: Numerical analysis of submerged flows for jet grouting. Proc. ICE - Ground Improve. **169**(1), 42–53 (2014)
16. Leach, S.J., Walker, G.L.: Some aspects of rock cutting by high speed water jets. Phil. Trans. Royal Soc. London. Series A **260**(1100), 295–308 (1966)
17. Abramovich, G.N.: The theory of turbulent jets. MIT Press Cambridge, Massachusetts (1963)
18. Davidson, P.A.: Turbulence. Oxford University Press, Oxford (2004)
19. Chu, E.H.: Turbulent fluid jet excavation in cohesive soil with particular application to jet grouting. PhD thesis, Massachusetts Institute of Technology
20. Heng, J.: Physical modelling of jet grouting process. PhD thesis, University of Cambridge (2005)
21. Modoni, G., Croce, P., Mongiovì, L.: Theoretical modelling of jet grouting. Géotechnique **56**(5), 335–347 (2006)
22. Modoni, G., Croce, P., Mongiovì, L.: Theoretical modelling of jet grouting – closure. Géotechnique **58**(6), 533–535 (2008)
23. Flora, A., Modoni, G., Lirer, S., Croce, P.: The diameter of single-, double-and triple-fluid jet grouting columns: Prediction method and field trial results. Géotechnique **63**(11), 934–945 (2013)
24. ASTM D1586-11 Standard Test Method for Standard Penetration Test (SPT) and Split-Barrel Sampling of Soils. ASTM International, West Conshokocken (2022)
25. ASTM D3441-04 Standard Method of Deep Quasi-Static Cone and Friction-Cone Penetration Tests of Soil; ASTM International, West Conshohocken (2016)
26. Bell, A.L.: Jet grouting, in Ground Improvement. M. P. Moseley, Blackie Ed, pp. 149–174 (1993)
27. Ochmański, M., Modoni, G., Bzòwka, J.: Prediction of the diameter of jet grouting columns with artificial neural networks. Soils Found. **55**(2), 425–436 (2015)
28. Spagnoli, G., Modoni, G., Arciero, M., Salvatore, E.: Improving the hydrodynamic performance of jet grouting with chemical additives. Int. J. Geosynthetics Ground Eng. **8**(1), 1–8 (2022)
29. Modoni, G., Bzówka, J.: Design of jet grouting for foundation. J. Geotech. Geoenviron. Eng. **138**(12), 1442–1454 (2012)
30. Hinze, J.O.: Turbulence, 1st edn. McGraw-Hill, New York (1948)

Practical and Theoretical Study of Abrasive Particle Velocities Observation in the High-Speed Water Jet Using Optical PTV–LIF Method and CFD Numerical Computation

Michal Zeleňák[1](✉), Zdeněk Říha[2], Kamil Souček[3], Fernando Kevin Miranda[1], and Vladimír Foldyna[1]

[1] Fluid Jets Department, Institute of Geonics of the CAS, Studentská 1768, Ostrava, Czechia
michal.zelenak@ugn.cas.cz

[2] Fluid Jets Department, Institute of Geonics of the CAS, Drobného 38, Brno 60200, Czechia

[3] Department of Geomechanics and Mining Research, Institute of Geonics of the CAS, Studentska 1768, Ostrava, Czechia

Abstract. This paper deals with the observation and determination of abrasive particle velocities in high-speed water jet using optical PTV-LIF methods. A commercial injection cutting head was used for the generation of multiphase turbulent AWJ. High-pressure flow monitoring, PTV-LIF methods and computational fluid dynamics (CFD) were used to study the multiphase flow. The main emphasis of this work is on the comparison of the results obtained by direct observation of abrasive particle velocities and from high-pressure water flow measurements with the results obtained by numerically modelling 3D multiphase turbulent flows. For this purpose, four levels of water pressure and one level of abrasive flow are compared. The abrasive size distribution parameter is considered for the experimental part and the numerical computation. The results obtained from the experiments and the numerical simulation of the abrasive water jet show very good agreement. The details of the performed studies are presented in this paper.

Keywords: Abrasive Injection Cutting Head · High-Speed Abrasive Water Jet · Numerical Flow Modelling · Particle Tracking Velocimetry · Laser Induced Fluorescence

Nomenclatures

AWIJ	Abrasive water injection jet
AWJ	Abrasive water jet
AWSJ	Abrasive water suspension jet
CAD	Computer aided design
CFD	Computational fluid dynamics
CT	Computed tomography
DPM	Discrete particle method

L. Sitek et al. (Eds.): Advances in Water Jetting II, LNME, pp. 19–32, 2025.
https://doi.org/10.1007/978-3-031-72778-8_2

LDA	Laser doppler anemometry
LTA	Laser transit anemometer
LIF	Laser induced fluorescence
LVT	Laser transit velocimetry
Nd:YAG	Neodymium-doped yttrium aluminium garnet
PTV	Particle tracing velocimetry
RANS	Reynolds-averaged Navier–Stokes
RhB	Rhodamine B
SST	Shear stress transport
STL	Stereolithography
t	Time
x_i	Mixture velocity component
u_{mi}	Volume fraction of the mixture
ρ_k	Density of the mixture
α_k	Volume fraction
μ_k	Viscosity of the mixture
m_{pk}	Mass transfer from the phase p to k
m_{kp}	Mass transfer from the phase k to p
K	Bulk modulus of water
ρ_{ref}	Reference density
p_{ref}	Reference pressure
p	Pressure
Y_d	Volume fraction of particles
$\overline{d}$	Mean diameter of the particles
n	Spreading exponent

1 Introduction

Erosion of material occurring due to the collision of a high-speed abrasive water jet with a body surface is the basic principle of abrasive water jet (AWJ) machining. AWJ machining is a process that uses a high-speed stream of water mixed with abrasive particles and air to cut various materials and it offers many advantages over other technologies, such as high flexibility, low environmental impact, and no thermal distortion. Abrasive water injection jet (AWIJ) and abrasive water suspension jet (AWSJ) technologies were developed in the 70 and 80s last century [1].

In the case of AWIJ a high-speed water jet is generated by a high-pressure liquid passing through a water nozzle (with an internal diameter of 0.05–0.5 mm). In the nozzle smallest cross-section, the pressure energy is transformed into kinetic energy and part of it is dissipated. The axial high-speed water jet generates low pressure inside the mixing chamber causing the suction of abrasive particles and air mixture into the mixing chamber through a side hole connected to the inlet duct as shown in Fig. 1. The mixture is then accelerated due to the kinetic energy imparted by the high-speed water jet, generating in this way AWJ in the mixing chamber and the focusing tube before leaving the cutting tool [2].

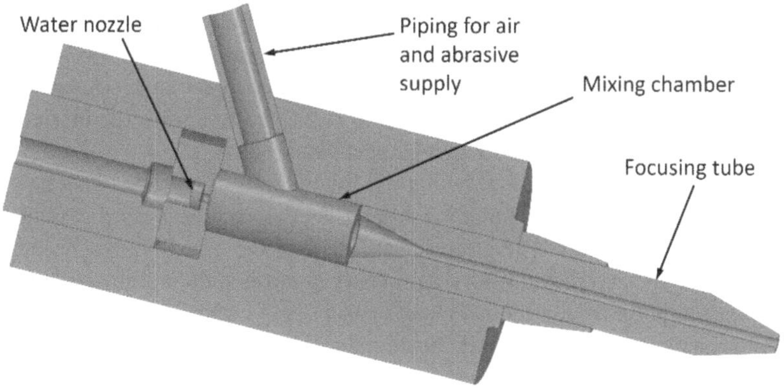

Fig. 1. CAD model of the injection cutting head.

The first papers introducing the observation of abrasive particles velocities were published in the late nineteen nineties, where non-optical methods were used for particles observation. In 1987 Swanson et al. [3] tested an inductive method using magnetic coils. This method consists in spreading a small amount of ferritic tracer particles over a volume of mineral grains. It was demonstrated that different density of ferritic-mineral abrasive does not influence significantly in the acceleration of particles. Lo et al. [4] proved this method in 2019 and presented a very complex study with phenomenological models mapping the AWIJ process generation with respect to the operation conditions. Another mechanical method developed for the determination of abrasive particles speed make use of high-speed rotary discs. The phase shift of two rotating discs and the length of the erosion trace caused by AWJ is recalculated on the speed of AWIJ. This method was firstly introduced authors Isobe et al. [5]. In a similar study recently published by Haghbin et al. [6], the authors used a modified rotating dual disc anemometer and compared their results with existing analytical models. Li et. al [7] also attempted the estimation of the jet dynamics employing an indirect forcing measurement. The disadvantage of the above-mentioned methods was the low validation of the calculated velocities for different mass ratios.

A very good progress in the field of measurement of abrasive particle velocities in high-speed water jet was achieved using optical methods. Laser doppler anemometry LDA and laser transit anemometers (LTA) were tested for the determination of average velocity fields of plain water (PW) and AWIJ. The authors in [8, 9] used LDA in their studies and reported the complexity of the determination of the boundary between water droplets and solid particles. Balz et al. [10] tested the LTA using He-Ne laser. LTA is based on the principle of laser beams hitting a small region with two focal points at a distance offset. The particles velocity is determined based on the offset between the focal points and the time delay caused by particle pass. Related problem of this method is low spatial resolution. The authors in [11, 12] found a solution to the measurement accuracy improvement of abrasive particle velocities using the laser-induced fluorescence (LIF) method combined with planar and stereo PTV. The crucial part for the successful measurement was the stability of fluorescent coating on the particle surface under test conditions, presented in the conclusion of the authors´ publications. Riha et al. [13] improved the coating procedure of fluorescent abrasive grains, which was successfully applied for the observation of abrasive particle velocities in abrasive suspension water jet. The method introduced by Jerman et al. [14] was also tested for the observation of cryogenic precooled ice particles in a high-speed water jet.

During the last 20 years, several numerical calculations of particle laden flows were carried out with the aim of understanding AWJ. Liu et al. [15] developed computational fluid dynamics (CFD) models for three-phase flow at the nozzle outlet based on 2D axisymmetric geometry. The authors significantly simplified the mixing chamber inner shape. The calculation of continuous phases (water, air) was performed as a first step, then the motion of abrasive particles in the domain was calculated. The 2D axisymmetric model was very simple and quick to compute. Such configuration allowed to calculate the distribution of water jet velocity along the domain axis with reasonable accuracy. Due to domain transition from 3D to 2D space, all the other results were inaccurate. It is valid for velocity distribution in the mixing chamber and the focusing tube, also for the shape of abrasive particle traces in the domain, the distribution of abrasive particle velocities, etc. Prisco et al. [16] described the calculation of two-phase flow in 3D geometry of the abrasive head. The flow phase interactions were described using the Euler-Euler model. This sophisticated model enabled the description of a wider range of individual flow phase interactions. The result of this study comprises pressure, velocity and density distribution with the absence of abrasive particles in the calculation. The computation domain did not consider the real geometry of the CAD model. The study published by the authors in [17] investigate three-phase flow in the abrasive cutting head. The inclination angle change of the inlet suction connector geometry is considered in their study. The models were solved as time-independent tasks using two-equation Reynolds-averaged Navier–Stokes (RANS) turbulent flow models. Their study results summarized velocity distribution inside the abrasive head and its outlet as well as the wear in the inner components. Moreover, theoretical computation of particle velocity distributions behind the focusing tube outlet is compared with experimental data showing a good agreement between calculated and measured values. Unfortunately, only a one-way coupling Discrete Particle Method (DPM) method was adopted i.e., abrasive particle motion was affected by the continuous phases (water and air), while the solid phase (abrasive particles) had no feedback on the continuous phase flow. The model used did not include a real nozzle geometry.

The present work focuses on the practical and theoretical study of abrasive particle velocities observation in the high-speed water jet using PTV-LIF method and CFD numerical computation of a standard injection cutting head. The main emphasis is on the mutual comparison of the results achieved by direct particle observation the ones computed using numerical models. The details of this study are summarized in the paper.

2 Material and Methods

A standard injection cutting head with one abrasive inlet and diamond orifice of 0.33 mm (0.013") combined with a focusing tube of 1.02 mm (0.04") were selected for the studies presented in this paper. Four levels of water pressure (105, 194, 406 and 406 MPa) and one mass flow rate (100 g/min) were used for the abrasive speed particle measurement and the numerical computation. Firstly, a real CAD model obtained by computed tomography (CT) scans was used as the water nozzle inner geometry for the CFD calculations. The 3D CAD model was acquired using a Nikon XT H225 CT scanner based on the acquisition of 2D radiographic pictures during the scan process of the nozzle. Different perspective views of the scanned object are obtained during rotation of the CT table presented in Fig. 2a. The distance between the X-Ray source and the flat detector panel determines the magnification of the scanned object and the number of 2D radiographic images improve the accuracy of the 3D model during the volumetric reconstruction. The 3D geometric model (Fig. 2b) was stored in a standard triangle language format (*.stl) using VG Studio Max.

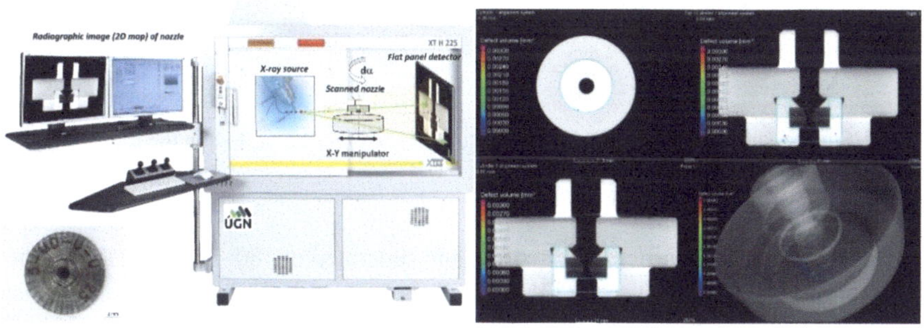

a) b)

Fig. 2. **a** Principle of the measurement using CT method; **b** 3D geometric reconstruction of the nozzle [18].

Secondly, the measuring assembly for the high-pressure flow diagnostic was built up with respect to the position of the cutting head (see Fig. 3). A high-pressure pump PTV Jets delivered pressurized water through a Polyflex hose into the HP inlet point. A high-pressure turbine flow meter KEM Küppers HM 006/TC-NS/V in combination with an ESI 5200 SB high-pressure sensor were used to collect experimental data. The sensors were installed as close as possible to the abrasive cutting head inlet in order to minimize the hydraulic losses. DAQ interface based on the LabView interface presented in publications [19, 20] allow to acquire information about the flow rate, the water pressure, the hydraulic power and the fluid temperature in time.

The abrasive particles velocity was measured at the focusing tube outlet using the planar laser induced fluorescence PLIF method. Due to the low density of the AWIJ seeding particles, the 2D particle tracing velocimetry (PTV) (Fig. 4) well known as low density particle image velocimetry (PIV) was used.

AWIJ represents a multiphase turbulent flow therefore, the attempt to detect abrasive particles within the water-air turbulent jet requires the usage of a LIF method (Fig. 5b). The process of abrasive particle detection in a high-speed jet requires the visibility of abrasive grains. For this purpose, an Australian garnet of 150–400 μm size was coated by Rhodamine B (RhB) fluorescent matter mixed inside acrylate lacquer (Fig. 5a)

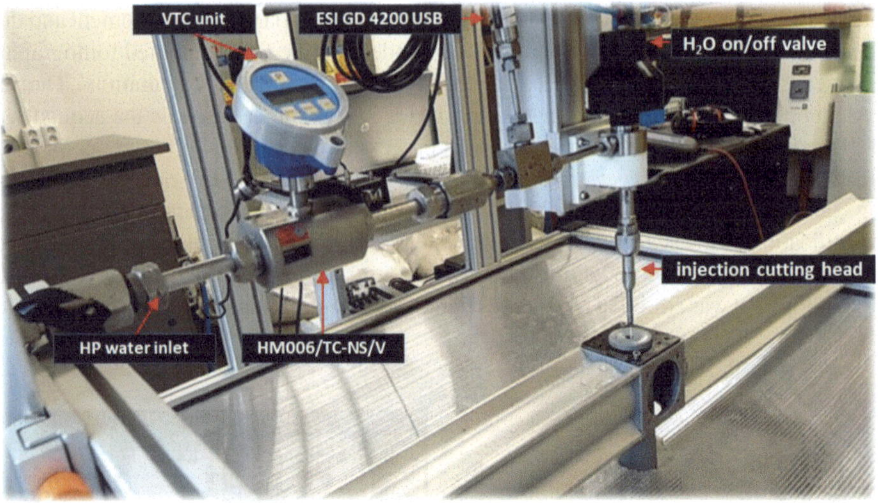

Fig. 3. High pressure flow diagnostics sensors for data acquisition of flow in the injection cutting head.

[13]. Subsequently, the particles were dried and separated in 180–250 μm size fractions using vibration sieving shakers. PTV optical setup for particles observation consisted of a 14-bit CCD camera with macro lens and long pass LIF filter. A standard Nd:YAG double-pulse laser 135 mJ x 15 Hz operated in double pulse mode was employed as illumination light source. The length of light pulse was approx. 6 ns. A divergent lens mounted on the laser output allows the generation of 15° light sheet with a minimal of 0.2 mm thickness in the place of the jet interaction. A short camera exposure time (1 μs) and relatively short dt time 110–200 ns allow the high-speed capture of grains (up to 1000 m/s). The area of interest (AOI) was adjusted to 19 x 14 mm. For each of the mentioned combination of water pressures (105, 194, 406 and 406 MPa) and mass flow rate (100 g/min), 100 experimental images operated in double frame mode were acquired. To improve the PTV cross-correlation method quality (Fig. 5f), the image sources were pre-processed using image subtraction (Fig. 5c), normalization (Fig. 5d) and intensity average functions (Fig. 5e). The main purpose of the motioned function is the reduction of the image background noise and the uniformization of the light intensity due to laser light interaction with non-spherical grains. The correlation matching of the grains based on PTV (Fig. 5g) considers vectors of particle sizes between 40–250 μm with overlap of interrogation windows of particles. The output of one recording sequence is schematically presented in a 2D vector map, which statistically represents one series of the measurement under the operated conditions.

In order to reach a better accuracy of the numerical computations with respect to the experimental results, an inner abrasive head geometry identical to the experimental one was used. The 3D geometry used for the CFD calculations (Fig. 1) combined the CAD geometry of the abrasive head provided by the supplier and the accurate CAD model of the water nozzle from a CT scan. Therefore, the computational domain of the abrasive cutting head including the water nozzle geometry was not simplified. This

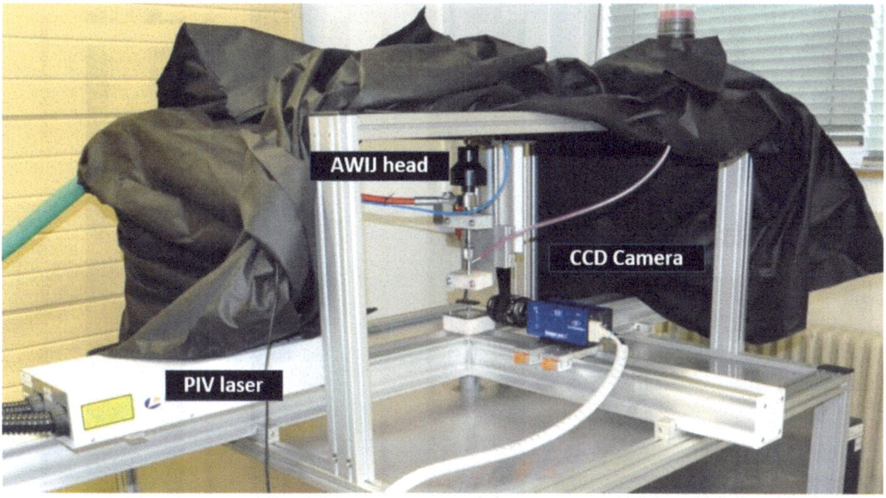

Fig. 4. A PLIF optical setup for the abrasive particle velocity measurement.

procedure ruled out a possible source of errors and discrepancies between calculated results and measured values. The computational mesh was mainly built using hexahedral elements. In the transition region between the mixing chamber and inlet abrasive duct, tetrahedral elements were considered. The rest was filled with pentahedral elements. The computational domain contained 1217562 elements in total. The CFD geometry and the boundary conditions are schematically depicted in Fig. 6.

A steady state turbulent multiphase fluid flow was solved in the domain. The mixture model [21] was selected for the multiphase numerical calculation (water and air). Neglecting the slip velocity between the two phases, the continuity equation and the volume fraction α_k for the secondary component could be expressed by Eq. (1):

$$\frac{\partial}{\partial t}(\alpha_k \rho_k) + \frac{\partial}{\partial x_i}\left(\alpha_k \rho_k u_{m_i}\right) = \sum_{p=1}^{n} \left(\dot{m}_{pk} - \dot{m}_{kp}\right)$$

$$\sum_{k=1}^{n} \alpha_k = 1 \tag{1}$$

The turbulent multiphase flow (water and air) was described by the two-equation RANS shear stress transport (SST) k-ω model [21]. In order to get the correct velocity profile shape at the water nozzle outlet, the compressibility of the water was considered in the solution. The water density was solved in CFD model using Eq. (2):

$$\rho = \frac{\rho_{ref}}{\left(1 - \frac{K}{p - p_{ref}}\right)} \tag{2}$$

The solid particles motion in the continuous phases of water and air was described by the discrete particle model (DPM). The abrasive particles diameter variable was

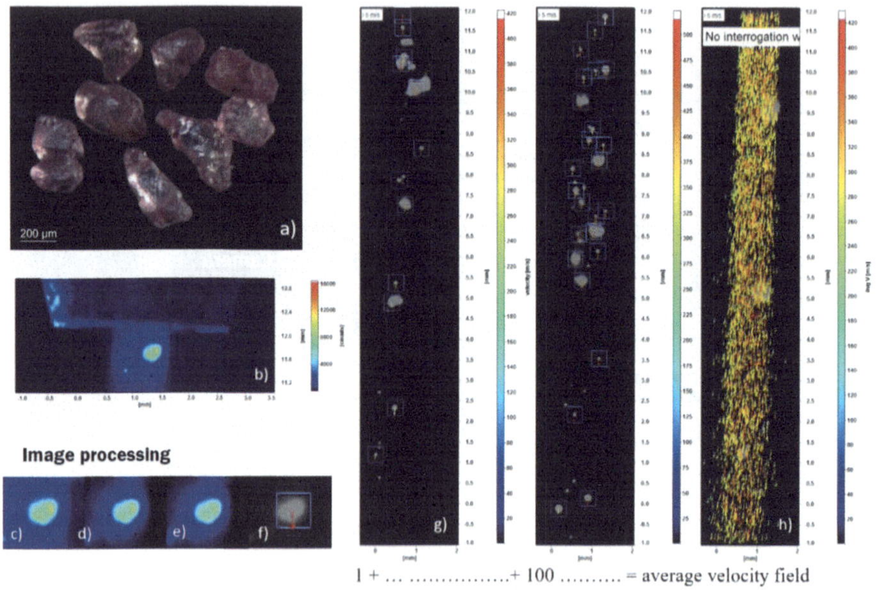

Fig. 5. **a** abrasive grains coated by RhB; image processing function: **b** original experimental image of abrasive grain at focusing tube outlet **c** source image-particle detail, **d** background subtraction, **e** averaging of pixels **f** PTV tracking of individual grains from double frame image, **g** example of vector map from one double frame image, **h** statistical vector map of particle velocities.

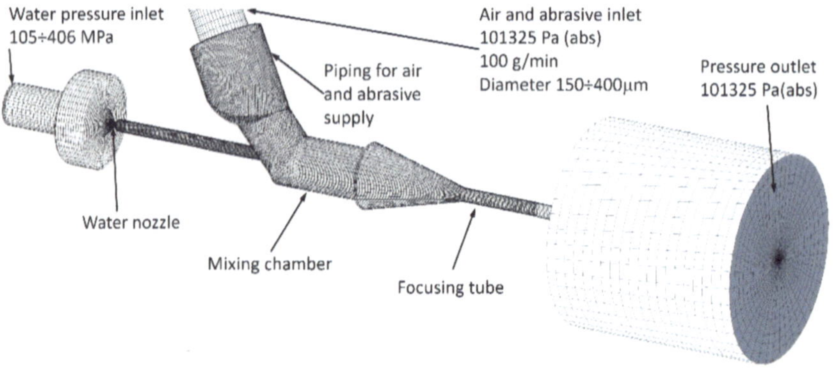

Fig. 6. The final mesh of the computational domain with description of the boundary conditions.

considered by the Rosin - Rammler distribution function (Eq. 3) in the computational model. This setup allowed to reach better accuracy in the velocity shape calculation of the continuous fluid flow filed including the discrete particles distribution.

$$Y_d = e^{-\left(d/\bar{d}\right)} \tag{3}$$

For this purpose, the measurement of the abrasive particle diameter distribution was performed using a particle size analyzer. It was observed that the mean particle diameter was 281 μm as shown in Fig. 7. The size of the initial abrasive particle fraction was in the interval of 150–400 μm. The mentioned Rosin - Ramller distribution was observed to fit for n = 5.9, also plotted in Fig. 7.

The behaviours of the mentioned 3 phase flow differ for optimal spherical and non-spherical grains reported by authors [22, 23]. It is possible to set particle sphericity parameter in the calculation. This parameter represents the ratio between the smallest and the largest dimension of the particle. For a perfect sphere, the particle sphericity is equal to 1. Another important parameter is the restitution parameter (RP), which defines the energy losses due to the collision of abrasive grains with the inner walls of the cutting head. In this study the particle sphericity was set to 0.65 and the RP was set 0.7 for both, tangential and normal, particle impact direction. The mentioned parameters setup corresponds to the setup presented in the publication [24].

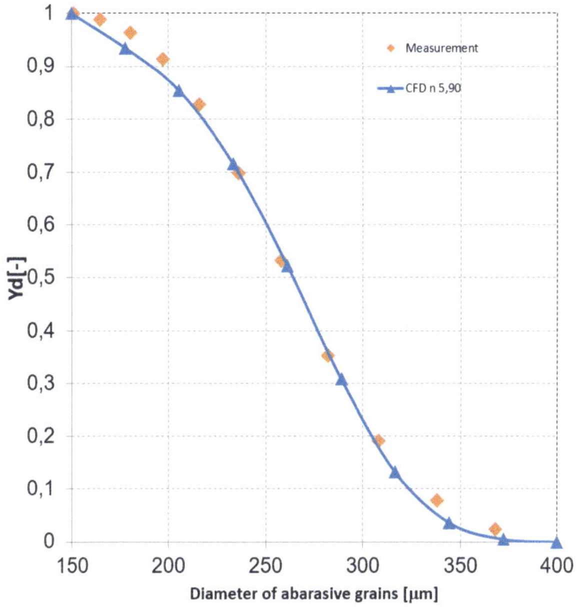

Fig. 7. Distribution of abrasive grains particles.

3 Results

The results of the water flow rate monitoring at the injection cutting head for the selected water pump pressure levels are presented in Fig. 8 and in Table 1. The figure presents the results of the numerical computation and shows a clear dependence of the flow rate on the water pressure and on the hydraulic power as well. These dependences are very close each other. The results given by the numerical computation and the practical measurement reached a very good agreement. The deviations between the measured and the calculated values are in the range of 1%.

Table 1. Results of measured pressures, volume flow rates and the power distribution for the water nozzle p_a: mean value of the water pressure, $\dot{V}$: mean value of the volume flow rate, P_a: average value of the power, σ: standard deviation, σ_M: standard error of mean.

p_a [MPa]	σ [MPa]	σ_M [MPa]	$\dot{V}$ [l/min]	σ [l/min]	σ_M [l/min]	P_a [kW]	σ [kW]	σ_M [kW]
105.558	1.680	0.215	1.576	0.012	0.001	2.773	0.052	0.006
194.464	15.236	1.950	2.077	0.079	0.010	6.750	0.746	0.095
302.722	2.126	0.272	2.529	0.022	0.001	12.762	0.156	0.019
406.165	3.224	0.412	2.860	0.017	0.002	19.363	0.172	0.002

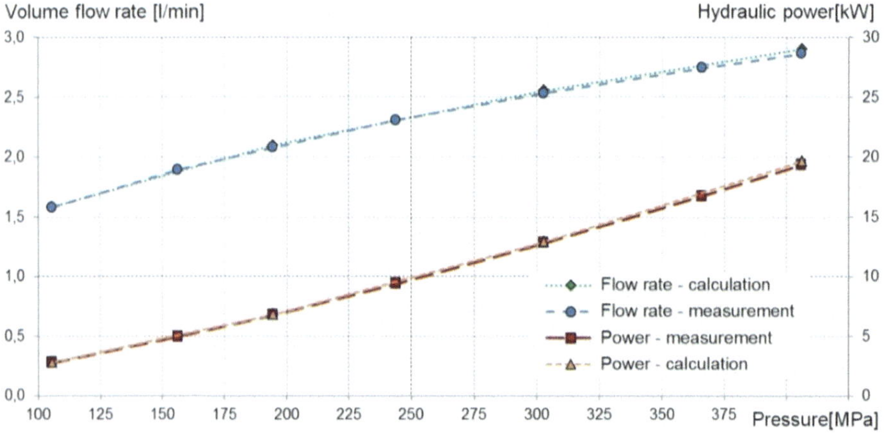

Fig. 8. Measured and calculated volume flow rates and hydraulic powers at the water nozzle inlet.

The results of the abrasive particle velocities at the outlet of the focusing tube calculated using PTV measurements and CFD models and are compared in Table 2 and the histograms are presented in Fig. 9.

It is obvious from the charts in Fig. 9 that a higher level of input water pressure accelerates the abrasive grains to higher velocities at the outlet of the focusing tube. The same trend is valid for the mean values of abrasive grains velocities listed in Table 2. The mentioned trend between the input water pressure and the abrasive grains velocities is valid for both measurements and CFD calculations. The measured values are distributed in a wider interval of velocities against the calculated values. Gauss distributions of the abrasive particle's velocities are obvious from the charts.

The results from the CFD numerical studies are presented for a selected combination of boundary conditions, 105 MPa, 194 MPa, 302 MPa, 406 MPa water input pressure and 100 g/min abrasive particles mass flow rate. Pressure conditions were selected based on the results of the flow and the pressure monitoring. The liquid velocity profile and the pressure distribution along the axis of the focusing tube are presented in Fig. 10. It is clear that the velocity and pressure distribution of the curves have a similar trend for

Table 2. Comparison of the mean values of abrasive particle velocities from PTV and CFD measurement p_a: mean value of the water pressure, N: number of abrasive grains, v_{avg}: mean velocity of abrasive grains, σ_M: standard deviation and σ_M standard error of the mean, 95% CI: 95% confidence interval of values, UL: upper limit, and LL: lower limit.

p_a [MPa]	PTV measurement					CFD calculations				
	N	v_{avg} [m/s]	σ [m/s]	σ_M [m/s]	UL LL [m/s]	N	v_{avg} [m/s]	σ [m/s]	σ_M [m/s]	UL LL [m/s]
105.558	1565	330.893	58.240	1.472	333.779 328.008	3057	296.885	87.974	1.591	300.004 293.766
194.464	2650	457.075	67.683	1.314	459.652 459.498	3060	395.734	118.067	2.134	399.918 391.550
302.722	1489	583.325	70.497	1.826	586.906 579.744	3054	473.679	143.181	2.591	478.758 468.600
406.165	2421	692.75	64.169	1.304	695.314 690.201	3054	562.359	166.590	3.014	568.268 556.449

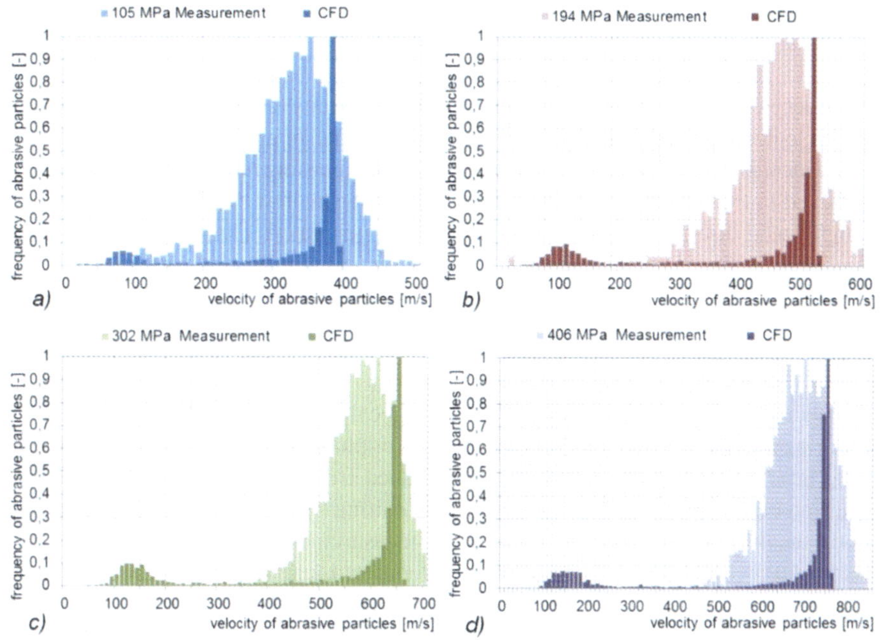

Fig. 9. Abrasive particle velocities at the focusing tube outlet, measurement vs. computational models for the given water pressure levels at 100 g/min abrasive particles mass flow rate: **a** 105 MPa; **b** 194 MPa; **c** 302 MPa; **d** 406 MPa.

all the tested pressures. The pressure increases over the length of the mixing chamber for all the tested levels.

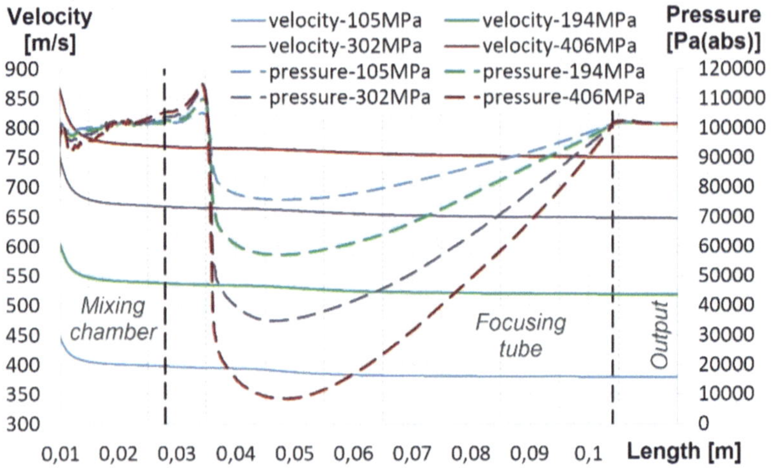

Fig. 10. Calculated velocity and pressure distribution along the axis of the focusing tube, 100 g/min abrasive.

The acceleration of abrasive particles causes the velocity magnitude of the high-speed water jet to decrease, while the velocity magnitude of the air and the abrasives increased with the increasing length of the cylindrical part of the focusing tube. The highest velocity is achieved in the water nozzle, more precisely in the smallest cylindrical space, where the high-speed liquid jet (at more than 800 m/s) is generated. The reason is the kinetic energy transfer from high-speed water jet to the air and the abrasive particles.

4 Conclusions

This paper presents a comparison between CFD calculations with high-pressure flow monitoring and PTV results in the study of the multiphase flow behaviour at the abrasive cutting head and at its outlet. The results obtained from the 3D numerical modelling of multiphase turbulent flow inside an abrasive cutting head are contrasted side by side with the results of the high-pressure water flow measurements. Four levels of input water pressure (105–406 MPa) were numerically computed and verified with experiments. The comparison of abrasive particle velocities based on numerical models and PTV measurements was performed for a constant flow rate of abrasive particles at 100 g/min and for the interval of particle diameters of 150–400 μm.

A very good agreement of the volume flow rates and of the input pressures given by measurement and CFD calculations was observed at the input of the pressurized water into the cutting tool. The numerical simulation of the flow provides information about the specific behaviour of multiphase mixture in the computational domain of the abrasive head. A water jet with high velocity is generated at the water nozzle. The velocity of the water jet gradually decreases while the length increases due to kinetic energy transfer to the air and to the abrasive particles (see Fig. 10). The water jet flowing through the axis of the cutting tool causes vacuum in the mixing chamber and at the inlet duct of air-abrasives mixture. Then, the automatic suction of abrasive particles with air is possible. In the mixing chamber and focusing tube, the water jet accelerates the mixture. The abrasive water jet is then generated. The data calculated by the presented CFD simulations validate this well-known cutting tool behaviour. The calculated shape of the high-speed water jet does not significantly change with an increasing length, even in the focusing tube, in contrast to the experiment presented in Fig. 7 and the presumptions mentioned by Osman et al. [24]. The calculated pressure distribution in the axis of the cylindrical part of the focusing tube corresponds to the pressure distribution [24]. The pressure increases with increment of the focusing tube length. Measurement of the input and output parameters combined with the numerical results of flow in the abrasive head, provides a unique information on the behaviour of the mixture flowing in the abrasive cutting head. The created numerical model can be further used for tool geometry optimization.

The next research will focus on the elimination of the abrasive particle degradation caused by their collisions with the inner walls of the cutting tool. In other words, the contact between abrasive particles and the surrounding walls should be avoided. It is expected that the elimination of this negative effect will positively influence in the cutting performance and in the life of the abrasive head.

Acknowledgement. This work was supported by the Czech Science Foundation under grant project 23-07415S.

References

1. Valicek, J., Cep, J. R., Rokosz, K., Lukianowicz, C., Kozak, D., Zelenak, M., Kostial, P., Hloch, S., Harnicarova, M., Hlavacek, P., Haluzikova, B.: New way to take control of a structural grain size in the formation of nanomaterials by extrusion. Materialwissenschaft Und Werkstofftechnik (2012). https://doi.org/10.1002/mawe.201200973
2. Hreha, P., Hloch, S., Monka, P., Monková, K., Knapčíková, L., Hlaváček, P., Zeleňák, M., Samardžić, I., Kozak, D.: Investigation of sandwich material surface created by abrasive water jet (AWJ) via vibration emission. Metalurgija **53** (2014)
3. Swanson, R.K., Kilman, M., Cerwin, S., Tarver, W., Wellman, R.: Study of particle velocities in water driven abrasive jet cutting (1987)
4. Lo, M., Henning, A., Hay, K., Miles, P.: Particle velocity in abrasive waterjets. In: Lecture Notes in Mechanical Engineering (2021). https://doi.org/10.1007/978-3-030-53491-2_15
5. Isobe, T., Yoshida, H., Nishi, K.: Distribution of abrasive particles in abrasive water jet and acceleration mechanism. In: Proceedings of 9th international symposium on jet cutting technology, pp. 217–238 (1988)
6. Haghbin, N., Khakpour, A., Schwartzentruber, J., Papini, M.: Measurement of abrasive particle velocity and size distribution in high pressure abrasive slurry and water micro-jets using a modified dual disc anemometer. J. Mater. Process. Technol. (2019). https://doi.org/10.1016/j.jmatprotec.2018.08.014
7. Li, H.Y., Geskin, E.S., Chen, W.L.: Investigation of forces exerted by an abrasive water jet on a workpiece (1989)

8. Neusen, F.K., Gores, T.J., Amano, R.S.: Axial variation of particle and drop velocities downstream from an abrasive water jet mixing tube. In: Proceedings of the 12th International Conference on Jet Cutting Technology (1994)
9. Neusen, K.F., Gores, T.J., Labus, T.J.: Measurement of particle and drop velocities in a mixed abrasive water jet using a forward-scatter LDV system (1992). https://doi.org/10.1007/978-94-011-2678-6_5
10. Balz, R., Heiniger, K.C.: Determination of spatial velocity distributions of abrasive particles in abrasive water jets using laser-induced fluorescence under real conditions. In: Proceedings of 16th WJTA-IMCA Conference and Expo (2011)
11. Roth, P., Looser, H., Heiniger, K.C., Bühler, S.: Determination of abrasive particle velocity using laser-induced fluorescence and particle tracking methods in abrasive water jets. In: WTJA American Waterjet Conference (2005)
12. Zeleňák, M., Foldyna, J., Linde, M., Pude, F., Rentsch, T., Fernolendt, J., Poort, H.U.: Measurement and analysis of abrasive particles velocities in AWSJ. In: Procedia Engineering (2016). https://doi.org/10.1016/j.proeng.2016.06.641
13. Riha, Z., Zelenak, M., Soucek, K., Hlavacek, P.: Flow field analysis inside and at the outlet of the abrasive head. Materials **14** (2021). https://doi.org/10.3390/ma14143919
14. Jerman, M., Zeleňák, M., Lebar, A., Foldyna, V., Foldyna, J., Valentinčič, J.: Observation of cryogenically cooled ice particles inside the high-speed water jet. J. Mater. Process. Technol. 116947 (2020). https://doi.org/10.1016/j.jmatprotec.2020.116947
15. Liu, H., Wang, J., Kelson, N., Brown, R.J.: A study of abrasive waterjet characteristics by CFD simulation. J. Mater. Process. Technol. 153–154 (2004). https://doi.org/10.1016/j.jmatprotec.2004.04.037
16. Prisco, U., D'Onofrio, M.C.: Three-dimensional CFD simulation of two-phase flow inside the abrasive water jet cutting head. Int. J. Comput. Methods Eng. Sci. Mech. **9** (2008). https://doi.org/10.1080/15502280802225598
17. Yanaida, K.: Flow characteristics of water jets. In: Proceedings of the Second International Symposium on Jet Cutting Technology. BHRA Fluid Engineering, Cambridge (1974)
18. Zelenak, M., Riha, Z., Soucek, K., Pude, F.: Analysis of micro continuous water jet based on numerical modelling and flow monitoring (2019). https://doi.org/10.1007/978-3-319-99353-9_16
19. Zelenak, M., Riha, Z., Fech, A., Soucek, K., Pude, F., Hlavacek, P., Kevin Miranda, F.: Theoretical and practical research on micro hydrodynamic nozzles for modulated water jet generation as an innovation to replace continuous water jet in soft materials erosion. Measurement **221**, 113530 (2023). https://doi.org/10.1016/j.measurement.2023.
20. Riha, Z., Zelenak, M., Nag, A., Poloprudsky, J., Kruml, T., Hloch, S.: A study of the erosion characteristics of an EN AE-6060 aluminium alloy processed using middle and high power continuous and modulated water jets, 205154 (2024). https://doi.org/10.1016/j.wear.2023.205154
21. Menter, F.R.: Two-equation eddy-viscosity turbulence models for engineering applications. AIAA J. **32** (1994). https://doi.org/10.2514/3.12149
22. Long, X., Liu, Q., Chen, Z., Xue, S.: Numerical study of the particle movement in the abrasive entrained waterjet nozzle at 300 MPa. In: 14th Asian International Conference on Fluid Machinery (2017)
23. Long, X., Ruan, X., Liu, Q., Chen, Z., Xue, S., Wu, Z.: Numerical investigation on the internal flow and the particle movement in the abrasive waterjet nozzle. Powder Technol. **314** (2017). https://doi.org/10.1016/j.powtec.2016.09.089
24. Osman, A.H, Mabrouki, T., Théry, B., Buisine, D.: Experimental analysis of high-speed air-water jet flow in an abrasive water jet mixing tube. Flow Meas. Instrum. **15** (2004). https://doi.org/10.1016/j.flowmeasinst.2003.08.001

The Effect of Multiple Impacts of Water Droplets on the Integrity of CFRP Composite

Alice Chlupová[1](✉), Jakub Poloprudský[1], Akash Nag[2], and Sergej Hloch[2,3]

[1] Institute of Physics of Materials, Czech Academy of Sciences, Brno, Czechia
chlupova@ipm.cz
[2] VŠB—Technical University of Ostrava, Ostrava, Czechia
[3] Institute of Geonics, Czech Academy of Sciences, Ostrava, Czechia

Abstract. The paper deals with the surface erosion of a carbon fiber reinforced polymer (CFRP) due to the multiple impacts of water droplets. The innovative technology of pulsating water jet (PWJ) was used for the study, where a continuous water jet was split into discrete clusters of water droplets. Increasing exposure time was used to analyze the evolution of erosion over time. Erosion exposure increased from 2 s to 15 s in 1-s increments. The hydraulic parameters were as follows: pressure 20 MPa, water jet speed 184 m/s, nozzle diameter of 0.2 mm. Resulting clusters of water droplets hit the surface of the composite with a frequency of 40 kHz. The depth of the erosion craters was assessed using a confocal microscope and the mechanism of erosion was observed using a scanning electron microscope. The work aimed to carry out preliminary research dealing with the erosion development of erosion and determining the erosion resistance of the CFRP composite.

Keywords: Erosion · Pulsating Water Jet · CFRP · Confocal Microscope · Scanning Electron Microscope

1 Introduction

The erosion of materials caused by the repeated impact of water drops still represents a practical and theoretical problem. The issue has been intensively studied since the first rotary blades were used in turbochargers. Even engineered multi-component materials with high strength-to-weight ratios, such as CFRP composites, used as wind turbine blades, are subject to erosion damage. It significantly damages the aerodynamic shape of the blades, i.e., their leading edges and reduces the performance of the wind power plant [1]. To eliminate energy losses, it is very important to investigate this issue, i.e., to observe the erosion footprints and the prevailing deformation mechanisms in the form of impact pressure and subsequent lateral flow.

The issue of erosion of wind turbine blades is gaining importance as the pressure on the use of renewable resources and the efficiency demands increase. Higher efficiency is often related to the extension of turbine blades and increasing rotational speed. The result is a higher speed of falling drops, which leads to accelerated erosion. Droplet

L. Sitek et al. (Eds.): Advances in Water Jetting II, LNME, pp. 33–42, 2025.
https://doi.org/10.1007/978-3-031-72778-8_3

velocity as a function of blade speed at different rain intensities was modelled by Verma et al. [2]. The intensity of the rain was divided into 4 categories from light rain, through average and heavy rain, to very heavy rain with a raindrop speed from 4.9 m/s through 6.4 and 7.1 m/s to 7.7 m /s. Combined with a blade speed of 140 m/s, the resulting drop impact speed can reach up to 148 m/s in very heavy rain.

The effect of droplet impact on carbon fibre-reinforced polymer (CFRP) composites was reported in the work of N. L. Hancox [3], where the erosion was divided into several stages as "fibre removal", "the appearance of small erosion craters" and "larger erosion craters".

One of the methods suitable for the formation of discrete water droplets is the pulsating water jet (PWJ) method described by Foldyna [4]. In this technique, a continuous water jet is interrupted using a sonotrode causing a hydrodynamic impingement [5]. The periodic segmentation of the water jet causes increased pressure on the surface of the solid at the moment of initial contact with the liquid phase. The resulting pressure is several times higher than the pressure created by a continuous water jet [6]. The stagnation pressure $\boldsymbol{p_s}$ of the continuous jet is given by Bernoulli's equation:

$$p_S = \frac{1}{2}\rho v^2 \tag{1}$$

where ρ is the fluid density, and v is the impact velocity. In contrast, the impact pressure p_{WH} known as the "water hammer effect" can be written as:

$$p_{WH} = \rho C v \tag{2}$$

where C is the speed of sound in the liquid. The radius of the affected area r can be calculated as follows:

$$r = Rv/C \tag{3}$$

where R is the diameter of the interacting water droplet. The duration of impact pressure T is then:

$$T = \frac{R}{v}\left[1 - \sqrt{\left(1 - \frac{v^2}{C^2}\right)}\right] \tag{4}$$

and when the impact pressure duration ends, the stagnation pressure stabilizes.

In this paper, the integrity of CFRP composite exposed to water droplets impingement was studied using a technique of pulsating water jet. Preliminary results obtained during the simulation of weather conditions leading to erosion of the composite used for rotor blades of wind turbines are presented. Based on microscopic observations, the influence of multiple impacts of water droplets on the depth and variability of the erosion stages depending on exposure time is identified. Erosion mechanisms of CFRP will also be discussed.

2 Experiment and Methods

The experimental material was a symmetric CFRP composite with eight layers, where the carbon fibres were laid at angles in the sequence [45, 0, 90, −45] (see Fig. 1a).

The surface of the CFRP composite was exposed to the periodic impact of water clusters using an ultrasonically modulated water jet (see Fig. 1b, c). Pressurised water controlled by a HAMMELMANN HDP 253 hydraulic pump (pressure up to 250 MPa and flow rate 40 l/min) enters the acoustic chamber. The sonotrode connected to the ECOSON WJ-UG-630–40 ultrasonic generator oscillates in the acoustic chamber with a frequency of 40 kHz. It causes pressure waves in a compressed liquid. As the pressurized liquid passes through the exit nozzle, the pressure waves transform into velocity waves. Inhomogeneity in the water velocity profile leads to the disintegration of the jet into discrete clusters of water [7]. The movement of the nozzle head over the sample was performed using a robotic arm of the ABB IRB 6640–180/2.55 robot.

As an initial approximation of the erosive effect of rain, the impact velocity in the PWJ experiments was set to 184 m/s, which is worse than heavy rain [2] to be sure that certain erosion will be reached. The impact velocity in future experiments may be reduced to values corresponding to the droplet impact velocity in operational weather conditions. Increasing exposure time was used to analyse the evolution of erosion damage with the increasing number of impingements. The exposure time can be converted to the number of water clusters hitting the surface as a function of frequency.

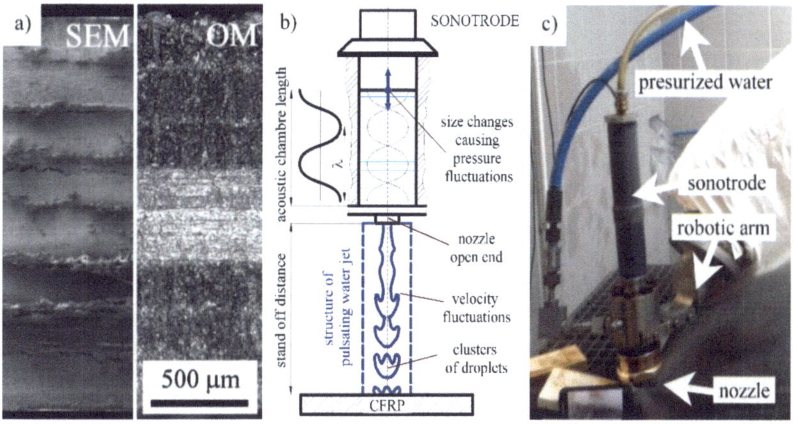

Fig. 1. Experimental setup: **a** micrographs of CFRP cross-section of, **b** diagram of sonotrode generating water clusters, **c** experimental setup during PWJ interaction with the surface of the composite.

Technological and hydraulic parameters of the process are shown in Table 1. During the experiment, all parameters were set constant except for time exposure. It varied in steps with an interval of 1s. Each step was repeated 3 times, i.e., 3 eroded areas were obtained at each condition. The distance between individual erosion areas was at least 8 mm.

The appearance of erosion stages was observed, and the erosion mechanism was evaluated using a Tescan LYRA 3 XMH FEG/SEMxFIB scanning electron microscope (SEM) working at a low accelerating voltage of 2kV.

The shape of erosion, i.e., depth and profiles of selected craters, were evaluated using OLYMPUS DSX1000 optical microscope (OM). The maximum width and depth were evaluated at the centre of the crater in a direction perpendicular to the direction of fibres in the first layer of the composite.

Table 1. Technological and hydraulic parameters of the PWJ experiment.

Technological parameters		Hydraulic parameters	
Repetition	3 [-]	Pressure	20 [MPa]
Working distance (SOD)	11 [mm]	Nozzle diameter	0.2 [mm]
Acoustic chamber length	6 [mm]	Liquid density	1007 [kg.m^{-3}]
Sonotrode frequency	40 [kHz]	Flow*	$5.66 \cdot 10^{-6}$ [m^3.s^{-1}]
Sonotrode power	112 [W]	Jet speed	184 [m.s^{-1}]
Exposure time	2–15[s]	Hydraulic power	116 [W]
Time increment	1 [s]	Pulse volume	0.145 [mm^3]

*Nozzle efficiency 0.92

3 Results and Discussion

An experimental series of areas of the eroded upper surface of CFRP were created with a gradually increasing exposure time from 2 s to 15 s in 1 s increments. Exposure time from 2 s to 15 s can be, depending on the frequency, converted to the number of water clusters hitting the surface ranging from 80,840 impacts to 606,300 impacts.

Micrographs in Fig. 2 show examples of erosion observation using SEM. The shape of the erosion kerf was formed in a direction parallel to the direction of the upper layer fibres. A significant material removal was recorded after only 2 s of exposure, corresponding to 80,840 impacts (see Fig. 2a). The cover polymer layer with a chess-board pattern (an imprint from the material´s manufacture) was eroded, creating a relatively narrow but deep erosion kerf, leaving many intact fibres uncovered. The application of PWJ for 5 s resulted in the removal of a wider area of a protective polymer layer. Many fibres were broken and carried away by water flow, and a narrow area of a second fibre layer appeared at the bottom of the erosion crater (Fig. 2b). After 10 s, the width of the erosion kerf increased (Fig. 2c). The area of uncoated carbon fibres in a first layer was relatively flat as well as the increased area of intact fibres from the second layer of fibres. Many broken fibres are present at the long ends of the crater. Exposure to PWJ for 15 s (Fig. 2d) resulted in further lengthening, deepening and widening of the crater, i.e., an extensive erosion pit was formed. A fracture of the upper polymer layer was visible along one edge of the crater (right side of the crater) and steep deepening of the crater from the first to the second fibre layer was observed.

Details of the CFRP damage mechanism at different magnifications and various PWJ exposure times are in Fig. 3.

Figure 3a shows the initial stages of CFRP erosion when the polymer surface covering the first layer of fibres was removed (2 s exposure). Unlike the erosion of metallic materials published in work [8], the initial stage reached in our experiment after 2 s PWJ exposure is already significantly larger than the incubation stage defined in [5].

Exposure for 4 s (see Fig. 3b) resulted in more extensive fracture damage of carbon fibres. The fracture failure mechanism of fibres with a relatively straight fracture surface indicating brittle fracture of high-strength carbon fibres loaded mainly in flexure is captured there. When carbon fibres are subjected to tensile loading [9] the fracture surface usually contains the origin of the crack (initiation site) and then diverging lines resulting from crack propagation on the fibre fracture surface.

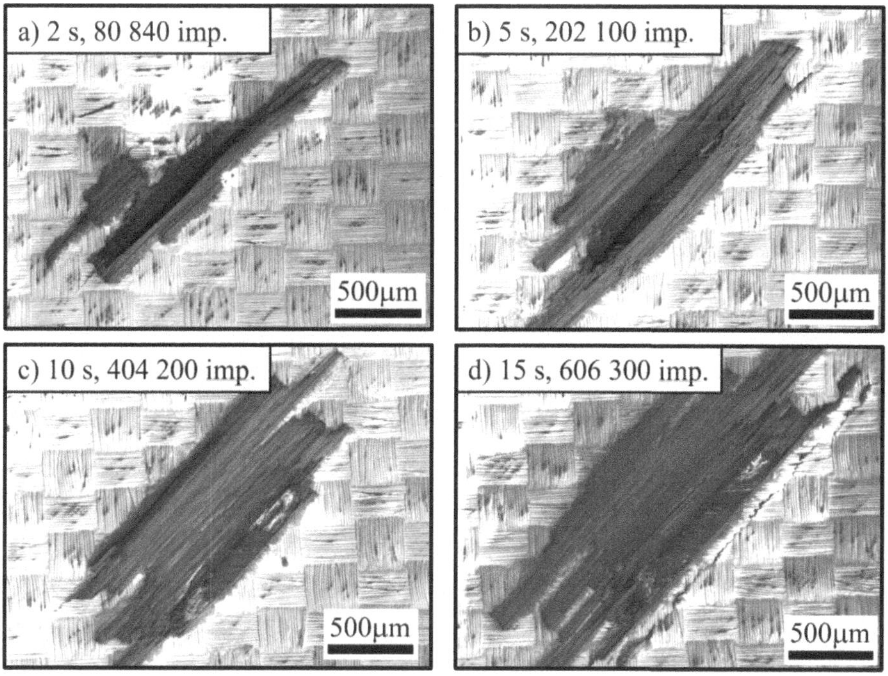

Fig. 2. SEM micrographs of the eroded surface of CFRP after exposition to PWJ for 2 s, 5 s, 10 s and 15 s.

The quality of the interface between fibres and polymer matrix can be assessed, similarly, as in [10]. In Fig. 3a and b is possible to observe many fragments of the epoxy matrix still attached to the fibres, indicating a perfect adhesion between fibres and the matrix. Moreover, the rough fracture surface of the polymer matrix is free of signs typical for brittle fracture (flat shiny areas with river lines [11]).

Figure 3c shows a crack propagating in a thin layer of polymer covering the surface of CFRP composite from the impact centre to kerf ends in the fibre direction. This effect could be observed from 4 s exposure to PWJ.

Application of PWJ for approximately 5 s led to a deepening of erosion up to the second fibre layer, as seen in Fig. 3d, where a small area of fibres running horizontally was detected.

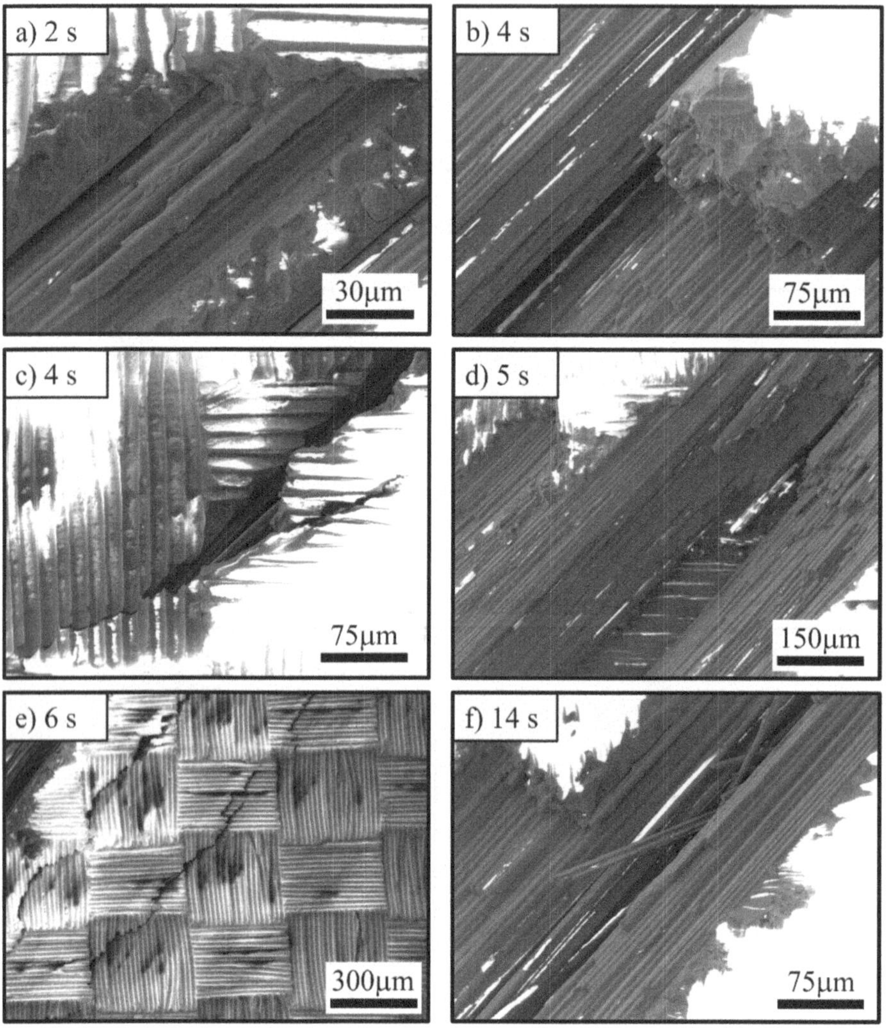

Fig. 3. SEM micrographs of the details of the mechanism of CFRP damage at various PWJ exposure times: **a** removal of the polymer surface, **b** fracture of fibres, **c** growth of the crack in polymer, **d** appearance of the second layer of fibres, **e** incorporation of broken fibres into erosion kerf.

The surface in the vicinity of the erosion crater after exposure up to 5 s showed mild stochastically distributed damage, but at higher exposure times massive cracking of the polymer surface also appeared (see Fig. 3e).

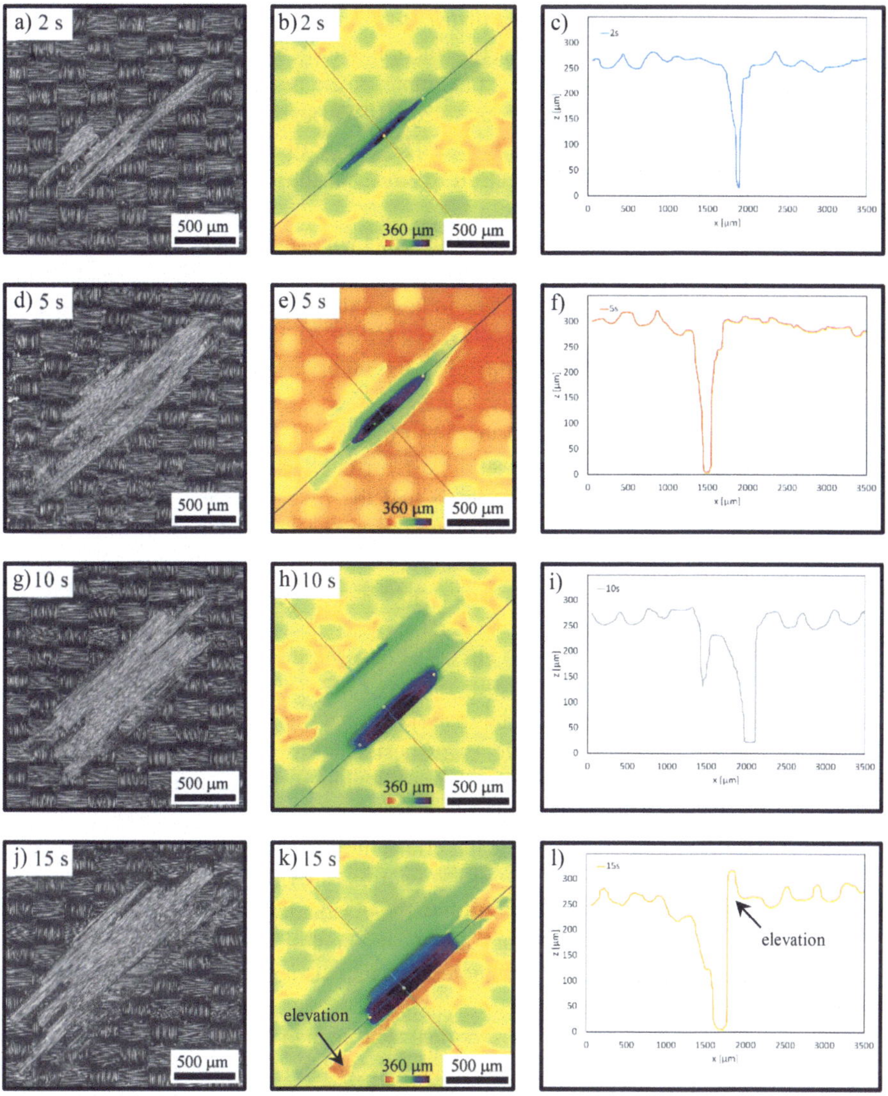

Fig. 4. Results of OM observation and evaluation of the eroded surface of CFRP after exposition to PWJ for 2 s (a, b, c), 5 s (d, e, f), 10 s (g, h, i) and 15 s (j, k, l). In the first column are OM micrographs, the second column shows heat maps as 2D reconstruction of erosion depth, and the third column presents results of measured kerf profiles.

The massive devastation of CFRP during the application of PWJ for longer exposure time (14 s) is demonstrated by the incorporation of broken fibres into erosion kerf (see Fig. 3f). Fibres embedded in the cracks can further accelerate the crack propagation and enlarge mass loss of CFRP composites [10].

CFRP erosion and surface damage distributed in the vicinity of the kerf were also observed using OM, allowing depth reconstruction and measurement of depth profiles. The data were further processed and an example of the results of OM observation of erosion after exposition to PWJ for four selected PWJ exposure times (2 s, 5 s, 10 s and 15 s, respectively) is shown in Fig. 4. The shape of erosion kerf exhibited not only directionality related to the direction of fibres in the first layer but a very steep ascent/descent angle of was also notable.

In the first column in Fig. 4 are OM micrographs with a large depth of focus (Fig. 4a, d, g, j). The second column (Fig. 4b, e, h, k) shows a 2D reconstruction of erosion depth for selected exposure times. The blue line indicates the direction of the erosion kerf in relation to the fibre direction in the first layer of CFRP. The red line represents the line for evaluation of the depth profile. In the third column, there are individual depth profiles of erosion kerfs (Fig. 4c, f, i, l).

Significant erosion with a maximum depth of around 250–300 μm was achieved after only 2 s exposure. As the exposure time increased, the maximum crater depth remained approximately the same. It means that the first fibre layer with a covering layer of polymer was eroded, and the bottom of the narrow kerf reached the second layer of fibres. Increasing exposure, nevertheless, affected the width of the erosion kerf and occasionally, next to the primary erosion kerf also secondary kerf appeared (see Fig. 4h and i for exposure to PWJ 10 s).

Depth profiles gave us information not only concerning the kerf depth and width but also shape. In addition to the wavy profile of the CFRP surface, it was also possible to observe a certain lifting (pile-up) of the material above the level given by the plane of the original material surface. This effect was best seen in Fig. 4k where the upheaval of material is depicted by the red band on the right side of the erosion kerf (marked by arrow). The elevation of material on the right side of the profile is visible also in Fig. 4l. In the case of PWJ treatment of metallic materials, the material pile-up is usually related to plastic deformation [11]. In the case of PWJ action of CFRP composite, the situation needs to be studied more deeply. In addition to the plastic deformation, the other effects can also act in CFRP composites as delamination or polymer swelling [10] which can be related to the infiltration of water molecules increasing the distance between the molecules of polymer.

The plot in Fig. 5 shows the cross-sectional (CS) area evaluated from depth profiles of the crater below the basal plane of CFRP for selected PWJ exposures. The data show an increase in the CS area with increasing exposure time, however, at higher exposure the trend shows saturation. It can be assumed that the saturation is related to reaching the second layer of fibres and a further increase in CS area will occur if the erosion crater deepens into the next layer of fibres.

4 Conclusions

The following conclusions can be drawn from the results of the research dealing with the erosion development and determination of the erosion resistance of the CFRP exposed to a PWJ technology:

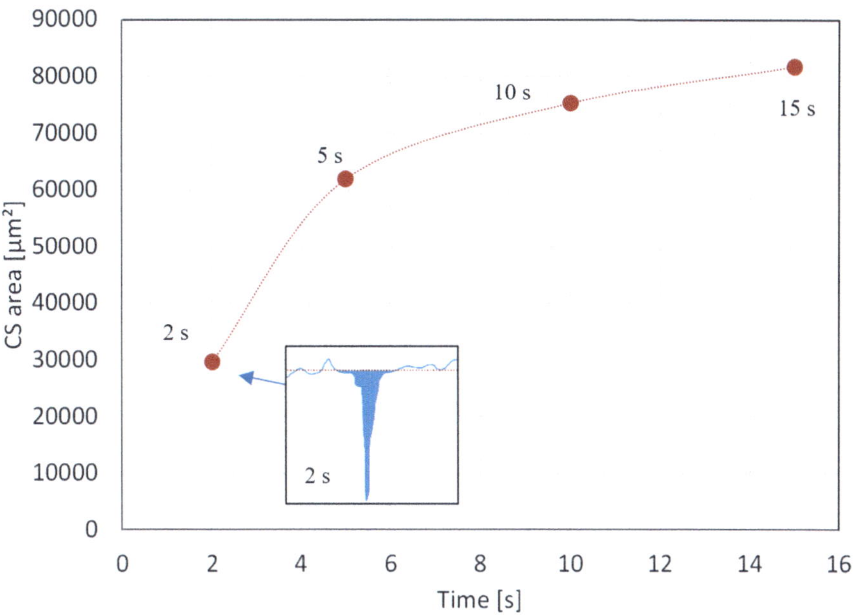

Fig. 5. Dependence of cross-sectional area on exposure time.

- During the experiment with a flow speed of 184 m/s and a frequency of 40 kHz, the speed of raindrop impacts on wind turbine blades, higher than very heavy rain, was simulated. The technology appears suitable for accelerated erosion tests.
- The experimentally applied conditions resulted in erosion damage of both the covering polymer and first fibre layer just after only 2 s of exposure, i.e., after 80,840 impacts reaching the maximum depth of around 250–300 μm.
- The shape of erosion is related to the direction of the first fibre layer.
- The increasing exposure time resulted in crater widening and deepening to further fibre layers.

Acknowledgement. The work was created with the financial support of the GAČR project 23-05372S.

References

1. Keegan, M.H., et al.: Modelling rain drop impact on offshore wind turbine blades. ASME Turbo Expo 2012, Article GT 2012–69175 (2012)
2. Verma, A.S., et al.: Numerical investigation of rain droplet impact on offshore wind turbine blades under different rainfall conditions: A parametric study. Compos. Struct. **241**, 112096 (2020)
3. Hancox, N.L.: The erosion of carbon fibre reinforced plastic by repeated liquid impact. Wear **23**, 71–81 (1973)

4. Foldyna, J., et al.: Utilization of ultrasound to enhance high-speed water jet effects. Ultrason. Sonochem. **11**, 131–137 (2004)
5. Hloch, S., et al.: Effect of pressure of pulsating water jet moving along stair trajectory on erosion depth, surface morphology and microhardness. Wear **452**, 203278 (2020)
6. Říha, Z., et al.: Comparison of the disintegration abilities of modulated and continuous water jets. Wear **478**, 203891 (2021)
7. Zelenak, M., et al.: Visualisation and measurement of high-speed pulsating and continuous water jets. Measurement **72**, 1–8 (2015)
8. Poloprudský, J., et al.: Surface and subsurface analysis of stainless steel and titanium alloys exposed to ultrasonic pulsating water jet. Materials **14**, 5212 (2021)
9. Srinivasa, V., et al.: Fracture morphology of carbon fiber reinforced plastic composite laminates. Mater. Res. **13**, 417–424 (2010)
10. Fang, M., et al.: Effects of hydrothermal aging of carbon fiber reinforced polycarbonate composites on mechanical performance and sand erosion resistance. Polymers **12**, 2453 (2020)
11. Cholake, S.T., et al.: Improved Mode I fracture resistance of CFRP composites by reinforcing epoxy matrix with recycled short milled carbon fibre. Constr. Build. Mater. **111**, 399–407 (2016)

Measurement of Impact Pressure of the Water Jet

Josef Foldyna(✉) and Michal Zeleňák

Institute of Geonics of the Czech Academy of Sciences, Studentská1768, 708 00 Ostrava, Czechia

josef.foldyna@ugn.cas.cz

Abstract. Assessing the quality and behavior of the water jet produced by the nozzle can provide indirectly information about the quality and condition of the nozzle. There are many different ways of selecting the best nozzle for a given application or comparing the performance of several nozzles. A special method for measuring the impact pressure of a high velocity water jet was developed at the Institute of Geonics of the Czech Academy of Sciences. The impact pressure of the jet generated by the nozzle is measured using a specially designed sensor. The paper presents results of measurements of impact pressure of the jet generated by two types of nozzles - a cylindrical nozzle and a flat jet nozzle – using two modes of measurement: in lines and point-by-point. The performed experiments and results showed that both methods are fully consistent and suitable for assessing the quality and condition of a given nozzle. The in-line measurement method is preferable for routine nozzle quality determination. It is less time-consuming and provides more detailed information on the distribution of the impact pressure across the jet. The point-by-point method allows a more accurate analysis of the dynamic behavior of the jet at individual points in the cross-section of the jet.

Keywords: Impact Pressure · High-Speed Water Jet · Round Jet · Flat Jet

1 Introduction

The advent of high-speed waterjet technology has revolutionized many industrial processes, offering unprecedented precision and efficiency. Despite its widespread adoption, a comprehensive understanding of the factors influencing the quality of high-speed waterjet cutting remains elusive. This paper aims to fill this knowledge gap by presenting some of the results of a systematic investigation into measuring the quality of high-speed waterjets, which is a critical determinant of their performance and effectiveness, affecting the accuracy, speed and overall efficiency of operations [1–3].

Nozzles used to produce high velocity water jets are widely used in cutting, cleaning, deburring and trimming applications. In all of these applications, a good understanding of the characteristics of the specific nozzle is essential to designing the optimum toolpath. Determining the quality of a nozzle in high-speed waterjet systems involves several methods, each focusing on different aspects of the nozzle's performance. Visual

L. Sitek et al. (Eds.): Advances in Water Jetting II, LNME, pp. 43–59, 2025.
https://doi.org/10.1007/978-3-031-72778-8_4

inspection is the simplest method, where the nozzle is examined for visible signs of wear or damage, checking for cracks, deformation or changes in orifice size. By measuring the flow rate of water through the nozzle, one can infer its condition - a lower than expected flow rate could indicate a blockage or wear in the nozzle. Some advanced methods involve analyzing the sound produced by the jet, where changes in the acoustic signature can indicate changes in nozzle condition.

Assessing the quality and behavior of the water jet produced by the nozzle can indirectly provide information about the quality and condition of the nozzle. By choosing the right nozzle, it is possible to avoid significant energy losses in the conversion of high pressure water into a jet, achieve higher equipment performance and thus contribute to better economic indicators in the use of waterjets in a wide variety of applications.

The paper presents a method for measuring the impact pressure of high-speed water jets generated by different types of nozzles. The measurement results provide information about the nozzle quality and the dynamic behavior of the high-speed water jet. This could potentially lead to a significant improvement in performance and quality in the implementation of specific work with high-speed water jets, as well as to a better understanding of the processes of jet formation.

2 Background

Over the years, several methods have been developed to select the best nozzle for a given application or to compare the performance of several nozzles. They differ in cost, efficiency or time required to implement them.

The simplest way is to perform test cuts in standard material at constant jet parameters. This is a comparative evaluation, comparing depths of cut and/or removed volumes. The advantage of this method is its simplicity and the possibility of a direct comparison between two different nozzles. The disadvantage is that it does not provide qualitative information about the nozzle (it does not allow determining the reason why one nozzle is better or worse than another).

Another method is use of a jet visualization. However, this method requires an illumination device with sufficient intensity and a very short flash time (only a few ns) for the visualization results to be sufficiently informative. A correctly performed jet visualization will allow the determination of the jet decay length and coherence, factors that significantly affect its performance [4]. The disadvantages of this method are the impossibility of comparing the performance of individual jets and relatively high cost of the measurement.

The measurement of the pressure or force effects of the jet at the point of impact on the material represents another method for assessing the quality and behavior of the water jet. The impact pressure of a waterjet is a function of several parameters, including the velocity of the jet, the density of the fluid, and the angle at which the jet strikes the surface. However, the complex interplay of these factors makes the measurement of impact pressure a non-trivial task. Traditional methods of pressure measurement, such as the use of pressure transducers, often fall short in accurately capturing the dynamic nature of waterjet impact pressure.

Wright [5] conducted testing to measure the impact forces produced by the jet and determine the rate of deterioration of the impact force with increasing standoff distance using impact plate mounted to a load cell, at which a high-speed waterjet was directed.

Leach & Walker used a block of hardened silver steel with a hole 0.3 mm diameter and approximately 2.5 mm deep, behind which a pressure transducer was mounted (see Fig. 1) to determine the stagnation pressure on the jet axis as a function of standoff distance [6]. Barker & Selberg used the method to measure the transverse pressure profile of the jet at various distances from the nozzle [7].

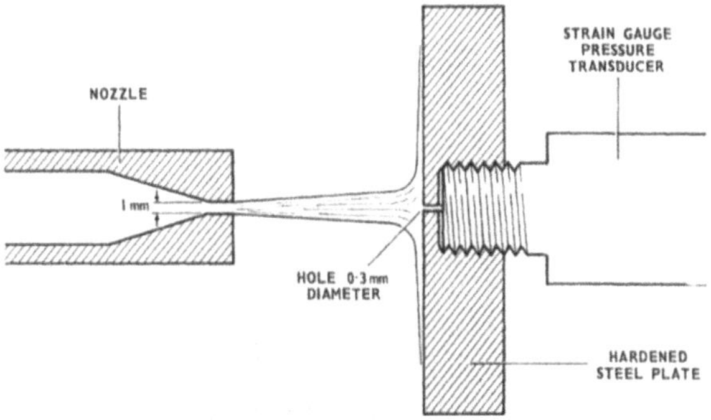

Fig. 1. Apparatus for the measurement of stagnation pressure of the jet [6].

In recent years, advancements in sensor technology and computational fluid dynamics (CFD) have opened up new avenues for the measurement of waterjet impact pressure. Therefore, an improved method of Leach and Walker for measuring the impact pressure of a high velocity water jet was developed at the Institute of Geonics of the Czech Academy of Sciences.

3 Experimental Procedure and Equipment

The impact pressure of the jet generated by the nozzle is measured using a specially designed sensor (see Fig. 2). The sensor is equipped with a piezoelectric pressure transducer Kistler 6229A. Basic technical data of the transducer are given in Table 1. The charge signal is amplified and converted into proportional voltage signal by a charge amplifier. The impacting jet acts on the pressure transducer through the orifice with a diameter of 0.2 mm. In order to scan the entire transverse jet profile at a given standoff distance, the nozzle moves over the sensor according to the diagram in Fig. 3a).

The experimental facility used for measuring the impact pressure of the jet consists of a high-pressure water supply system, a robot for manipulating the nozzle during tests, and the measurement equipment.

High-pressure water is supplied to the nozzle by a Hammelmann HDP 253 high-pressure pump capable of delivering up to 67 l/min of water at an operating pressure up to 160 MPa. The nozzle is manipulated using the ABB IRB 6640-180/2.55 Master robot.

Table 1. Basic technical data of quartz high-pressure sensor Kistler 6229A.

Pressure range	0 … 500 MPa
Sensitivity	-0.25 pC/MPa
Natural frequency	>200 kHz
Rise time	1 μs
Linearity (for all ranges)	≤ ±1%FSO
Operating temperature range	-50 … 200 °C

Fig. 2. Sensor for measuring the impact pressure of the jet.

Acquisition and processing of data from the sensor are performed using an USB-6251 multifunction I/O device (National Instruments) and a LabVIEW graphical programming environment (National Instruments). Graphical outputs are processed also in MountainsSPIP software.

The paper presents results of measurements of two types of nozzles - a cylindrical nozzle with a diameter of 2.08 mm (StoneAge Attack Tip 082) and a flat jet nozzle with an equivalent diameter of 2.05 mm and a spray angle of 15° (Lechler 15-08).

Two modes of measurement were used in experiments – measurement in lines and measurement point-by-point.

3.1 Measurement in Lines

Operating pressure was set to 30 MPa, the nozzle moved over the sensor at standoff distances of 10, 30 and 50 mm. The traversing velocity of the nozzle in the x-direction was set to 1 mm/s. The length of movement in the x-axis direction was set individually for each standoff distance to be approximately twice the jet footprint. After each pass, the nozzle moved 0.25 mm in the y-direction. Since the measurements were taken every second pass, the actual distance (offset) between measurements was 0.5 mm (see Fig. 3b). The measurement was terminated after the entire cross section of the jet has been scanned at given standoff distance. During the measurement, the signal from the pressure sensor was recorded at a sampling rate of 200 ksamples/s, which corresponds to 20 ksamples/mm.

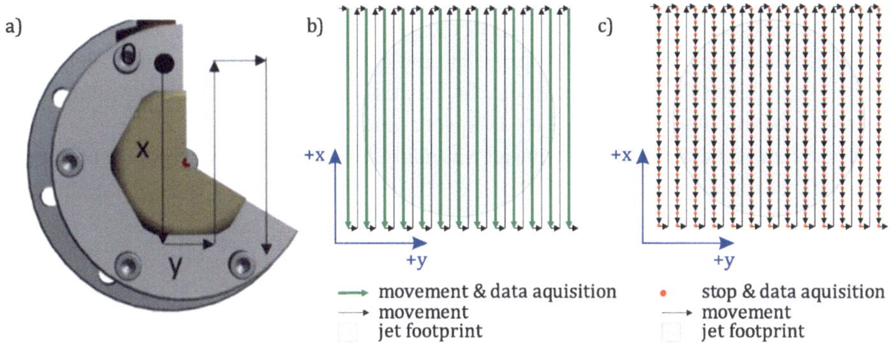

Fig. 3. a) Scheme of the movement of the waterjet over the sensor. b) Measurement in lines. c) Point-by-point measurement.

The measured data were processed in the graphical programming environment NI Lab-VIEW using a user-developed program. Because the high sampling rate of the pressure sensor signal led to acquiring a large number of data points for each measured line, the data points had to be compressed into a smaller number of points using Sample Compression Express VI (VI stands for virtual instrument) to be able to render the recordings as 3D surface plots and intensity graphs. The reduction factor was set to 100, i.e. 100 points at the input to the VI is reduced to one point in the resulting signal using the mean value of the points in each segment. Subsequently, the processed data were transferred to MountainsSPIP software, where 3D profiles of the impact pressure levels, plots of an impact pressure intensity in a plane perpendicular to the water jet axis, and impact pressure profiles in two mutually perpendicular slices passing through the jet axis were generated. While the impact pressure profile in the x-axis direction is represented by processed data from a single measured line, the impact pressure profile in the y-axis direction consists only of individual points spaced 0.5 mm apart (the number of points corresponds to the number of measured lines).

In order to be able to evaluate the dynamic behavior of the waterjet at different distances from the nozzle, selected samples of "rough" data 0.25 s long were extracted from recordings of measured impact pressure and the dynamic nature of the impact pressure was analyzed using Fast Fourier transformation (FFT) with Hanning window.

3.2 Point-by-Point Measurement

Point-by-point measurements were performed under identical operating pressure of 30 MPa as the measurements in lines. The nozzle traversed over the sensor at a standoff distance of 10 mm with constant step of 0.5 mm in the x-axis direction. The impact pressure was measured for 1 s at a sampling rate of 200 ksample/s after each step, then the nozzle was moved again by 0.5 mm in the x-axis direction. The number of steps in one cycle was set so that the total length of motion in the x-axis direction was approximately twice the jet footprint. The offset of points in y-axis direction was 0.5 mm (see Fig. 3c). The whole procedure repeated until the entire cross-sectional profile of jet had been measured.

The measured data were processed in the graphical programming environment NI Lab-VIEW using a user-developed program and exported to MountainsSPIP software, where 3D profiles of the impact pressure levels and plots of impact pressure intensity in a plane perpendicular to the water jet axis were generated.

The dynamic behavior of the waterjet was analyzed using FFT from records of impact pressure measured at selected points.

4 Results and Discussion

4.1 Results of Measurement in Lines

Results of measurements of impact pressure of the jet in lines are presented graphically in the form of 3D graphs in Figs. 4, 5, and 6 for the StoneAge Attack Tip 082 nozzle and in Figs. 7, 8, and 9 for the Lechler 15-08 nozzle. It should be noted that the 3D representation of the impact pressure measurements essentially corresponds to the velocity profile in the measured jet cross-section, because the amplitude of the impact pressure at a given point of the jet cross-section at a specific standoff distance from the nozzle corresponds to the flow velocity at that point.

Cylindrical Nozzle StoneAge Attack Tip 082. As can be seen in Fig. 4, the impact pressure distribution of the water jet generated by the StoneAge Attack Tip cylindrical nozzle at a standoff distance of 10 mm from the nozzle is nearly symmetrically distributed around the axis of the jet. The impact pressure distribution is bell-shaped with an almost circular base. The maximum magnitude of the impact pressure is close to 30 MPa

At a standoff distance of 30 mm from the nozzle orifice, there is a more pronounced change in the impact pressure distribution, see Fig. 5. Also, the impact pressure profiles in the x- and y-axis direction are no longer axisymmetric and the maximum level of impact pressure magnitude has dropped to 28 MPa.

The above-mentioned changes in the impact pressure distribution are further intensified at a standoff distance of 50 mm from the nozzle exit, as can be seen in Fig. 6. This indicates the fact that the initially compact core of the jet starts to split into two streams at some standoff distance from the nozzle orifice. The maximum level of the magnitude of the impact pressure has dropped to 19 MPa.

The above changes in the distribution of the impact pressure as a function of standoff distance from the nozzle exit indicate certain imperfections in the nozzle geometry (probably caused during manufacturing) that affect negatively the quality of the generated high-speed water jet quite significantly, especially at greater standoff distances from the nozzle exit.

Flat jet nozzle Lechler 15-08. Of course, the distribution of the impact pressure generated by the flat Lechler nozzle differs significantly from the distribution of the impact pressure generated by the cylindrical StoneAge Attack Tip nozzle.

As can be seen in Fig. 7, the shape of the impact pressure distribution at a standoff distance of 10 mm from the nozzle exit is distinctly oval, elongated in the y-axis direction.

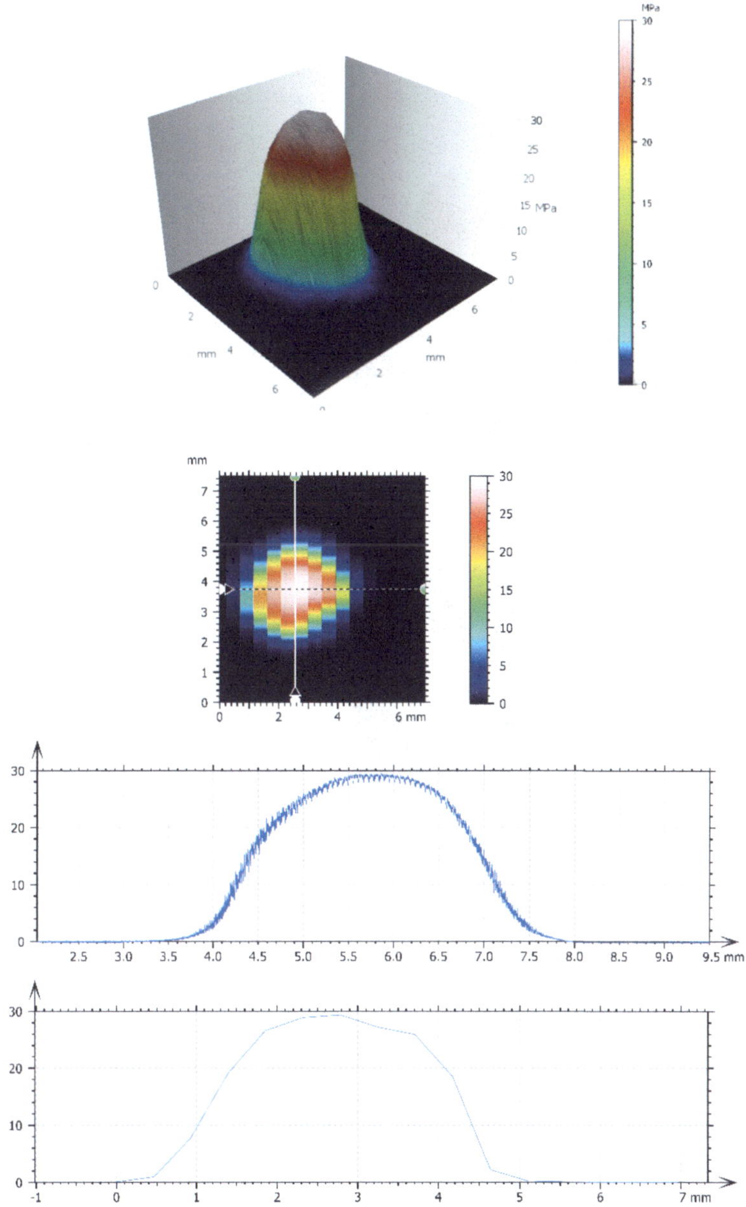

Fig. 4. Impact pressure generated by StoneAge Attack Tip 082 nozzle at operating pressure of 30 MPa at standoff distance of 10 mm. From top to bottom: 3D graph, intensity chart, impact pressure profiles at the jet axis in x-direction and y-direction.

The impact pressure profile in the x-axis direction is symmetrical and bell-shaped. The distribution of the impingement pressure in the y-axis direction is relatively uniform and

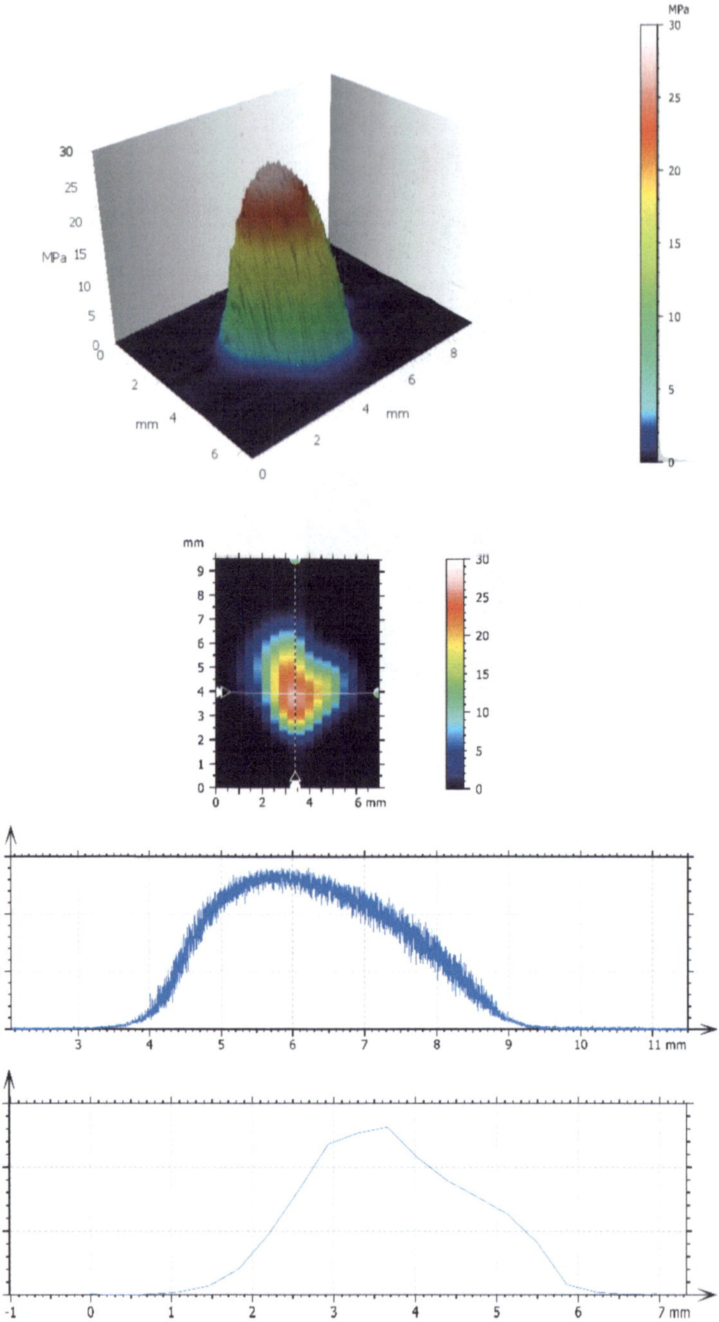

Fig. 5. Impact pressure generated by StoneAge Attack Tip 082 nozzle at operating pressure of 30 MPa at standoff distance of 30 mm. From top to bottom: 3D graph, intensity chart, impact pressure profiles at the jet axis in x-direction and y-direction.

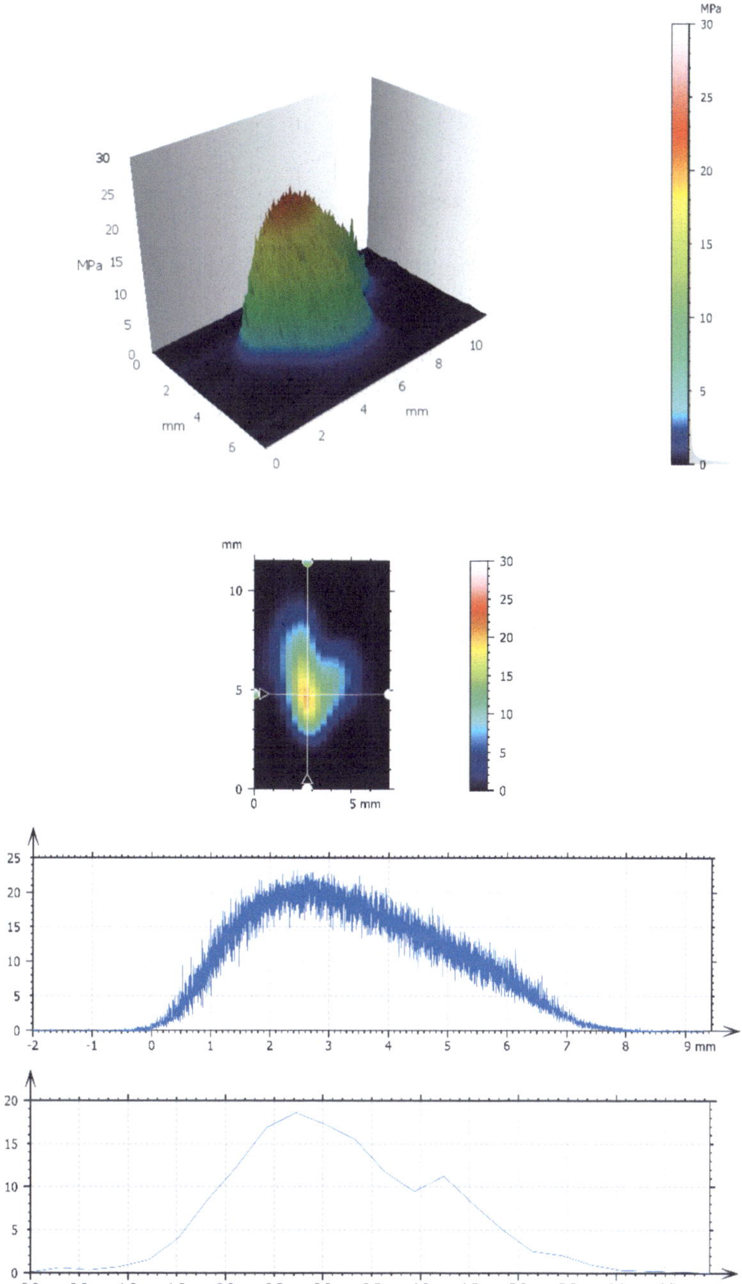

Fig. 6. Impact pressure generated by StoneAge Attack Tip 082 nozzle at operating pressure of 30 MPa at standoff distance of 50 mm. From top to bottom: 3D graph, intensity chart, impact pressure profiles at the jet axis in x-direction and y-direction.

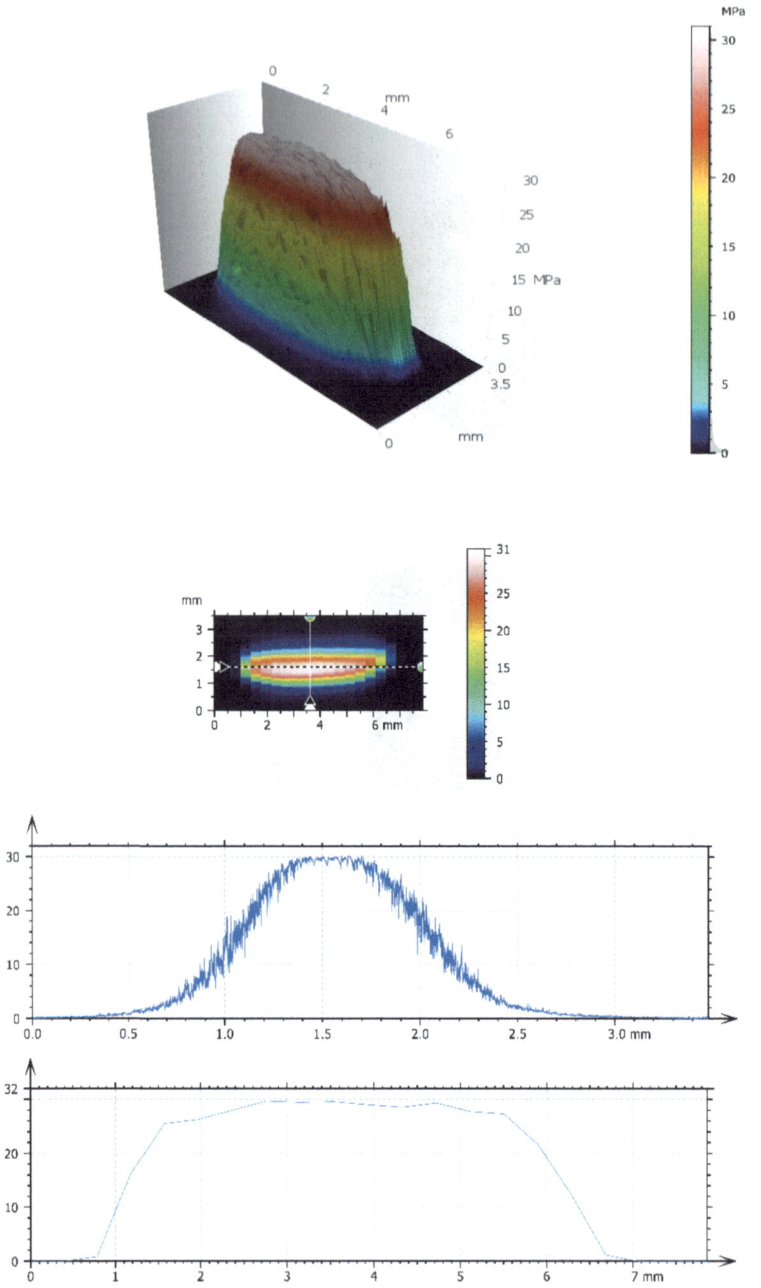

Fig. 7. Impact pressure generated by Lechler 15-08 nozzle at operating pressure of 30 MPa at standoff distance of 10 mm. From top to bottom: 3D graph, intensity chart, impact pressure profiles at the jet axis in x-direction and y-direction.

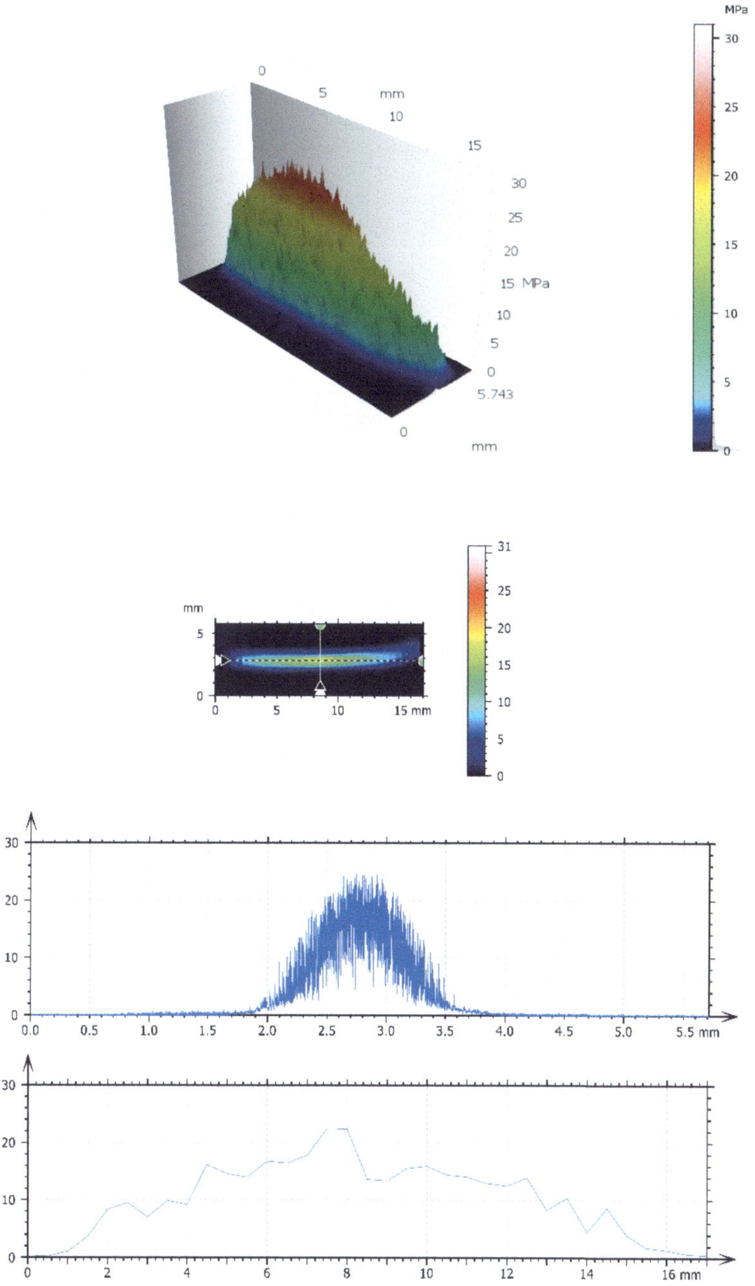

Fig. 8. Impact pressure generated by Lechler 15-08 nozzle at operating pressure of 30 MPa at standoff distance of 30 mm. From top to bottom: 3D graph, intensity chart, impact pressure profiles at the jet axis in x-direction and y-direction.

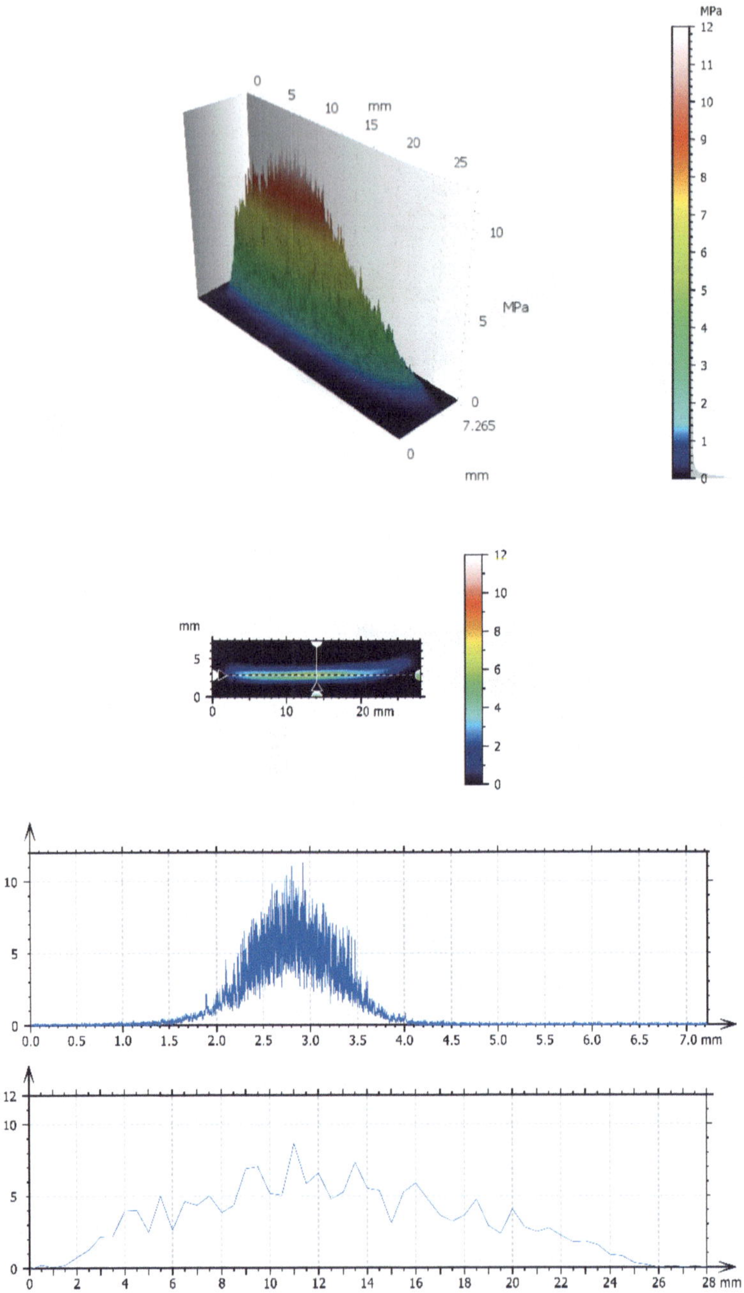

Fig. 9. Impact pressure generated by Lechler 15-08 nozzle at operating pressure of 30 MPa at standoff distance of 50 mm. From top to bottom: 3D graph, intensity chart, impact pressure profiles at the jet axis in x-direction and y-direction.

the profile is trapezoidal. The maximum level of the magnitude of the impact pressure is almost 30 MPa.

The impact pressure profile at a standoff distance of 30 mm from the nozzle exit starts to change, see Fig. 8. While the impact pressure in the x-axis direction is still bell-shaped, but with significantly higher dynamic behavior, the impact pressure profile in the y-axis direction starts to become irregular with a relatively pronounced peak at about the center of the profile. The maximum level of impact pressure magnitude at this standoff distance from the nozzle exit drops to approximately 22 MPa.

At a standoff distance of 50 mm from the nozzle exit, a slight distortion of the impact pressure distribution can be observed at the very right edge (see Fig. 9), which is probably caused by a minor inaccuracy in the nozzle manufacturing. The profile of the impact pressure in the x-axis direction is still bell-shaped, but the high dynamics of the measured signal indicates a significant breakup of the jet into smaller clumps and droplets of water. The impact pressure profile in the y-axis direction is quite markedly non-uniform. The maximum level of the magnitude of the impact pressure dropped below 10 MPa.

Presented results of the measurement of the impact pressure generated by the flat Lechler nozzle show that the impact pressure distribution at a small standoff distance from the nozzle orifice has an almost ideal character. However, the impact pressure distribution changes significantly at greater standoff distances from the nozzle exit as the jet starts to break up into smaller water clusters and droplets quite quickly. This may be the result of some minor inaccuracies in the manufacturing of the nozzle or due to its faulty design.

Analysis of dynamic behavior of the waterjet. The results of the analysis of the dynamic behavior of the jet generated by the cylindrical nozzle StoneAge Attack Tip performed by frequency analysis of recorded unprocessed impact pressure recordings using FFT are presented in plots in Figs. 10 and 11. Following three sections were selected and analyzed from the time domain of the impact pressure recording taken in the jet axis, namely in the jet axis and at a distance of 1 mm in the x-axis direction to either side of the jet axis.

The analysis of the dynamic behavior of the jet at a distance of 10 mm from the nozzle exit shows that the dynamics of the impact pressure in the jet axis shows a distinct peak around 11 kHz in the frequency domain, while the amplitudes of the other frequencies are small. However, at a distance of 1 mm from the beam axis, this peak disappears from the signal and, on the contrary, the impact pressure amplitudes increase with lower frequencies.

The dynamic behavior of the jet changed significantly at a standoff distance of 50 mm from the nozzle outlet, which is documented by the fact that the frequency peak around 11 kHz completely disappears from the signal measured in the jet axis and, on the contrary, the amplitudes of the impact pressure increase significantly with frequencies below 7 kHz. At a distance of 1 mm from the jet axis there is then an increase in impact pressure amplitudes with frequencies below 5 kHz. Such dynamic manifestations of the impact pressure indicate a gradual natural breakup of the jet with increasing standoff distance from the nozzle outlet.

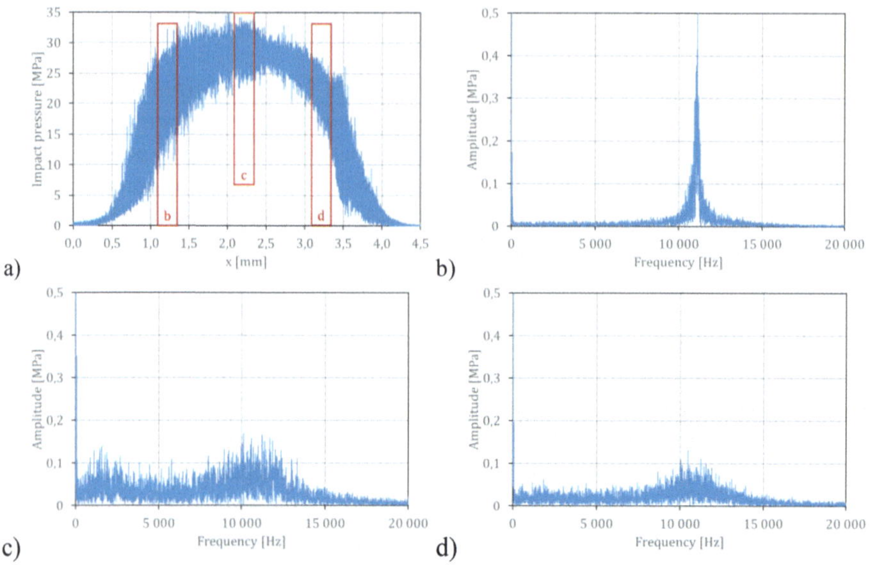

Fig. 10. Spatial distribution and frequency domains of impact pressure of the jet generated by StoneAge Attack Tip 082 nozzle at operating pressure of 30 MPa at standoff distance of 10 mm: a) spatial distribution of impact pressure at the jet axis in x-direction, b) to d) frequency domains determined from portions of the signal as indicated by red rectangles in a).

It should be mentioned that the reason for the presence of a distinct frequency peak in the impact pressure amplitude around 11 kHz in the jet axis is not clear and will be the subject of further investigation.

4.2 Results of Measurement Point-by-Point

An example of the result from a jet impact pressure measurement by point-by-point method is shown in Fig. 12. The impact pressure of the waterjet generated by the cylindrical nozzle StoneAge Attack Tip at a standoff distance of 10 mm from the nozzle outlet is almost symmetrically distributed around the axis of the jet. The distribution and profiles of the impingement pressure were not analyzed in this case, as only a few values in the direction of both the x and y axes would be involved. The maximum level of the magnitude of the impact pressure is close to 30 MPa.

The analysis of the dynamic behavior of the jet generated by the cylindrical nozzle StoneAge Attack Tip was again carried out using FFT frequency analysis of the unprocessed signals measured at each point. As an example, the results of the analysis of the time domain of the signal recorded in the jet axis (Fig. 13) and at a distance of 1 mm from the jet axis in the x-axis direction (Fig. 14) are presented.

The analysis of the dynamic behavior of the jet at a standoff distance of 10 mm from the nozzle outlet measured by the point-by-point method also confirmed that the impact pressure in the jet axis shows a significant peak in the frequency domain around 11 kHz, while the other frequencies are not significant. At a distance of 1 mm from the jet axis,

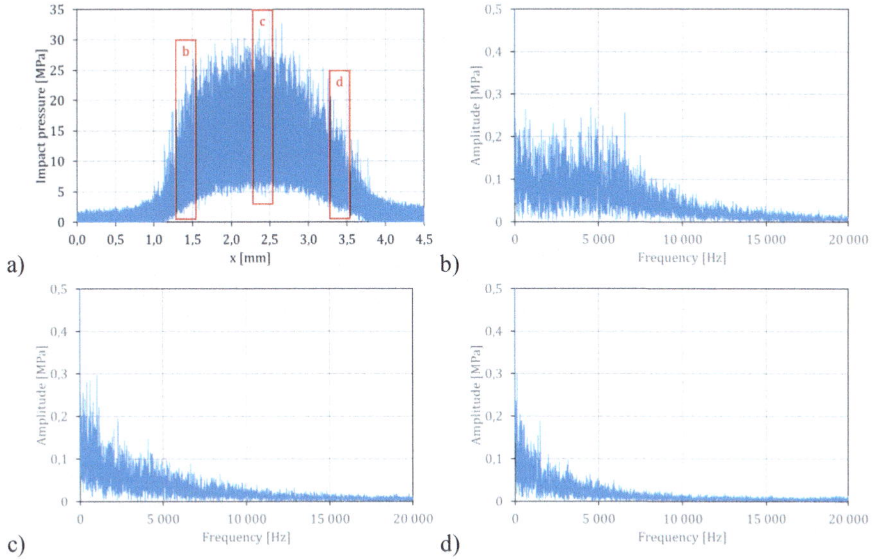

Fig. 11. Spatial distribution and frequency domains of impact pressure of the jet generated by StoneAge Attack Tip 082 nozzle at operating pressure of 30 MPa at standoff distance of 50 mm: a) spatial distribution of impact pressure at the jet axis in x-direction, b) to d) frequency domains determined from portions of the signal as indicated by red rectangles in a).

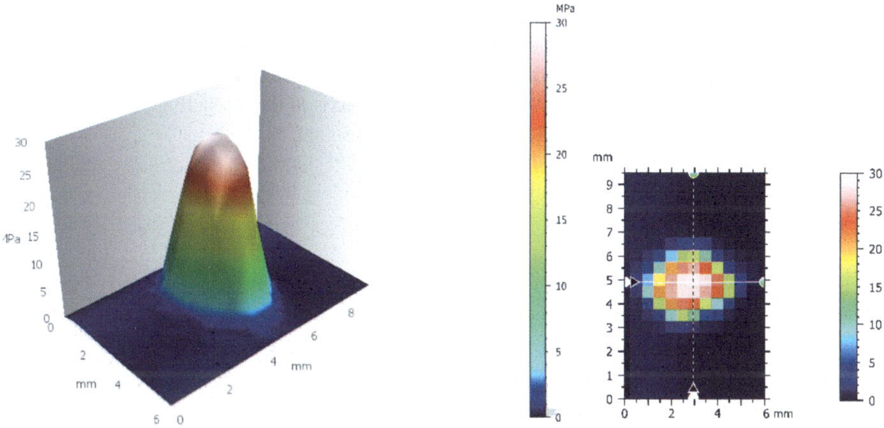

Fig. 12. Impact pressure generated by StoneAge Attack Tip 082 nozzle at operating pressure of 30 MPa at standoff distance of 10 mm: 3D graph (left), intensity chart (right).

this peak starts to disappear from the signal and, on the contrary, the amplitudes of the impact pressure increase mainly at lower frequencies.

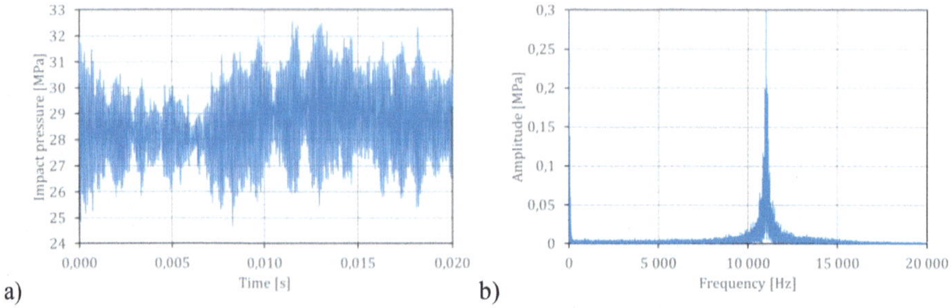

Fig. 13. Time and frequency domains of impact pressure of the jet generated by StoneAge Attack Tip 082 nozzle at operating pressure of 30 MPa at standoff distance of 10 mm: a) time domain of impact pressure measured at the axis of the impacting jet, b) frequency domain of impact pressure determined from the measured signal.

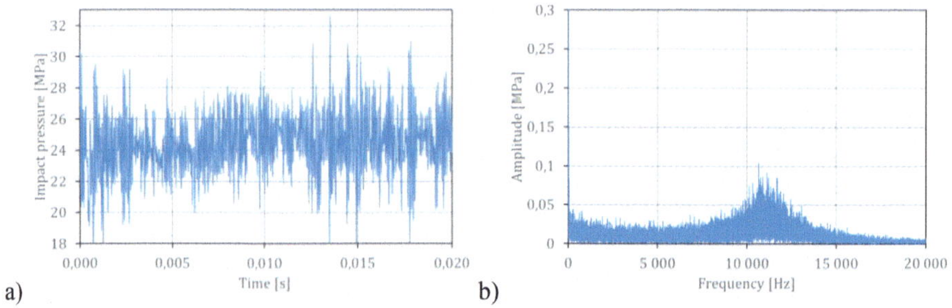

Fig. 14. Time and frequency domains of impact pressure of the jet generated by StoneAge Attack Tip 082 nozzle at operating pressure of 30 MPa at standoff distance of 10 mm: a) time domain of impact pressure measured 1 mm in x-direction from the axis of the impacting jet, b) frequency domain of impact pressure determined from the measured signal.

5 Conclusion

The paper presents two methods of measuring impact pressure using a specially designed impact pressure sensor. The proposed methods of measurement allows gaining a better knowledge of the distribution of the impact pressure (and therefore the jet velocity) in the measured cross-section. It is therefore suitable for assessing the quality and condition of a given nozzle.

The aim of the paper was to present possibilities of the proposed methods for the evaluation of nozzle quality and dynamic behavior of high-speed water jets. Therefore, only selected partial results with two different nozzle types are presented here, illustrating the possibilities of proposed measurement procedures. A detailed comparison of the different nozzle types with each other was not the focus of this work.

It can be concluded that the measurement results obtained by both methods are fully consistent. However, for routine nozzle quality determination, the in-line measurement method is preferable because it is less time consuming to implement and provides more detailed information about the impact pressure distribution across the jet cross-section.

On the other hand, the point-by-point method allows a more accurate analysis of the dynamic behavior of the jet at individual points in the cross-section of the jet because single-point measurements are performed without nozzle movement.

Frequency analysis of the impact pressure recordings revealed a distinct peak at a frequency of about 11 kHz. It was detected only in the beam axis and only at a small distance from the nozzle exit. The cause of the presence of this frequency peak has not yet been clarified and will be the subject of further research.

Acknowledgements. The work was supported by grants of the Czech Science Foundation (23-05235S, 23-05372S, 23-07415S) and by project for the long-term conceptual development of research organizations RVO: 68145535. The authors are very thankful for the support.

References

1. Liu, X., Liang, Z., Wen, G., et al.: Waterjet machining and research developments: a review. Int. J. Adv. Manuf. Technol. **102**, 1257–1335 (2019)
2. Pan, Z., Ni, Y.: Scaling of on-board characteristics of waterjet propulsion based on bench thrust testing data and loop test results. Adv. Mech. Eng. **13**(9) (2021)
3. Yang, G.L.: Simulation of the dynamic characteristics of high-speed waterjet using SIMULINK. Int. J. Heat Technol. **38**(2), 371–376 (2020)
4. Zeleňák, M., et al.: Methods for the behaviour analysis of continuous flat water jet structures. Measurement **202**, 111886 (2022)
5. Wright, D.: Impact force of high pressure waterjets. In: Proceeding of WJTA-IMCA Conference and Expo. Houston, Texas (2013)
6. Leach, S.J., Walker, G.L.: The application of High Speed Liquid Jets to Cutting. In: Philosophical Transaction, pp. 295–308. Royal Society London (1966)
7. Barker, C.R., Selberg, B.P.: Water jet nozzle performance tests. In: Proceeding of 4th International Symposium on Jet Cutting Technology: Paper A1. Canterbury, UK (1978)

Relation of the AWJ Machine Construction and the Amplitude of Vibrations Measured on the Sample

Jakub Gřunděl(✉), Libor M. Hlaváč, and Petr Pětroš

VSB – Technical University of Ostrava, 17. Listopadu 2172/15, Ostrava, Czechia
jakub.grundel@vsb.cz

Abstract. Using abrasive water jet for micromachining requires the highest possible accuracy. One of the problems that need to be overcome are vibrations of the sample. When cutting using identical or similar settings of the abrasive water jet, one might learn to know what to expect from the results. However, having a model predicting the vibrations based on the parameters and the structure of the machine would be more desirable. Models taking the parameters into account already exist, but none of them include settings that are unchangeable on the respective machine. These include the direction of the cutting relative to the ribs, the isolation of the machine from external conditions and the fixing of the arm with the cutting head. Experiments were performed on different workplaces with different conditions using the same relative traverse speeds derived from the respective settings of parameters. Results show that the isolation of the machine does not affect the results noticeably. The fixing of the arm at only one point reduces the amplitude of sample vibrations but it might increase the vibrations of the cutting head. Results for the direction of the cutting show mixed results so more research should be conducted.

Keywords: AWJ · Abrasive Water Jet · Vibrations · Vibration Measurement

1 Introduction

Abrasive water jet (AWJ) cutting is one of the recent machining methods making it a non-traditional machining process alongside with, e.g., laser beam machining, plasma beam machining or electrical discharge machining. These methods were created because sometimes the traditional ones cannot achieve desirable results. The most usual problems come from the shape that the material is supposed to be machined into and from the machinability of the material. Traditional machining tools need to be in touch with the material but their own size can make that difficult or even unattainable. Machining this way also works by removing mass from both the material and the tool. If the tool gets damaged in a much higher rate than that of the desired material, the huge financial losses can be expected. For that reason, the non-traditional machining processes can sometimes be better suited for the job than the traditional machining ones. Original application of the abrasive water jet was cutting but during the recent years some studies have been

L. Sitek et al. (Eds.): Advances in Water Jetting II, LNME, pp. 60–71, 2025.
https://doi.org/10.1007/978-3-031-72778-8_5

performed to figure out, if other machining applications have any advantages. Among these applications are machining operation such as milling [1], piercing [2], grinding [3], turning [4] and polishing [5]. To predict the outcome of the AWJ machining, one would have to use a very complex model, that considers all the necessary parameters.

Many models were prepared, some of them being the theoretical ones [6, 7] and some other based on regressions [8, 9]. The theoretical models are derived namely from the physical laws such as the laws of conservation, while the regression ones are calculated from measured data. Another way to distinguish the different types of models is through the type of material being cut. For ductile materials, such as aluminum or copper, are used other types of the models [10] than for brittle materials, like ceramics or glass [11]. These models usually do not consider the structure of the AWJ as something that should be taken into consideration. The theoretical models work with idealized versions that are not affected by the structure. On the other hand, the regression models were usually calculated based on measurements made on one particular machine and one kind of material, which means that the effect of the structure is already included in the model and transferring of this model to another machine or material can cause some errors. These errors could be insignificant unless one tries to use AWJ to cut small products, where even the slightest increase in vibrations can cause the quality of the product to be insufficient. Presented research is aimed at determining, how big the fluctuations in vibrations can be. To do that one must determine, what parts of the structure are changing.

The AWJ models usually work with the traverse speed of the jet, the pumping water pressure, the abrasive mass flow rate and many types of orifices and focusing tubes, which can be mostly selected identical on all machines. What cannot be changed up much is the cutting head, the supporting grid and the environment. The cutting head is mounted on a plotter which moves it in XYZ directions (sometimes having possibility to rotate also in two axes). This plotter is installed so that the cutting head is moving above the attenuating vessel (designed to absorb residual jet energy). The cutting head can be alternatively mounted also to a robotic hand firmly connected to the ground just at one point. In all cases the cutting head moves in all directions above the grid placed on the metal plates (the ribs) inside the attenuating vessel. When the cutting head is mounted to the plotter the connection of the plotter to the ground is in four points (Fig. 1, right hand side). However, some positioning devices are constructed so that the stiff L type hand is moving along one side of the attenuating vessel and the mounted cutting head is moving in the perpendicular direction along the arm protruding above the vessel (Fig. 1, left hand side). This design makes it much easier to place heavy, large pieces of material on the cutting surface of the attenuating vessel. The negative feature of this construction is that the arm with the cutting head is fixed only at one point to the ground (just on one side of the attenuating vessel), opposite to the classical plotter, where the arm with the cutting head is fixed in two points (on both sides of the attenuating vessel). So, it can be expected that vibrations of the system with one fixture are higher.

As mentioned in the previous text, the grid supporting the machined sample is usually placed upon the metal plates called ribs. When the straight lines are cut, one can choose to make the lines in perpendicular or parallel in relation to the ribs (Fig. 2). Regarding the environment of the machine, the most important thing is the isolation of the cutting place from the external influences, especially vibrations. If there are other machining tools producing some vibrations in the vicinity, it might be possible that these vibrations are induced into the cutting machine with the AWJ. Another important aspect is isolation from the outside world. Having the machine in an unstable environment might cause the rise of both the sample and machine vibrations which can lead to decrease in quality. With the increase of interest in the AWJ micromachining [12, 13], it is important to figure out all the possible issues that can occur to find an appropriate solution helping to minimize their effect.

Fig. 1. Attachment of cutting head (left side: one-point fixture; right side: two-point fixture).

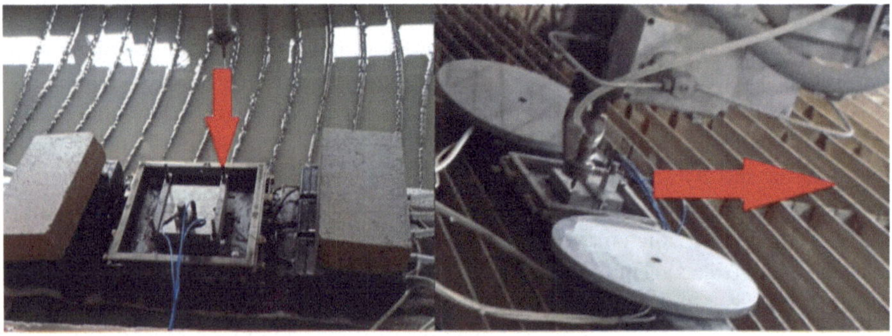

Fig. 2. Axis of the grid (left side: cutting parallel to the supporting ribs; right side: cutting perpendicular to the supporting ribs).

2 Performed Experiments

The main reason, why the relation of the structure of the AWJ and measured vibrations is not studied sufficiently, is because usually only one machine is available for obtaining the data. In other cases, data is taken from various sources but the sample and the measurement system are also different. To achieve the best results the presented research has been performed at five different workplaces, where each machine is different. The experiment includes making five cuts into a duralumin sample with dimensions 90 x 90 x 20 mm. This sample was screwed upon a metal plate which was then placed into a force measuring device which was used as a holder. This device was firmly secured to the supporting ribs using weights but the sample on a metal plate was freely embedded. The plate holder was used to assure that the sample doesn't move anywhere. However, it was loose enough to ensure that the measured vibrations were unaffected by the attachment. The cuts were made using identical traverse speeds relative to the limit traverse speed upon which the cuts stopped being throughout the material thickness in the whole path (see definition in [7]). Used relative traverse speeds (TSR) were 0.2, 0.4, 0.6, 0.8 and 1. The vibrations were measured using three 352C33 accelerometers from PCB Piezotronics. With the use of NI PXIe-4492 the signals were transferred to the computer, where they were recorded and saved using the program Signal Express. The cuts were performed always in the x-direction regarding the force sensor on which the sample was mounted. The first location is the VSB – Technical University of Ostrava (TU Ostrava), Laboratory of Liquid Jet at the Department of Physics (Fig. 3). At this location the ribs are perpendicular to the direction of the cut, machine is fairly well isolated from other machining tools but being in the same room as the water pump its isolation is not as good as at some other studied locations. The cutting head is attached to a plotter with two fixed points. In all experiments the maximum pressure 380 MPa was used.

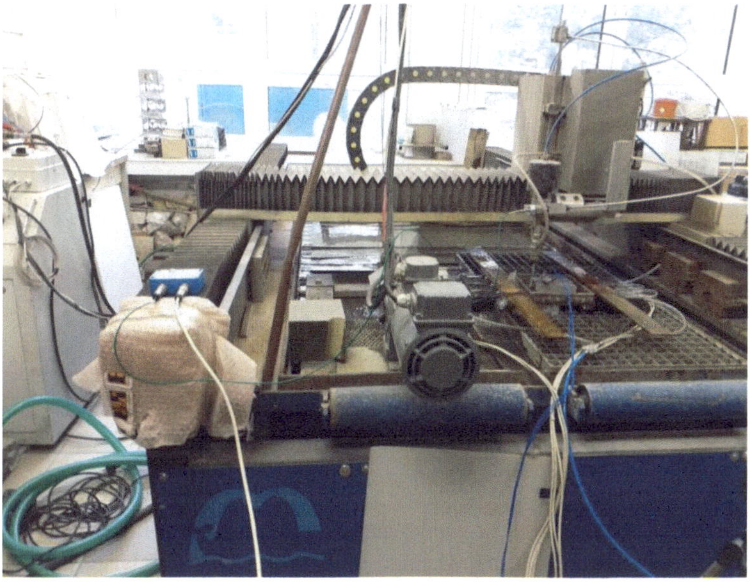

Fig. 3. Machine at the VSB – Technical University of Ostrava.

The second location, where the identical series of experiments using the same samples were performed, was the company Kámen Ostroměř (Fig. 4). Here the cuts were parallel to the grid. The machine was not well isolated, because there were other machining tools in the same hall, which affected the measured vibrations through the floor. The hall is also near the train station and train tracks and that causes spikes in vibrations when some train is passing by. The cutting head was mounted at the arm fixed in one point. The applied pressure was also 380 MPa.

The third location was the Faculty of Manufacturing Technology with a seat in Prešov (Fig. 5) that belongs to the Technical University of Košice (FMT, TU Košice). The cuts were parallel to the grid. Here the isolation was the very good because the machine was not near any other tool in fairly secluded location and the pump was in another room, which improved the measurement of vibrations. The cutting head was mounted to the plotter with two fixed points and the applied pressure was 400 MPa, because the water pressure pump has fixed setting. The difference from other workplaces is approximately 5.3%, which is a typical uncertainty of pressure in pumps at all measured workplaces.

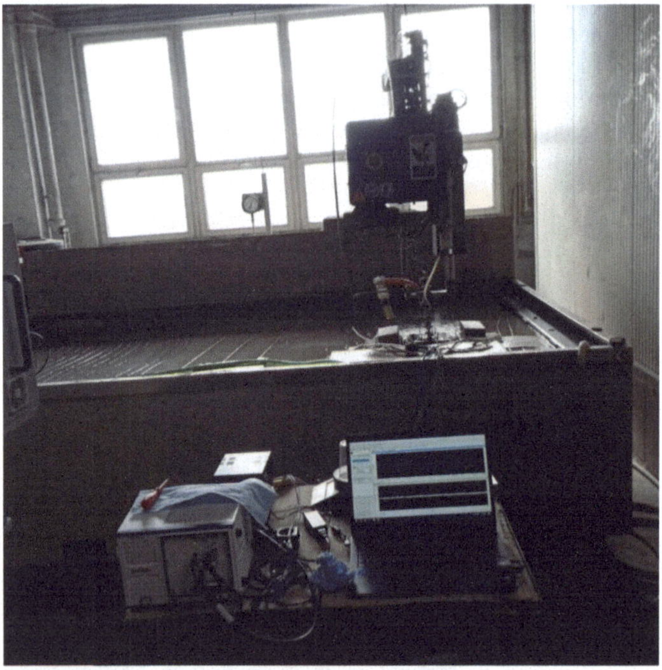

Fig. 4. Machine at the company Kámen Ostroměř.

The fourth location was the company Watting s.r.o. in Prešov (Fig. 6). At this location the cuts were perpendicular to the ribs and the isolation was similar to the one at the Faculty of Manufacturing Technology. The mounting of the cutting head again on the plotter with the two-point supporting system and the applied pressure was 380 MPa.

The last location was the Politecnico di Milano (Fig. 7), where the ribs were parallel to the cuts. The machine has an excellent isolation, which is achieved by having the machine in an isolated room without the pump. The cutting head is mounted to the plotter with two fixed points and the pressure used for experiments was 380 MPa.

The types and specifications of the machines are summarized in Table 1. Table 2 summarizes the fixed factors.

The traverse speeds for FMT, TU Košice were recalculated for respective parameters used at other locations using theoretical approach published in [7].

Fig. 5. Machine at the Faculty of Manufacturing Technology in Prešov.

Fig. 6. Machine in the company Watting s.r.o. in Prešov.

3 Results

The signals, recorded using the Signal Express, were processed in the program LabVIEW. Using this program the RMS of all the vibrations has been calculated to investigate the amplitudes. RMS of vibrations of all three axes were saved into a Microsoft

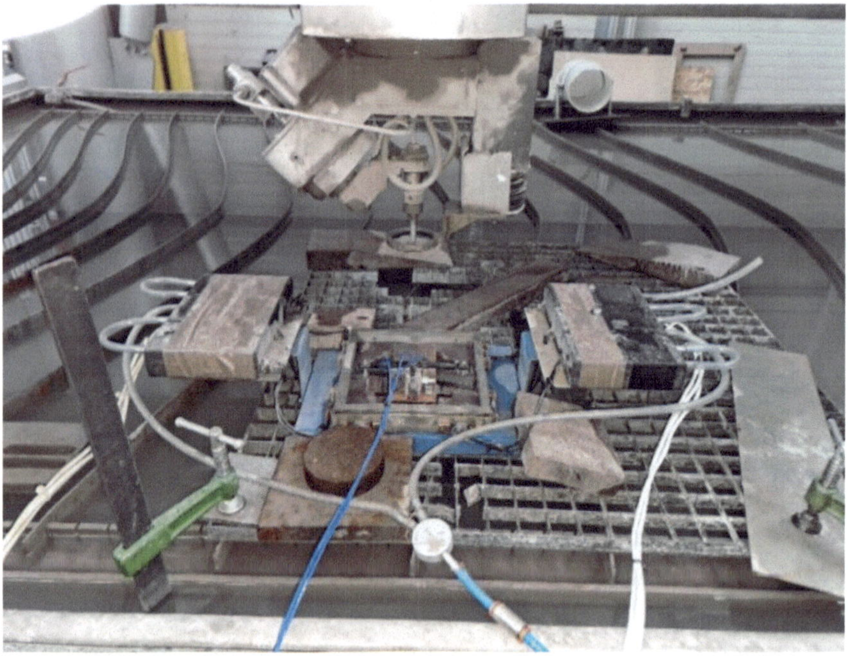

Fig. 7. Machine at the Politecnico di Milano.

Table 1. Characteristics of studied machines.

Location	Direction of cutting regarding the ribs	Isolation of the machine	Fixing of the arm with cutting head	Maximum pressure
TU Ostrava	Perpendicular	Good	2 points	380 MPa
Kámen Ostroměř	Parallel	Bad	1 point	380 MPa
FMT, TU Košice	Perpendicular	Good	2 points	400 MPa
Watting s.r.o	Perpendicular	Good	2 points	380 MPa
Politecnico di Milano	Parallel	Great	2 points	380 MPa

Excel file, where they were compared in relation to the relative traverse speed and the locations. Then they were put into three graphs for respective axes related to the sensors' locations. Figure 8 shows the signal in the program LabVIEW with the marked RMS of the amplitudes. The average amplitudes for identical material (3.2315) were determined were determined for all selected workstations.

The figures show graphs comparing average values of signals' amplitudes in x (Fig. 9), y (Fig. 10), and z (Fig. 11) axes.

Table 2. Fixed factors.

Location	d_0 [mm]	d_A [mm]	l_A [mm]	L [mm]	m_A [g/min]	Abrasive Type	Abrasive grain [mesh]
TU Ostrava	0.25	0.76	76.2	2	250	Australian garnet	GMA80
Kámen Ostroměř	0.25	0.76	76.2	2	250	Australian garnet	GMA80
FMT, TU Košice	0.33	1.02	76.2	2	250	Australian garnet	GMA80
Watting s.r.o	0.25	0.76	76.2	2	250	Australian garnet	GMA80
Politecnico di Milano	0.25	0.76	76.2	2	250	Australian garnet	GMA80

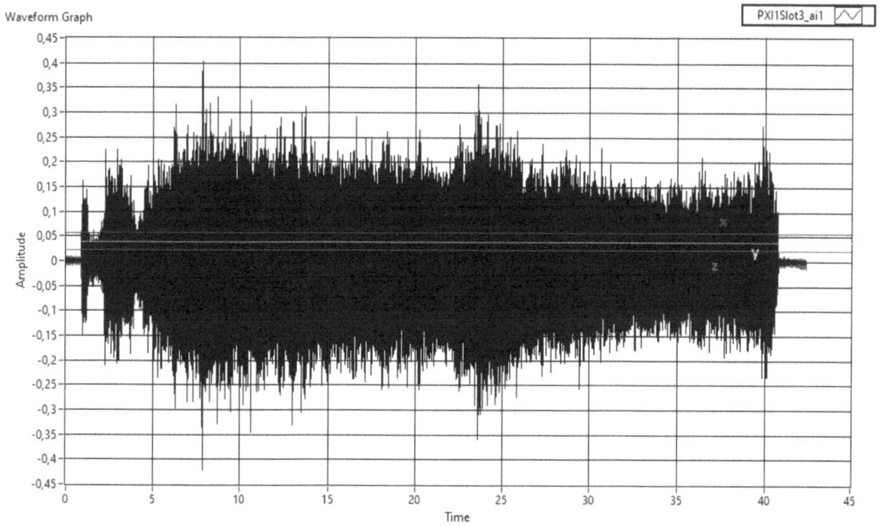

Fig. 8. Measured amplitude of vibrations in the x-axis.

4 Discussion of Results

It can be seen that the traverse speeds below 80% of the limit one cause very similar and quite low amplitudes of vibrations of the sample. When micromachining using the AWJ, using the lower traverse speeds can be expected and preferred. If the traverse speed increases, the amplitude of vibrations on almost all samples rise. The only amplitude of vibrations that does not rise significantly are those measured in the firm Kámen Ostroměř. That can be explained by the fact that cutting head was mounted at the arm fixed only at one point. Although it might seem that having less rigid cutting head can

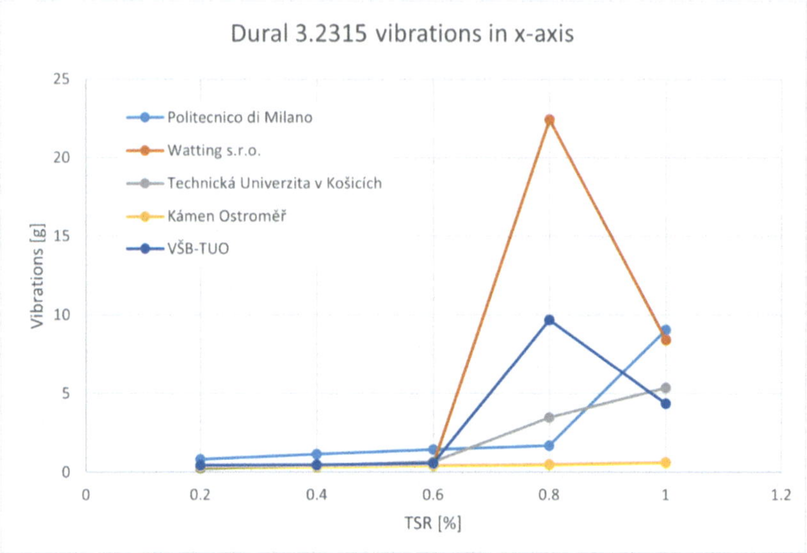

Fig. 9. Measured amplitude of vibrations in the x-axis.

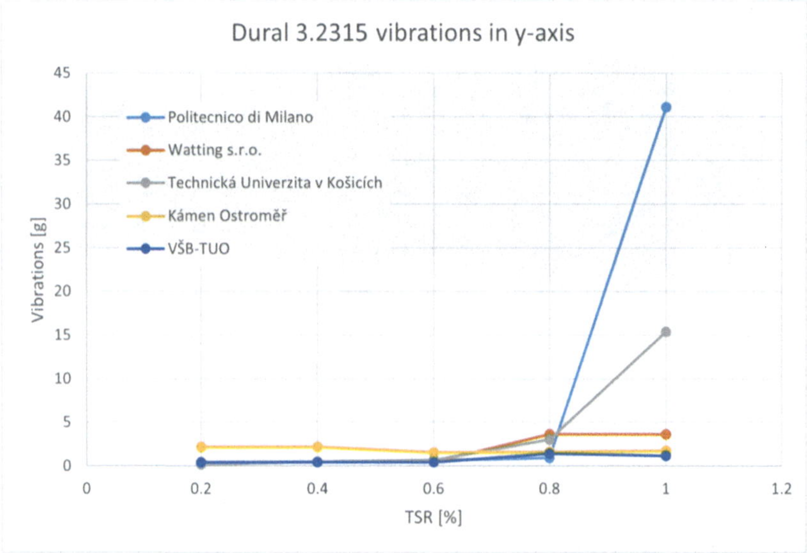

Fig. 10. Measured amplitude of vibrations in the y-axis.

cause higher amplitudes of vibrations, the result is opposite. This fact can be a proof that vibrations of the cutting system need not be transferred to the cut sample. Because the smallest amplitudes of vibrations are in the firm Kámen Ostroměř while the biggest ones are at the Politecnico di Milano, it can be assumed that the isolation of the machine

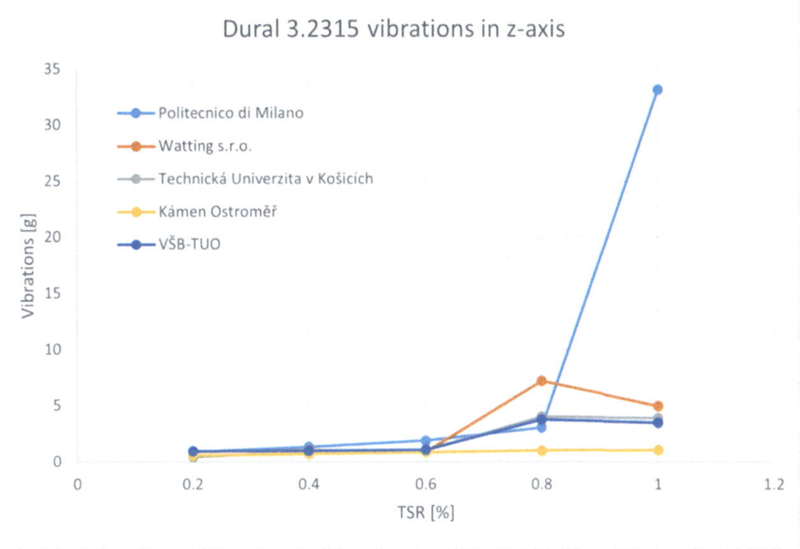

Fig. 11. Measured amplitude of vibrations in the z-axis.

does not have high effect on the measured vibrations. Similarly, the position of the ribs regarding the direction of the cut shows mixed results. The machines used at the VSB–Technical University of Ostrava and at the Faculty of Manufacturing Technology were the most similar ones. The amplitudes of vibrations measured at the VSB – Technical University of Ostrava were somewhat smaller. This can be explained by the fact that the arm with cutting head is shorter in Ostrava than in Prešov. Therefore, the deflection amplitude of this arm is smaller, which probably also reduces the amplitude of vibrations on the sample. However, more experiments should be performed to confirm or deny this assumption. It would be also a good idea to measure the amplitude of vibrations of the cutting head itself, which was unfortunately impossible during this research phase. The final conclusion can be made that up to the relative traverse speed of approximately 60% of the limit one all the studied devices cause similar vibrations of the sample. The unexpected significant increase in vibration amplitudes for some higher traverse speeds may be caused by other influences, for example significant waves of water in the attenuating vessel, resonance of the underlying ribs, etc. The frequencies of the measured signals were not included into this article, because the extent of the usable signal was up to 10 kHz (as stated by Nyquist–Shannon sampling theorem), because the measuring device was set to 20 kHz sampling frequency. No significant frequency dependence of the signal on the type of technological equipment was observed in the measured range. Therefore, only the vibration amplitude was chosen to evaluate the influence of the layout of the technological equipment on the signal measured on the cut sample.

5 Conclusions

Due to the fact that usual traverse speed ratio applied in practice is at most 60% during cutting processes, the difference between workstations is insignificant. Cutting at higher speeds it is impossible to assume that the results achieved using one machine will be similar to another machine even if the identical parameters are used. For that reason, it might be a good idea to monitor the amplitude of vibrations during the cutting process to ensure the quality of the cut. For a better insight to the problem more experiments should be conducted to provide more information.

Acknowledgement. Presented research was supported by project SP 2023/036 of the Ministry of Education, Youth and Sports of the Czech Republic.

References

1. Rabani, A., Madariaga, J., Bouvier, C., Axinte, D.: An approach for using iterative learning for controlling the jet penetration depth in abrasive waterjet milling. J. Manuf. Process. **22**, 99–107 (2016)
2. Schwartzentruber, J., Papini, M.: Abrasive waterjet micro-piercing of borosilicate glass. J. Mater. Process. Technol. **219**, 143–154 (2015)
3. Liang, Z.W., Xie, B.H., Liao, S.P., Zhou, J.H.: Concentration degree prediction of AWJ grinding effectiveness based on turbulence characteristics and the improved ANFIS. International Journal of Advanced Manufacturing Technologies **80**, 887–905 (2015)
4. Zohourkari, I., Zohoor, M., Annoni, M.: Investigation of the effects of machining parameters on material removal rate in abrasive waterjet turning. Advances in Mechanical Engineering 6 (2014)
5. Loc, P.H., Shiou, F.J.: Abrasive water jet polishing on Zr-based bulk metallic glass. In: Lin, Z.C., Huang, Y.M., Chen, C.C.A., Chen, L.K. (eds.) Advanced Materials Research, vol. 579, pp. 211–218. Trans Tech Publications Ltd., Bäch (2012)
6. Hashish, M.: A model for abrasive–Waterjet (AWJ) machining. J. Eng. Mater. Technol. **111**, 154–162 (1989)
7. Hlaváč, L. M.: Revised Model of Abrasive Water Jet Cutting for Industrial Use. Materials 14 (2021)
8. Aydin, G., Karakurt, I., Hamzacebi, C.: Artificial neural network and regression models for performance prediction of abrasive waterjet in rock cutting. The International Journal of Advanced Manufacturing Technology **75**, 1321–1330 (2014)
9. Gupta, V., Pandey, P.M., Garg, M.P., Khanna, R., Batra, N.K.: Minimization of kerf taper angle and kerf width using Taguchi's method in abrasive water jet machining of marble. Procedia Material Science **6**, 140–149 (2014)
10. Paul, S., Hoogstrate, A.M., van Luttervelt, C.A., Kals, H.J.J.: Analytical and Experimental Modeling of Abrasive Water Jet Cutting of Ductile Materials. J. Mater. Process. Technol. **73**, 189–199 (1998)
11. Paul, S., Hoogstrate, A.M., van Luttervelt, C.A., Kals, H.J.J.: Analytical Modeling of the Total Depth of Cut in Abrasive Water Jet Machining of Polycrystalline Brittle Materials. J. Mater. Process. Technol. **73**, 206–212 (1998)
12. Shi, L.P., Fang, Y., Dai, Q.W., Huang, W., Wang, X.L.: Surface texturing on SiC by multiphase jet machining with microdiamond abrasives. Mater. Manuf. Processes **33**, 1415–1421 (2018)

13. Mehta, K.M., Pandey, S.K., Shaikh, V.A.: Unconventional Machining of ceramic matrix Composites—A review. Materials Today: Proceedings **46**, 7661–7669 (2021)
14. Hou, R. G., Wang, T., Lu, Z., Liu, Y. Y.: Experimental Study of the Ultrasonic Vibration-Assisted Abrasive Waterjet Micromachining the Quartz Glass. Advances in Materials Science and Engineering (2018)

Study of Erosion Wear of PVD-Coated Focusing Tubes in AWJ Cutting Process

Petr Hlaváček(✉), Libor Sitek, and Michal Zeleňák

Institute of Geonics of the CAS, Studentská 1768 Ostrava, Czech Republic
petr.hlavacek@ugn.cas.cz

Abstract. This paper investigates the erosive wear of focusing tubes in abrasive water jet (AWJ) cutting processes, in particular the effect of physical vapor deposition (PVD) coating on the tube durability. Through controlled experiments, the wear rates of focusing tubes coated with different PVD multi-coatings are analyzed and compared with uncoated tubes. Initial results show that PVD coatings exhibit higher wear rates in the early stages due to their brittle nature, however they stabilize and ultimately enhance the overall lifetime of the focusing tubes. This study provides valuable insights into the mechanisms of wear and the effectiveness of PVD coatings in the operational life extension of focusing tubes, and also suggests potential improvements in coating techniques to optimize the tube performance.

Keywords: Abrasive water jet · Focusing tube wear · PVD coating

1 Introduction

Focusing tubes, also known as mixing tubes or focusing nozzles, are integral components of abrasive waterjet (AWJ) cutting systems. These tubes concentrate a high-speed water jet, which contains abrasive particles, onto the material to be cut, enabling precise and efficient cutting processes [1]. However, the focusing tubes are subjected to significant wear due to the high-velocity impact of the abrasive particles, leading to a gradual loss of material from the inner surface of the tube. This wear negatively affects the accuracy and efficiency of the cutting process, necessitating the replacement of the focusing tubes after tens of hours of use [2].

The wear rate of focusing tubes is influenced by several factors, including water pressure, focusing tube dimensions, water nozzle diameter, abrasive mass flow rate and the focusing tube material composition [3, 4].

However, one of the most critical factors affecting the lifetime of the focusing tube is the abrasive material used [5]. When olivine is used as abrasive, the lifetime of a standard focusing tube can exceed 400 h. Garnet based abrasives are most commonly used, with a typical focusing tube lifetime between 50 and 100 h. The shortest focusing tube lifetime is for corundum-based abrasives (lifetime less than 1 h), therefore corundum abrasives are only used for special applications [6–8].

L. Sitek et al. (Eds.): Advances in Water Jetting II, LNME, pp. 72–80, 2025.
https://doi.org/10.1007/978-3-031-72778-8_6

By understanding the factors that affect focusing tube wear and selecting the appropriate abrasive material, operators can optimize cutting performance and extend the lifetime of their focusing tubes, leading to increased efficiency and cost savings.

The focus of this study, is to investigate the wear characteristics of focusing tubes with different PVD coatings and to compare them to uncoated tubes. Specifically, we analyze the wear rates and mechanisms of focusing tubes coated with three different PVD multi-coating systems, each of them composed by layers of titanium carbonitride (Ti(CN)), aluminum oxide (Al_2O_3), and titanium nitride (TiN) in varying thicknesses. By conducting controlled wear tests and detailed analyses of the coated and uncoated focusing tubes, we aim to provide a comprehensive understanding of the effectiveness of PVD coatings in the lifetime extension of these critical components for AWJ cutting technology.

This research contributes to the ongoing efforts to improve the performance and durability of focusing tubes, ultimately enhancing the efficiency and cost-effectiveness of AWJ cutting processes. By optimizing coating techniques and understanding the wear mechanisms, more durable and reliable focusing tubes can be developed, leading to significant advancements in the field of advanced manufacturing technologies.

2 Method

The experimental setup consisted of a high-pressure water pump and a manipulator to move the cutting head. The high-pressure water was supplied to the water nozzle by a PTV75–60 multiplier pump capable of delivering up to 7.8 L of water per minute at water pressures up to 400 MPa. A commercially available Slice I cutting head with a 0.33 mm diameter diamond water nozzle was used to generate the abrasive water jet. Standard commercially available focusing tubes from the manufacturer Kennametal were used for testing. The focusing tubes were made of Roctec 100 material. The dimensions of the focusing tubes were as follows: length of 76.2 mm; outer diameter of 7.14 mm; inner diameter of 1.02 mm. Table 1 shows the basic parameters used to test the wear of the focusing tubes.

Table 1. Basic parameters of AWJ used in testing

Water pressure	400 MPa
Abrasive material	Australian garnet GMA 80 MESH
Abrasive mass flow rate	400 $g{\cdot}min^{-1}$
Water nozzle diameter	0.33 mm

Three different types of multilayer PVD coatings were applied to the focusing tubes and an uncoated focusing tube was used as a reference. The first and second PVD coatings (designated E-3053 and E-3067) consisted of three PVD layers applied sequentially to the surface of the focusing tubes. The first layer was titanium carbonitride (Ti(C, N)), the second layer was aluminum oxide (Al_2O_3), and the last layer was titanium nitride

(TiN). A PVD multi-coating consisting of two PVD layers titanium carbonitride (Ti(C, N)) and aluminium oxide (Al_2O_3) was deposited on the third focusing tube (designated F-3220). The individual coatings and their thicknesses are listed in Table 2.

Table 2. List of tested PVD multi-coatings and their thickness

Marking	Composition of individual layers	Thicknesses of layers	Total thickness
E-3053	Ti (C, N); Al_2O_3; TiN	9 μm; 2 μm; 1 μm	12 μm
E-3067	Ti (C, N); Al_2O_3; TiN	8 μm; 6 μm; 1 μm	15 μm
F-3220	Ti (C, N); Al_2O_3	7 μm; 9 μm	16 μm

The thickness of each layer was measured on control samples placed in the coating chamber. Only the E-3067 coating layers were measured on a cross-section of the focusing tube using an electron microscope, see Fig. 1 and Fig. 2. The thickness of individual PVD layers was measured on the outside of the focusing tube.

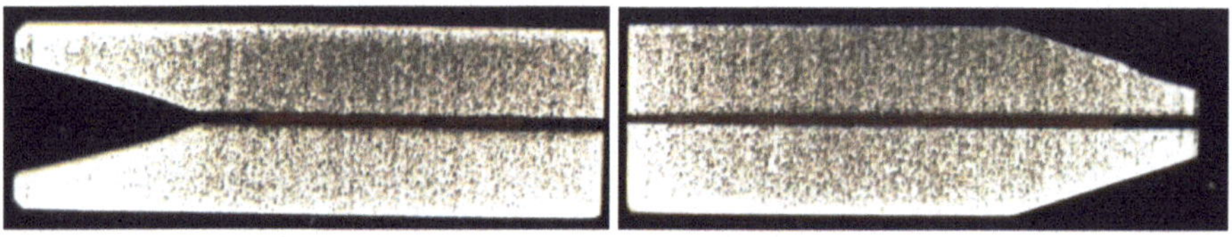

Fig. 1. Longitudinal cross-section of focusing tube with PVD multi-coating E3067 (left – input part of the focusing tube, right – output part of the focusing tube)

On the focusing tube with PVD coating E-3067, a control measurement of the chemical composition of individual layers was performed using energy dispersive X-ray analysis (EDX) along the line shown in Fig. 2. The results of the chemical analysis are shown in Fig. 3 and Table 3.

The elemental analysis confirms that the composition of the individual layers corresponds to the design specifications.

Figures 4 and 5 show the input and output diameter of new focusing tubes tested.

The measurement of focusing tube wear has been studied mainly by observing the changes in mass and calculating the focusing tube wear rate $\dot{w}$ from this equation:

$$\dot{w} = \frac{\Delta m}{t}$$

where Δm is mass loss in mg and t is running time in min.

Furthermore, the change in the output diameter of the focusing tubes was monitored using digital microscopy. The focusing tubes were measured every 10 min in the first hour of testing and then every 30 min.

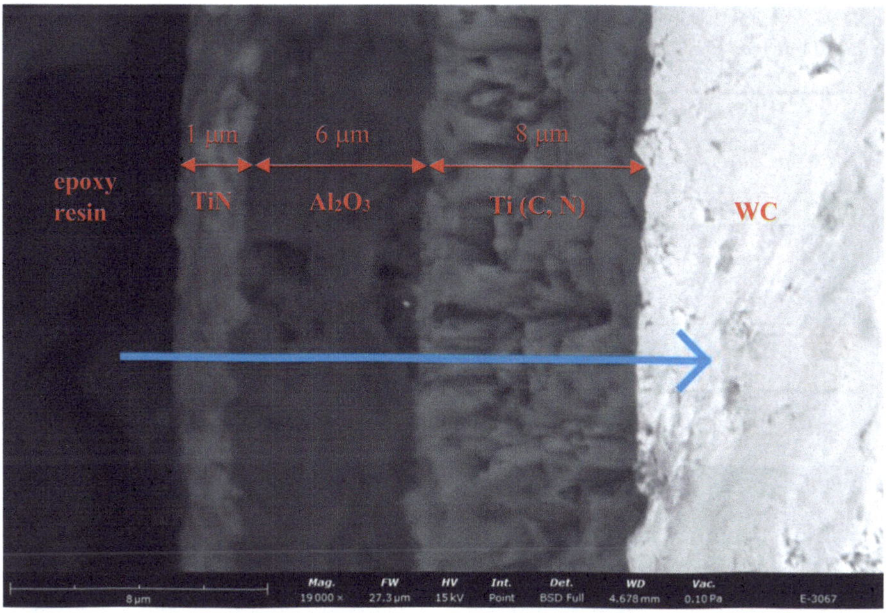

Fig. 2. SEM image of the cross-section of the PVD coating for the E-3067 process. The blue line marks the location of the control measurement of the chemical composition using energy dispersive X-ray (EDX) analysis

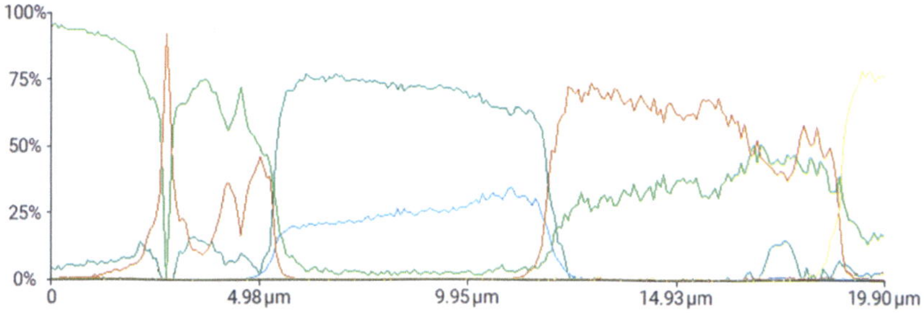

Fig. 3. Measured values of mass concentration of individual elements along the analyzed line (light green – carbon, dark green – oxygen, blue – aluminum, orange – titanium, yellow – tungsten).

3 Results

Figure 6 shows an example of damage of the PVD coating E-3067 near the exit hole of the focusing tube. After only ten minutes of use, peeling layers of the PVD coating are visible. After two hours of use, the damage is greater and there are also visible cracks in the coating, probably caused by reflected abrasive grains during cutting process.

For all focusing tubes, the highest wear rate was measured in the first ten minutes of use (see Fig. 7), in this time the lowest wear was measured in the reference focusing

Table 3. Measured values of atomic and weight concentration of individual elements along the analyzed line (see Fig. 3)

	Element number	Element symbol	Element name	Atomic concentration [%]	Weight concentration [%]
	6	C	Carbon	28.94	16.3
	8	O	Oxygen	43.44	32.6
	13	Al	Alumi-num	12.73	16.1
	22	Ti	Titanium	13.35	30
	74	W	Tungsten	0.41	3.5

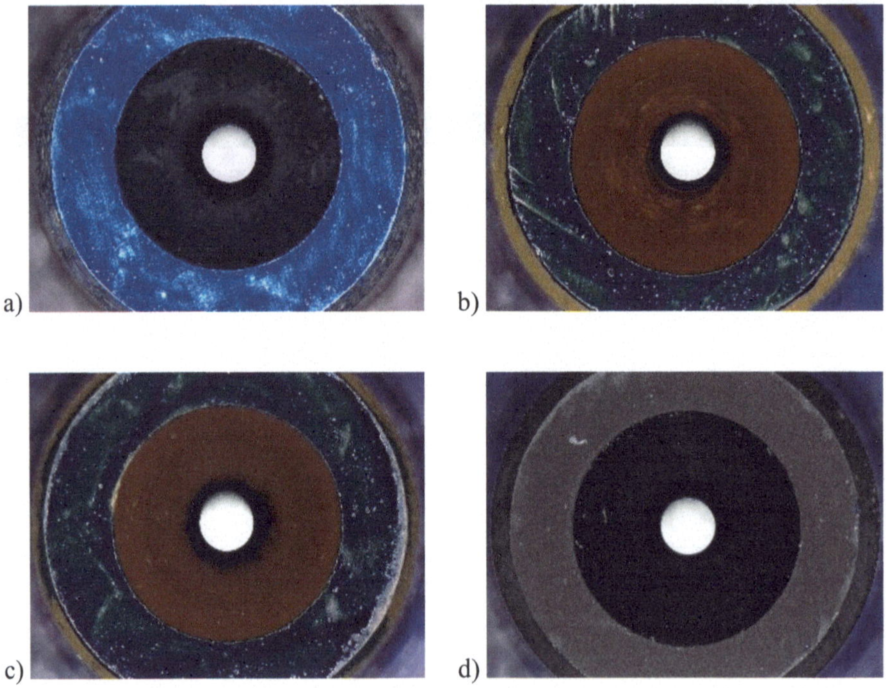

Fig. 4. Input diameter of focusing tubes tested (the blue color is only a matting of the surface due to the removal of reflections): **a** reference focusing tube; **b** E-3053 coating; **c** E-3067 coating; **d** F-3220 coating

tube and the highest one in the F-3220 coated focusing tube. For the other time intervals, the already coated focusing tubes showed lower values of wear than the reference.

Figure 8 shows the evolution of the focusing tube wear rate on the time interval of 1 to 4 h of use. It can be seen that the measured values for the E-3053 and E-3067 coatings are identical to the reference focusing tube after two and three hours of usage,

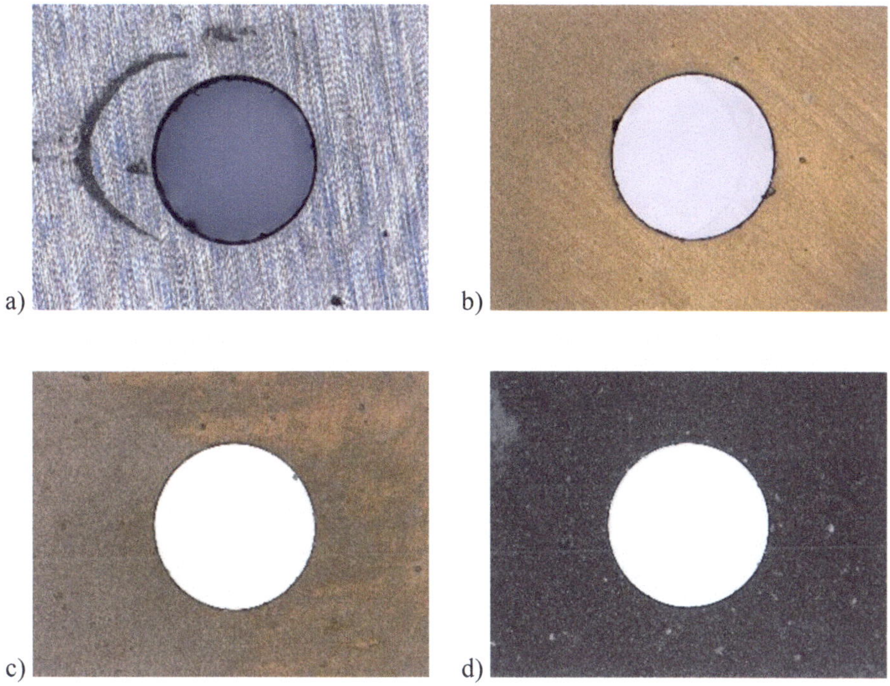

Fig. 5. Output diameter of focusing tubes tested: **a** reference focusing tube; **b** E-3053 coating; **c** E-3067 coating; **d** F-3220 coating

respectively. The best results were obtained with the F-3220 coating, which showed a wear rate of approximately 15% lower than the reference tube, even after 4 h of use.

4 Discussion

The results indicate that PVD coatings can influence the wear resistance of focusing tubes significantly. In the initial phase, the coated focusing tubes exhibit higher wear rates, probably due to the brittle nature of the PVD coatings, which can crack and peel off under abrasive action. However, after the initial phase, the wear rates stabilize and the coated focusing tubes show an improved performance compared to the uncoated reference tubes. This suggests that although PVD coatings have an initial vulnerability, their overall protective properties improve the lifetime of focusing tubes for longer periods of use.

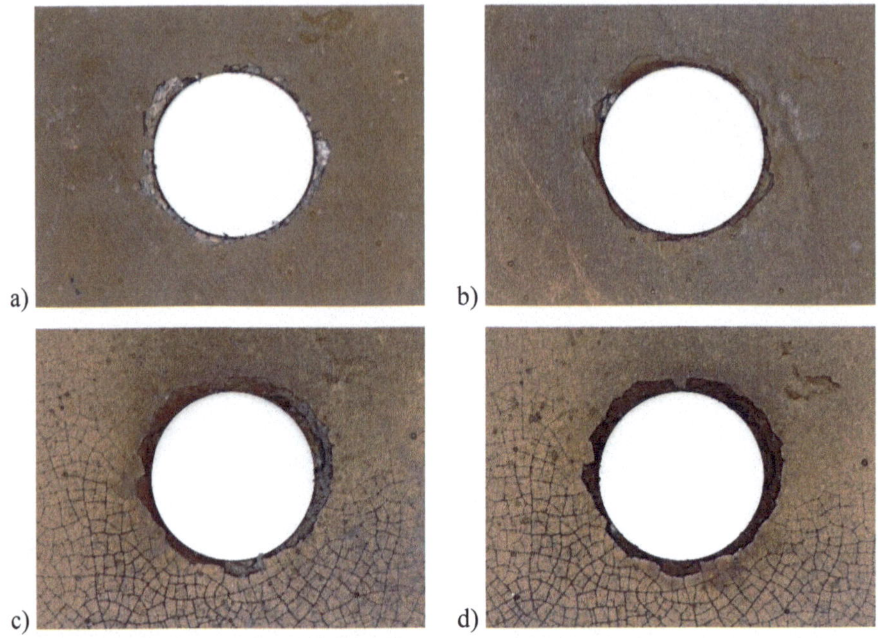

Fig. 6. Output diameter of focusing tubes with E-3067 coating - **a** after 10 min of using, **b** after 1 h of using, **c** after 2 h of using, **d** after 4 h of using

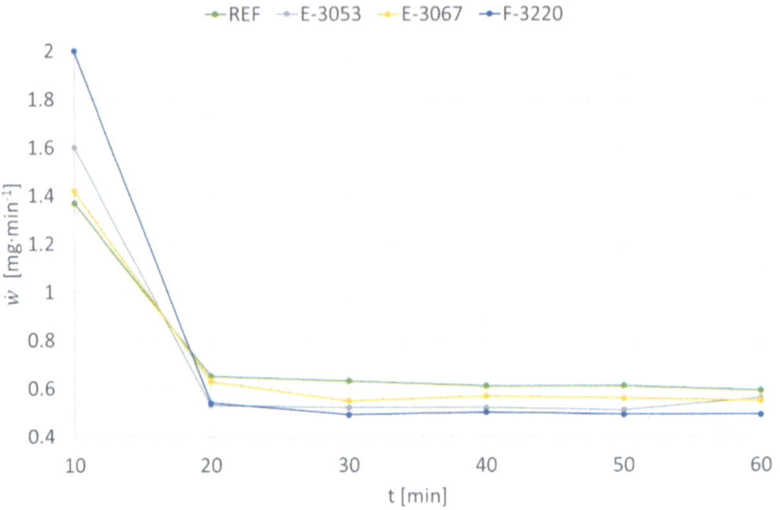

Fig. 7. Focusing tube wear rate $\dot{w}$ as a function of test time during the first hour of use

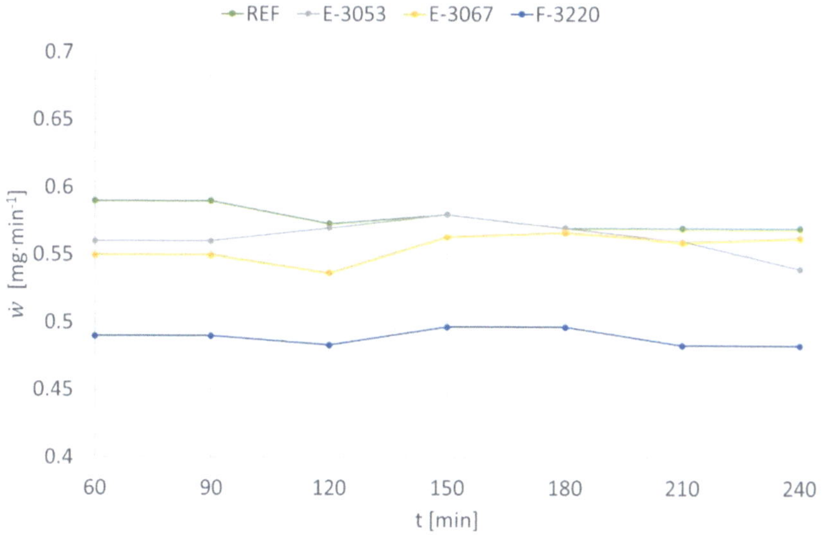

Fig. 8. Focusing tube wear rate $\dot{w}$ as a function of test time in intervals of 1 to 4 h of use

5 Conclusion

The study of PVD coated focusing tubes in the AWJ cutting process has revealed both challenges and benefits associated with the use of such coatings. The initial wear rate of the coated tubes is higher due to the brittleness of the PVD layers, but after this initial phase, the wear rate decreases, suggesting a stabilization of the wear process. In particular, the F-3220 coating showed the most significant improvement, achieving a wear rate of approximately 15% lower than the reference focusing tube even after four hours of continuous use. This suggests that PVD coatings, despite their initial vulnerability, can provide protective benefits and extend the lifetime of focusing tubes. Future work should focus on improving the adhesion and toughness of PVD coatings to minimize initial wear and improve their overall protective performance. The next challenge in PVD coating should focus on analyzing the thickness of the coating inside the focusing tube. Preliminary measurements show that the coating thickness inside the focusing tube is much lower than on the outer surface.

By understanding these mechanisms, we can develop more durable focusing tubes that will extend the service life, reduce maintenance costs, and improve the efficiency of AWJ cutting processes. This work contributes to the growing body of knowledge aimed at optimizing materials and coatings for advanced manufacturing technologies.

Acknowledgements. This work was supported by project for the long-term conceptual development of research organizations RVO: 68145535. Thanks also go to the industrial partners: company PTV, which provided focusing tubes for testing; and company Pramet tools, which applied the PVD coatings to the focusing tubes. The authors are very grateful for their help.

References

1. Summers, D.A.: Waterjetting technology. E & FN SPON, London (1995)
2. Nanduri, M., Taggart, D.G., Kim, T.J.: The effects of system and geometric parameters on abrasive water jet nozzle wear. Int. J. Mach. Tools Manuf **42**(5), 615–623 (2002)
3. Kim, T.J., Labus, T.J.: Influence of basic jet parameters and physics of abrasive water jet cutting. In: T.J. Labus (ed.) Fluid Jet Technology: fundamentals and applications, pp. 3.1–3.45. WJTA, St. Louis (1995)
4. Momber, A.W., Kovacevic, R.: Principles of Abrasive Water Jet Machining. Springer-Verlag, London Limited, London (1998)
5. Martinec, P., Foldyna, J., Sitek, L., Bílková, L., Barčová, K.: Olivine as industrial abrasive for high-speed water jet technology. In: P. Lake (ed.) Proceedings of the 16th International Conference on Water Jetting, pp. 287–297. BHR Group (2002)
6. Hlaváček, P., Sitek, L., Foldyna, J.: Evaluation of erosion performance of abrasive particles in abrasive water jet cutting process. MM Sci. J. 3869–3872 (2020)
7. Martinec, P., Foldyna, J., Sitek, L., Ščučka, J., Vašek, J.: Abrasives for AWJ cutting. Institute of Geonics of the CAS, Ostrava (2002)
8. Perec, A., Pude, F., Grigoryev, A., Kaufeld, M., Wegener, K.: A study of wear on focusing tubes exposed to corundum-based abrasives in the waterjet cutting process. Int. J. Adv. Manuf. Technol. **104**, 2415–2427 (2019)

Topaz as a Potential Abrasive for Cutting and Surface Treatment of Materials by AWJ Technology

Petr Martinec[1], Libor Sitek[1(✉)], František Veselovský[2], Petr Hlaváček[1], Josef Foldyna[1], Josef Klomínský[2], Roman Gabor[3], Lucie Němcová[1], and Lucie Ruppenthalová[1]

[1] Institute of Geonics of the CAS, Ostrava 708 00, Czechia
libor.sitek@ugn.cas.cz
[2] Czech Geological Survey, Prague 118 00, Czechia
[3] VŠB—Technical University of Ostrava, Ostrava 708 00, Czechia

Abstract. Rocks with an enrichment of topaz (quartz-topaz quartzolites) found in the Krkonoše-Jizera crystalline complex in northern Bohemia or a potential secondary raw material from the mining of cast mica from greisen (Czechia) can be a suitable raw material for obtaining topaz concentrate for its subsequent use in Abrasive Water Jet (AWJ) technology as an abrasive. Topaz from these rocks has good mineralogical parameters, density and hardness. The grains of the topaz concentrate are also less fractured. This article outlines the basic properties of topaz in relation to its use as a potential abrasive for AWJ, including basic testing of its cutting abilities compared to other abrasive concentrates. The cutting performance of topaz was found to achieve 90.5% of the cutting performance of almandine garnet GMA 80 when cutting AISI 304 steel, while exhibiting better cutting properties than andradite garnet or quartz sand of the same grain size.

Keywords: Abrasive Water Jet · Abrasive · Topaz · Garnet · Erosion Efficiency

1 Introduction

New finds of quartz-topaz quartzolites and the potential mining of lithionites (zinwaldite) from greisens in the Czech Republic [the locations of Cínovec and Krásno in the Krušné Hory (Ore Mountains / Erzgebirge)] enabled the current use of topaz as an accompanying secondary raw material. Topaz quartzolites found in the Krkonoše-Jizera orthogneiss in the western Sudetes consist only of quartz and topaz, with the quartz content being up to 65% and the topaz content about 32% (see [1, 2]). In the Czech part of the Krkonoše-Jizera crystalline complex, these rocks can be found, for example, near the settlements of Růžek and Vítkov in the vicinity of Chrastava [1]. In Poland, topaz quartzolites (greisens) are found in the same geological unit near the settlements of Rebiszow and Gierałtówek [3], where these rocks occur in a 15 km long stratified layer. Muscovite quartzite with primary topaz was found near Zdobnice in Orlické hory [4]. Other topaz deposits are also found in the Ore Mountains, where topaz is a part of granitoid massifs [5].

L. Sitek et al. (Eds.): Advances in Water Jetting II, LNME, pp. 81–92, 2025.
https://doi.org/10.1007/978-3-031-72778-8_7

The industrial use of topaz as a raw material is limited (e.g. for the production of aluminum, mullite and fluorine [6]). However, there is considerable potential of its use as an abrasive for Abrasive Water Jet (AWJ) cutting, blasting and treatment technologies.

2 Background

Topaz ($Al_2[SiO_4(F,OH)_2]$) is a mineral belonging to the group of silicates (nesosilicates). The symmetry is rhombic (Pbnm), the crystals have a prismatic habitus. This rock-forming mineral is colorless or yellowish, yellow-gray with a glassy luster. The mineral is easily cleavable according to {001}, the cleavage according to {101} and {011} is rare. The cleavage planes are conchoidal or uneven. The Mohs hardness of Topaz is 8, the Vickers hardness $HV = 1454$, the Knoop hardness $Hk = 1240$–1500 (for the recommended optimum load 981–1961 mN [7]), the Mohs hardness re-calculated according to Povarennykh [8] $H_{teor} = 7.83$–8.02 and the specific gravity 3.5 g·cm^{-3}. In addition to a short prismatic habitus (Fig. 1), it can occur as a *pycnite* in radially arranged in the aggregates. Topaz is mainly used as a gemstone valued for its distinctive color and hardness. The usual properties of topaz can be found in Table 1.

The use of topaz as a possible abrasive in AWJ technologies is based on its favorable density and high hardness. Compared to almandine garnet, i.e. an abrasive commonly used in AWJ technologies, topaz is harder, more cleavable and with lower density, which may, under certain conditions, cause difficulties in transporting the abrasive concentrate into the mixing chamber of an abrasive head. It would also be possible to use the fine fractions resulting from the cleavage of topaz grains after their disintegration for the treatment of material surfaces using AWJ.

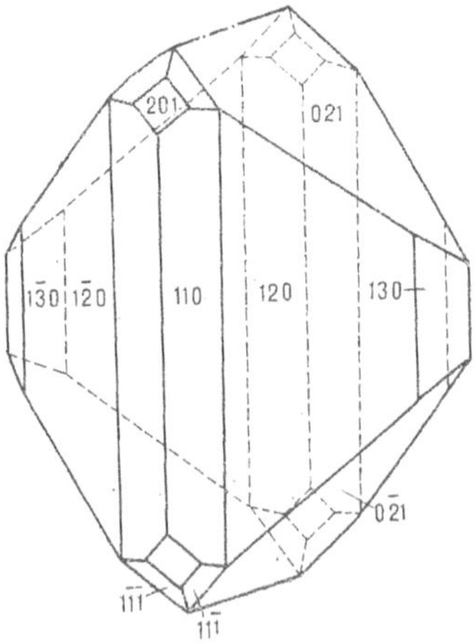

Fig. 1. An ideal short prismatic topaz crystal typical of greisens [9].

3 Materials in Experiment

For an experiment on the cutting performance of the topaz abrasive in the AWJ technology, **topaz concentrates** were prepared in the laboratories of the Czech Geological Survey (CGS) from crushed rock of quartz quartzolite from the Růžek locality and from Sn-W-Li greisen from the Cínovec locality. Magnetic separation and separation in heavy liquid (1,1,2,2-tetrabromoethane) were carried out to obtain concentrates with an enrichment of topaz (see Table 1).

Table 1. Basic properties of abrasives: garnet almandine, quartz, topaz, garnet andradite and garnet grossular.

Mineral	Crystal system	Composition	Specific density (He-method)	Cleavage
–	–	–	$[g.cm^{-3}]$	–
Garnet almandine	Cubic	$Fe_3Al_2[SiO_4]_3$	4.06–4.32*	{110}, bad or missing
Quartz	Trigonal	SiO_2	2.65*	Non-cleavable
Topaz	Rhombic	$Al_2[SiO_4(F,OH)_2]$	3.5–3.57	{001}, very good
Garnet andradite	Cubic	$Ca_3Fe_2[SiO_4]_3$	3.85	{110}, bad or missing
Garnet grossular	Cubic	$Ca_3Al_2[SiO_4]_2$	3.59	{110}, bad or missing

Mineral	Crackability of grains	Mohs hardness	Vickers hardness 100g	Cutting performance in steel
–	–	–	–	[% GMA]
Garnet almandine	Frequent to very frequent*	7–7.5	–	100
Quartz	Variously intensive	7	1135–1142*	70–86*
Topaz	Various according to genesis*	8	1454	90.5
Garnet andradite-grossular	Frequent*	6.5–7	950–1050	82.8

Note: *differs according to the mineral genesis depending on the geological conditions of mineral formation*

3.1 Separation of Topaz for Cutting Tests

The topaz concentrate was produced by crushing the rock fragments in a Retsch Jaw Crusher BB300 to a grain size of less than 4 mm. The product was then ground in a Retsch Hammer Mill SK-1. The fractions were separated in a heavy liquid (1,1,2,2-tetrabromoethane with a density of 2.967 g·cm^{-3}). By standardising the grain size of fractions and sieving, a fraction of 0.125–0.315 mm was prepared for the consequent mineralogical analyses and experimental cutting of materials using AWJ. The properties of the topaz concentrate prepared for the experiment are listed in Tables 1 and 2.

3.2 Mineralogical Analysis of Topaz and Topaz Concentrates Used in Experiment

The mineralogical analysis of the two topaz sources focused on the mineralogical identification of the topaz composition and the mineralogical purity of the topaz concentrate used in the experiment (Table 2). The shape analysis of topaz grains damaged by natural influences and by the treatment technology is also presented.

Table 2. Mineralogical parameters of topaz in the concentrate for the preparation of abrasive.

Sample	Locality and source	He-density	FTIR spectrum and X-Ray diffraction	Shape and cleavage of grains
[mm]	–	[g.cm^{-3}]	–	–
Topaz concentrate 0.125–0.3	Růžek quartzolite	3.486	Topaz	Polygonal {001}very good
Topaz concentrate 0.125–0.3	Cínovec Sn-W-Li greisen	3.468	Topaz	Polygonal {001}very good

Sample	Cell parameter a_0	Cell parameter b_0	Cell parameter c_0	Cell volume	Vickers Hardness 100g
[mm]	(Å)	(Å)	(Å)	(Å)	–
Topaz concentrate 0.125–0.3	4.6537	8.8073	8.3905	343.90	1450
Topaz concentrate 0.125–0.3	4.6493	8.7969	8.3925	343.25	1440

Topaz grains have an almost identical He-density of 3.47–3.49 cm^3.g^{-1}. The shape of the grains is polygonal, with distinct cleavage {001} (Figs. 2, 3 4, Table 2). The lattice parameters and the volume of the elementary cell are almost identical in both cases (Table 2). In both cases, the FTIR spectra (Figs. 5 and 6) confirm the high mineralogical quality of the topaz without the effect of mineral alterations.

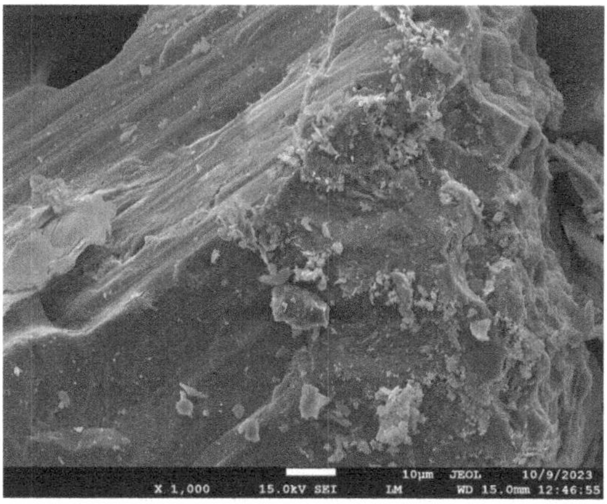

Fig. 2. Topaz grain of the quartz-topaz quartzolite from the Růžek locality. The topaz grain shows weak mechanical defects and inclusions. The cleavage planes are significant. The fracture perpendicular to the c-axis of the crystal is uneven and rough. SEM microscopy. Without metal plating. 1000x magnification, scale on image 10 μm.

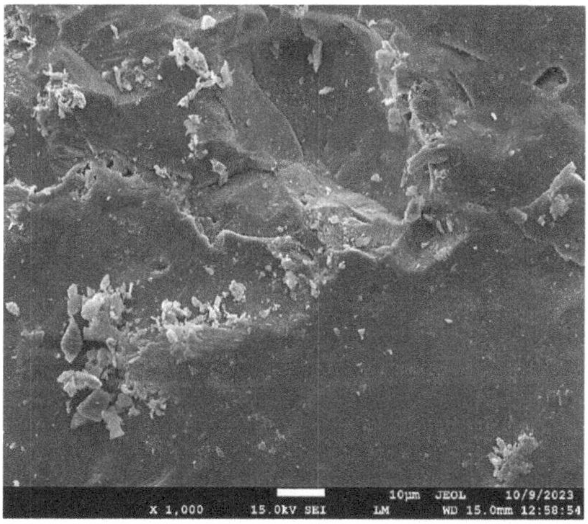

Fig. 3. Topaz grain of the quartz-topaz quartzolite from the Růžek locality. The topaz grain shows the cleavage plane {001} with very small cavities (gas-liquid inclusions), sporadic cracks and mineral inclusions. SEM microscopy. Without metal plating. 1000x magnification, scale on image 10 μm.

Quantitative X-ray powder diffraction (XRD) showed a high mineralogical purity of the topaz concentrate from both localities with mica and quartz impurities (Table 3). According to optical microscopy, impurities of other minerals occur in form of intergrowths and inclusions. Weak sericitisation (pycnite) in the topaz was also found.

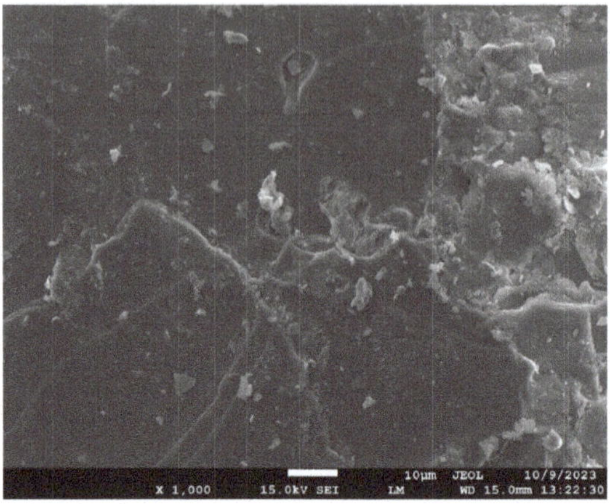

Fig. 4. Topaz grain of the Sn-W-Li greisen from the Cínovec locality. The topaz grain shows the cleavage plane {001} with very small cavities of gas-liquid inclusions and without mineral inclusions. SEM microscopy. Without metal plating. 1000x magnification, scale on image 10 μm.

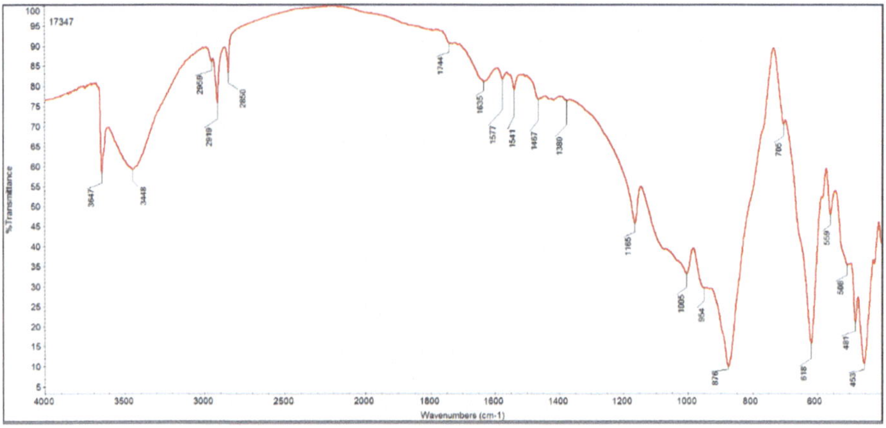

Fig. 5. Topaz concentrate from the Růžek locality, fraction of 0.125–0.3 mm, rock source: quartz-topaz quartzolite. Infrared spectroscopy (FTIR). Wavelength range of 400 - 4000 cm−*1*. KBr method, 2 mg sample weight, grain size of ~ 3–4 μm. *Interpretation: topaz dominance: 3647, 1165, 1085, 1005, 865, 950 (?), 619, 706, 559, 481, 454, 425 cm*$^{-1}$*; traces: carbonate: 1744, 1464 cm*$^{-1}$*, mica (overlaps): 1028 cm*$^{-1}$*, apatite (overlaps): 1005, 605, 575 (?) cm*$^{-1}$.

3.3 Fracture of Topaz Grains Determined by Surface Area Measurement and High-Pressure Adsorption of Nitrogen

Determining the fracture intensity of topaz grains is essential for assessing the stability of the abrasive grain in the process of its interaction with the mixing chamber, focusing tube and subsequently with the material to be disintegrated. Based on the experience with the evaluation of garnet grain fracture at the outlet and after recycling [10], the

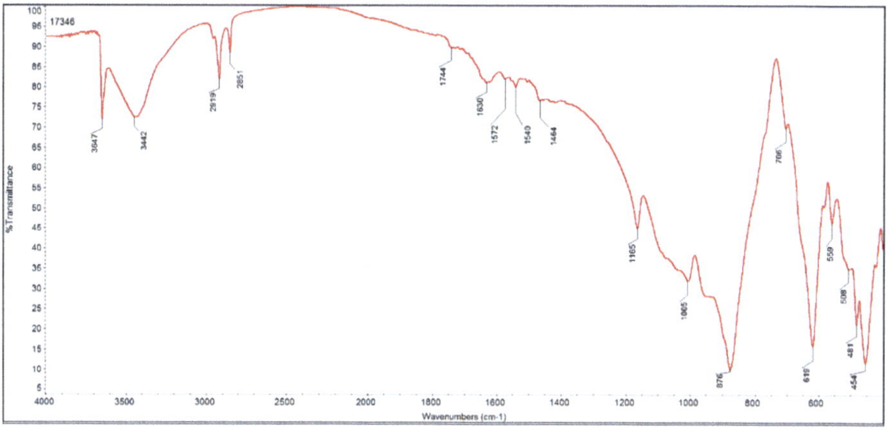

Fig. 6. Topaz concentrate from the Cínovec locality, fraction of 0.125–0.3 mm, source rock: greisen. Infrared spectroscopy (FTIR). Wavelength range of 400 - 4000 cm^{-1}. KBr method, 2 mg sample weight, grain size of ~ 3–4 μm. *Interpretation: topaz dominance: 3647, 1165, 1085, 1005, 865, 950 (?), 619, 706, 559, 481, 454, 425* cm^{-1}*; inclusion: carbonate: 1744, 1464* cm^{-1}*, traces: mica (overlaps):1028* cm^{-1}*, apatite (overlaps): 1005 a 605, 575 (?)* cm^{-1}.

Table 3. Mineralogical composition of laboratory-produced topaz concentrates for preparation of abrasives. XRD phase analysis (done in laboratory of CGS). The mineral content is given in percent weight.

Sample	Locality and source rock	Topaz	Apatite	Quartz	Mica
[mm]	–	[% weight]	[% weight]	[% weight]	[% weight]
Topaz concentrate 0.125–0.3	Růžek quartzolite	92.5	2.5	4	1
Topaz concentrate 0.125–0.3	Cínovec Sn-W-Li greisen	86	–	6	2

Sample	Locality and source rock	Fluorite	Kaolinite	Comments
[mm]	–	[% weight]	[% weight]	–
Topaz concentrate 0.125–0.3	Růžek quartzolite	–	–	Sericitisation of topaz
Topaz concentrate 0.125–0.3	Cínovec Sn-W-Li greisen	3	3	Small amount of chlorite and cassiterite (approx. 0.5% weight)

adsorption method was used to determine the specific surface area and internal fracture of a topaz abrasive grain according to the pore distribution (see methodology [11]).

The ASAP 2060 Micromeritics and the BET method [12] were used for measurements of the specific surface area and pore distribution by high pressure adsorption of nitrogen. The analysed sample with a grain size of 0.125–0.3 mm was dried and not further processed. The results are shown in Table 4 for a 20 g sample.

Table 4. Topaz concentrates from the Růžek and Cínovec localities. Surface area measured according to the BET method with specification of further parameters.

	Topaz concentrate from Růžek	topaz concentrate from Cínovec
	Quartz-topaz quartzolite	Sn-W-Li greisen
BET surface area [$m^2 \cdot g^{-1}$]	0.0484 ± 0.0006	0.2853 ± 0.0013
Y-intercept [$g \cdot cm^{-3}$ STP]	2.452168 ± 0.132014	0.231361 ± 0.009521
C	36.67	65.93
Q_m [$cm^3 \cdot g^{-1}$ STP]	0.0111	0.0656
Correlation coefficient	0.9997815	0.9999381

Explanations:
STP – standard conditions
BET surface area [$m^2 \cdot g^{-1}$] – grain surface determined according to the BET method
Y-intercept [$g \cdot cm^{-3}$ STP] – under standard conditions
C – BET constant
Q_m [$cm^3 \cdot g^{-1}$ STP] – adsorbed amount of gas

The graph in Fig. 7 shows that the samples of topaz grains (fraction of 0.125–0.3 mm) prepared by technological separation from rocks differ depending on the origin/genesis of the rock (Sn-W-Li greisen has different adsorption parameters than quartz-topaz quartzolite).

For the topaz sample from the Růžek locality, the course of the isotherm predicts that it is a non-porous material (i.e. the distribution of pores is not well developed; with a weak gas-mineral interaction). Isotherm type III. H4 type hysteresis curve (undeveloped slit-shaped pores) [12]. BET specific surface area of grains = 0.0484 ± 0.0006 $m^2 \cdot g^{-1}$, total pore volume of 0.000321 $cm^3 \cdot g^{-1}$.

In contrast, the topaz sample from the Cínovec locality shows signs of a fine-pored structure with mainly slit-shaped pores and probably also cavity pores of gas-liquid inclusions (Fig. 4). Isotherm type IV [12]. The BET specific surface area of grains = 0.2853 ± 0.0013 $m^2 \cdot g^{-1}$, the total pore volume of 0.001449 $cm^3 \cdot g^{-1}$ that is an order of magnitude higher than that of topaz from the Růžek locality (Table 4, Fig. 7).

As far as the grain fracture is concerned, the topaz grains from the Růžek locality are less fractured with the same technology of separation and enrichment of the topaz concentrate.

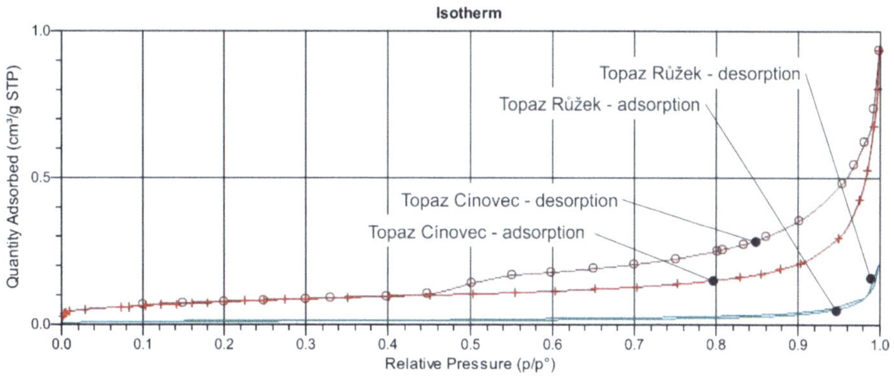

Fig. 7. Nitrogen adsorption isotherms of topaz grains (fraction of 0.125–0.3 mm). Topaz from Růžek and Cínovec localities.

4 Testing of Cutting Performance of Topaz as Abrasive in AWJ Technology

The cutting performance of the topaz abrasive for the AWJ technology was determined experimentally under defined conditions as the ratio of the weights of the collected test material and the reference material (GMA 80 garnet concentrate) after non-through cutting in samples of AISI 304 stainless steel [13]. This material was selected for its well-defined and homogeneous properties. The samples of material to be cut were processed into prisms of 100 mm × 50 mm and 10 mm thick. The tested abrasive was characterised by mineralogical type, designation of the industrial product, input grain size curve and shape properties [14–16]. The experiment was carried out in the Fluid Jets Department at the Institute of Geonics of the Academy of Sciences of the Czech Republic in Ostrava.

Due to the smaller amount of topaz concentrates from both localities that would be required to test cutting performance, the two topaz samples were mixed prior to testing and used as a single abrasive concentrate. The cutting performance of topaz was then compared with other abrasive concentrates used as abrasives in AWJ technology. The technological parameters of AWJ applied in the cutting performance testing are listed in Table 5. The observed relative cutting performance of topaz is given as a percentage of the cutting performance of the reference abrasive (GMA 80).

The cutting performance of the topaz abrasive is 90.5% of the cutting performance of the commercially available GMA 80 abrasive (=100%), which is a better result than when using a comparable andradite-grossular abrasive or quartz sand (Fig. 8).

Table 5. Technological parameters of AWJ.

Parameter/data	Value/type
Water pressure	400 MPa
Water nozzle diameter	0.33 mm
Inner diameter of focusing tube	1.02 mm
Length of focusing tube	76.2 mm
Stand-off distance	2 mm
Abrasive flow rate	300 $g{\cdot}min^{-1}$
Cut material	AISI 304
Abrasive cutting head	Slice I

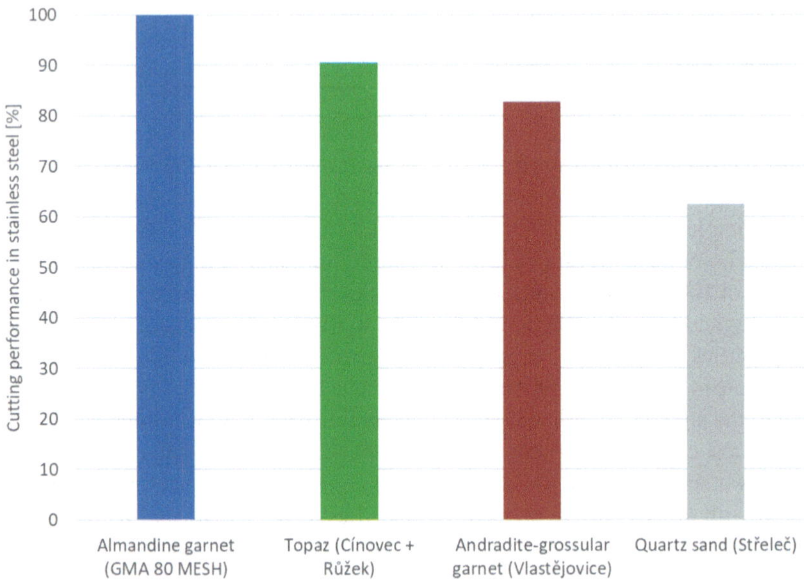

Fig. 8. Cutting performance of four different mineral abrasive concentrates when cutting the AISI 304 steel: almandine garnet GMA80, topaz (mixture of topaz from Cínovec and Růžek), andradite-grossular garnet (Vlastějovice) and quartz sand (Střeleč). All minerals come from the localities in Czechia.

5 Conclusion

Rock containing topaz appears to be a potential source of new mineral abrasive for AWJ technology. The properties of topaz (Tables 1 and 2) as an abrasive for the technology of cutting, machining and surface treatment using a high-speed abrasive water jet can be summarised as follows:

- *Favourable properties:* high hardness, higher specific weight, favourable morphology of grains, low grain fracture. The possibility of separation of larger grains is advantageous.
- *Unfavourable properties:* very good cleavage of topaz according to {001}, numerous micro-inclusions especially in topaz from Sn-W-Li greisen, potentially higher grain fracture due to cracks (natural fracture also during the treatment technology), relatively lower cutting performance in AISI 304 steel compared to almandine garnet.

In the experiments, the topaz abrasive showed a performance of 90.5% when cutting AISI 304 steel compared to the reference GMA 80 almandine garnet. Compared to andradite-grossular garnet concentrate or quartz sand, it showed better cutting performance.

When the water jet interacts with the abrasive particles or the particles hit the inner walls of the cutting head while passing through the mixing chamber and focusing tube, the topaz particles easily break up according to the cleavage surfaces into fine particles with high hardness and higher density. Due to its favourable properties, the waste slurry after cutting can be re-used for other technological operations, such as surface treatment of hard materials (metal, ceramics) by abrasive water jet blasting.

Deposits of topaz quartzolite and greisen with topaz enrichment offer good opportunities for obtaining a concentrate that can be used in AWJ technologies. In addition, this would mean the use of topaz as a secondary raw material after the mining of lithium containing mica.

However, the processing of the raw material into a commercial topaz concentrate as an abrasive for AWJ technology requires the development of an effective technology to process the raw material into a mono-mineral concentrate. In addition, the effect of topaz as an abrasive on the erosion of the internal spaces of the abrasive cutting head would need to be verified, which has not yet been done due to the small amount of abrasive concentrate used for testing. It would also be interesting to investigate the individual suitability of topaz for cutting other materials with AWJ (e.g. glass and glass derivatives) or, conversely, for treating material surfaces using fine fractions. The initial results indicate that this type of topaz abrasive can be a very promising abrasive material if its price is comparative with abrasives with similar performance.

Acknowledgements. This research was funded by Czech Science Foundation, grant number 23-07415S. The authors are very thankful for this support. The authors would also like to express their special thanks to Lucie Gurková for patience and significant benevolence during translation of this manuscript and Catherine Kraina for final linguistic revision by native speaker.

References

1. Klomínský, J., Fediuk, F., Veselovský, F., Schovánek, P., Jarchovský, T., Táborský, Z.: Topaz quartzolite (greisen) near Chrastava in northern Bohemia as a metallogenetic indicator of W - Sn mineralization (in Czech). Geoscience Research Reports in 2003, 112–114, Prague (2004)
2. Martinec, P., Hlaváček, P., Sitek, L., Foldyna, J., Klomínský, J., Veselovský, F., †Fediuk, F.: Topaz quartzolite (greisen) in North Bohemia – potential source of the natural abrasive. Geoscience Research Reports 56(2), 119–127 (2023)

3. Budkiewicz, M.: Quartz-topaz rock from Kameń in Lower Silesia (in Polish). Biur. Państ. Inst. Geol. 58, Warszawa (1949)
4. Opletal, M., Sokol, A.: Topaz-bearing quartzites at the contact with granitoid orthogneisses near Zdobnice in the Orlické hory. J. Czech geol. Soc. **39**(1), 81 (1994)
5. Hejtman, B.: Petrography of the igneous rocks of the Bohemian Massif (part 1. Bohemia - in Czech), Charles University, Prague, Intrusive igneous rocks of W. and N (1984)
6. Anonymous: Torrigton–Tungsten and Topaz. Green Critical Minerals Ltd. (ASX: GCM) (2015)
7. Szymański, A., Szymański, J.M.: Hardness estimation of minerals rocks and ceramics material. Materials Science Monographs 49, Elsevier, Amsterdam-Oxford-NY and PWN Polish Scientific Publishers, Warszawa (1989)
8. Povarennykh, A. S.: Mineral hardness (in Russian). Inz. AN USSR, Kyiv (1963)
9. Godovikov, A. A.: Mineralogia (in Russian). Nedra Moscow (1983)
10. Ruppenthalová, L., Martinec, P., Gurková, L.: Evaluation of Damage of Almandine Garnet Grains by N2 Adsorption Method. In: Hloch, S. et al. (eds.) Advances in Manufacturing Engineering and Materials - Proc. of ICMEM 2018, LNME, pp. 567–574. Springer Nature Switzerland (2019)
11. Pechoušek, J.: Measurement of the surface area of solid substances and determination of their porosity by the gas sorption method. Palacký University in Olomouc, Department of Experimental Physics. Version from 1/25/2010 (2010)
12. ISO 9257:1995 (E): Determination of the specific surface of solids by gas adsorption using the BET method
13. Hlaváček, P., Sitek, L., Foldyna, J.: Evaluation of erosion performance of abrasive particles in abrasive water jet cutting process. MM Science Journal, 3869–3872, (2020)
14. Martinec P.: A contribution to the characteristics of mineral abrasives for WJC. In: Rakowski (ed.) Geomechanics 93 (Proceedings of the International Conference Geomechanics 93, Hradec/Ostrava/Czech Republic/28–30 September 1993), pp. 361–365. A. A. Balkema Publishers, Rotterdam, Netherlands (1994)
15. Foldyna, J., Martinec, P., Sitek, L.: Testing of individual types of abrasives for abrasive water cutting. In: Proc. of 6th pacific Rim Int. Conf. On Water Jet Technology, pp. 127–131. Sydney, Australia (2000)
16. Martinec, J., Foldyna, J., Sitek, L., Ščučka, J., Vašek, J.: Abrasives for AWJ cutting. Academy of Sciences of the Czech Republic, Ostrava, Institute of Geonics (2003)

Numerical and Experimental Study of Oscillating Water Jets Injected in Submerged Conditions

Fernando Kevin Miranda[1](✉), Michal Zeleňák[1], Zdeněk Říha[2], Akash Nag[1], and Kamil Souček[3]

[1] Fluid Jets Department, Institute of Geonics of the CAS, Studentská 1768, Ostrava, Czechia
kevin.miranda@ugn.cas.cz

[2] Fluid Jets Department, Institute of Geonics of the CAS, Drobného 38, Brno, Czechia

[3] Department of Geomechanics and Mining Research, Institute of Geonics of the CAS, Studentská 1768, Ostrava, Czechia

Abstract. This paper studies the bulk effects generated during the underwater injection of self-excited oscillating water jets using a designed and manufactured in-house fluidic nozzle. As a first step to predict and analyze the process, transient tridimensional turbulent flows simulations were carried out, wherein two classic turbulent models (RANS and LES), with mesh refinement, were compared regarding their detailed description of the jets-induced vorticity and velocity fields. The submerged jet oscillation frequency was computed through CFD post-processing tools and it was found to be in a good match with the frequency value experimentally estimated by means of a pressure sensor at the desired input pressure level. A liquid-vapor phase change model allowed also the numerical computation of the cavitation field evolution during the jet injection, results of which correlated qualitatively well with the cavitation clouds footage recorded through the stroboscopic shadowgraph technique. Numerical and experimental results evidenced the important presence of cavitation and vorticity as constant effects of the studied process. Furthermore, submerged jet impingement to a soft plastic surface was performed to capture the added destructive potential of these effects and the final volumetric imprints were analyzed using digital microscope and computed tomography.

Keywords: Oscillating water jets · Submerged jets · Water jet soft erosion · Coherent structures submerged

Nomenclature

SEOJ	Self-Excited Oscillating Jets
NS	Navier-Stokes
LES	Large Eddy Simulations
SOD	Stand-Off Distance
CFD	Computational Fluid Dynamics
CT	Computed Tomography
FFT	Fast Fourier Transform

L. Sitek et al. (Eds.): Advances in Water Jetting II, LNME, pp. 93–109, 2025.
https://doi.org/10.1007/978-3-031-72778-8_8

PIV Particle Image Velocimetry
RANS Reynolds-averaged Navier-Stokes

1 Introduction

Fluidic oscillators, also called fluidic nozzles, are devices with the ability to induce self-excited oscillating jets (SEOJ), based on fluid dynamic interactions leading to flow instabilities inside a mixing chamber, without the use of moving parts. These devices have their roots in the field of fluid logic, the goal of which was to investigate different concepts and designs to achieve non-electronic analogic and digital operations – one of the best achievements being the assembly of fluid-based computers [1]. One of the first patents of this kind of device was made by Nikola Tesla in 1920 under the name of *valvular conduit*.

Fluidic oscillators can be divided into two representative types regarding the feedback channel (see Fig. 1): i) feedback-free oscillators, which depend solely on the collision of two jets inside the mixing chamber to generate periodic motions in the output stream [2]; and ii) two feedback channels oscillators, which are used in a way that each channel captures, one at a time, a portion of the main stream, guiding it back to the mixing chamber inlet. Once released, it pushes the main stream to the other side [3]. This process is periodically repeated. Depending on the application, this sort of device can be designed to generate continuous or discrete jets at different oscillation frequency ranging from $O(10^0)$ Hz to $O(10^4)$ Hz [4–6].

SEOJ are currently used in the industry for irrigation purposes, windshield cleaners, surface coating and surface treatment, and in general they show great efficacy over conventional methods [7]. SEOJ allows the applicability extension of WJ technology to new fields like medicine, propulsion, fuel injection and flow control, to mention a few [8–12]. Submerged jets injection with the right input parameters, could improve the material surface treatment and also the disintegration abilities of water jets [13–16], this open potential applications for reefs removal from oil rigs and subsea pipelines, cleaning of dam walls, underwater descaling and scuba archeology to mention a few. Since the information on underwater operation of SEOJ is still very scarce, this work intends to numerically and experimentally analyze the fluid interaction generated by the submerged operation of oscillating water jets and their effects imprint in soft materials.

2 Materials and Methods

2.1 Fluidic Nozzle Description

The selected fluidic oscillator used in this study (referred to as nozzle herein after) has a feedback-free mixing chamber type based on a pressure-jets interaction fluidic oscillator [17]. A wall-bounded thin cavity (1 mm from wall to wall) works as the mixing chamber. It has two pressurized inlets and one rectangular output slot. The outlet rectangular slot has an area of 2.5 mm^2 (detailed oscillator design can be provided upon request). The input power jets are directed in such angle that they collide and interact at the centerline

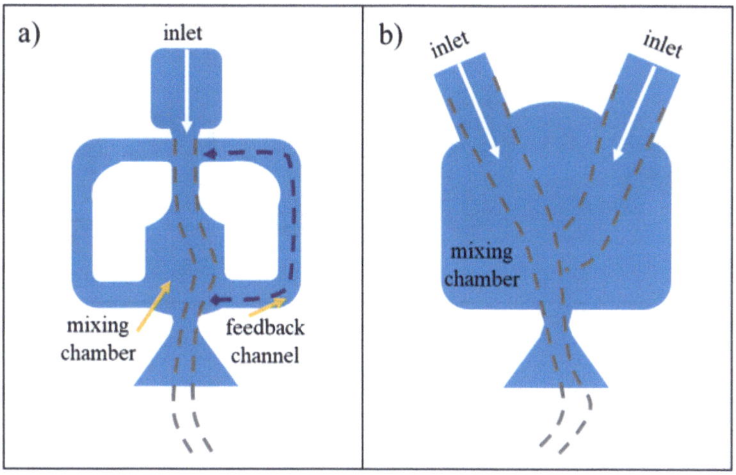

Fig. 1. Fluidic oscillator types: a) wall-attachment oscillator and b) jets interaction oscillator.

of the mixing chamber [18]. Fluid flow instabilities are then achieved, amplified and directed towards the output slot, where the fluid is accelerated to high-speed jets. The result of the fluid interactions inside the mixing chamber and of the fluid acceleration is a periodically oscillating water jet. The stainless-steel manufactured nozzle comprises several pieces, which are later assembled as shown in panels "a" and "b" of Fig. 2. Panel "c" shows important measured operational characteristics of the nozzle in question, such as flowrate and the hydraulic power for input pressure values ranging from 0.5 bars to 41 bars.

2.2 Numerical Methods

Computational Fluid Dynamics (CFD) is a powerful tool which helps, nowadays, with the research of complex fluid processes providing accurate description. Technology and new implementations allow CFD to simulate complex fluid interactions, e.g. multiphase-fluid interactions, fluid phase change, fluid structure interactions and chemical reactions in fluid processes.

In this study, the numerical simulation of SEOJ discharged underwater is provided. *OpenFOAM* platform is used to carry out the simulations, considering transient and turbulent fluid flows simulations for a fluid with water properties at room temperature (density of 997 $kg{\cdot}m^{-3}$ and dynamic viscosity of 8.9×10^{-9} Pa·s). The Finite Volume Method is employed to compute the Navier-Stokes (NS) partial differential equations to describe the fluid behavior. Turbulence modeling approaches are taken into account to avoid expensive computation effort. Before performing the case study simulations, two turbulent models are tested: i) Large Eddy Simulations (LES), a filtering method with the main focus of the detailed calculation of larger length scales, using sub-grid models to calculate smaller length scales; and ii) Reynolds-averaged Navier-Stokes (RANS), which uses time-averaging methods based on the Reynolds decomposition and is demonstrated to be less computationally expensive than LES.

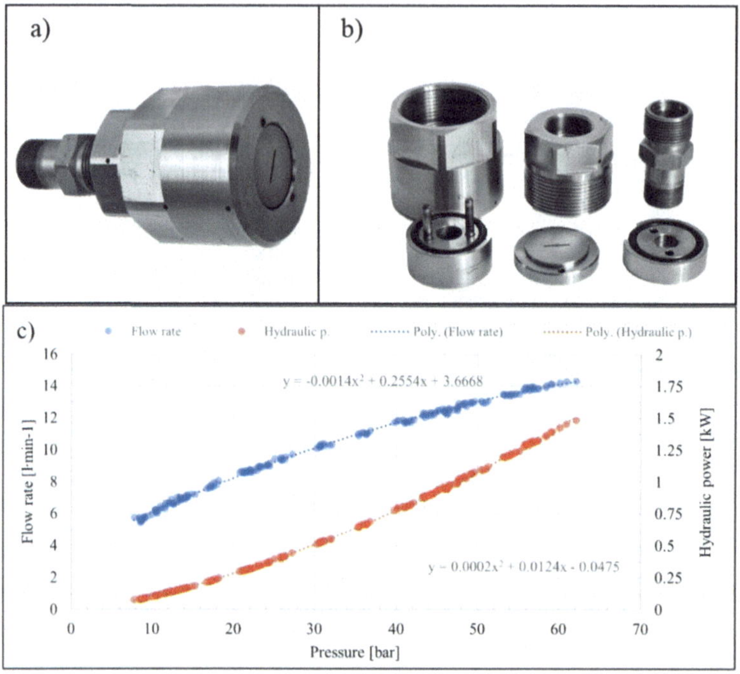

Fig. 2. Stainless-steel-built nozzle is shown fully assembled in panel "a" and before its assembly in panel "b". Panel "c" shows the nozzle operation flowrate (blue colored) and the hydraulic power (red colored) in function of the input pressure.

Figure 3 shows the comparison of results obtained using Smagorinsky (LES) and k-omega-SST (RANS) turbulent models in two different analyses: the velocity field contour and coherent vortical structures of an oscillating jet injected in a fluid. The Q-criterion (proposed by Hunt et. al [19] and redefined in terms of instantaneous velocity gradient tensor by Chong et. al [20]) was employed in this study in order to capture the vorticity coherent structures. During the calculations, RANS demands lower computational effort, but as evidenced in the figure, its results lack the multiscale details provided by LES model, which allows more reliable description of cavitation-vortex interactions.

The strong presence of vorticity due to the turbulent jets induces low-pressure zones, which depending on the fluid temperature reach the saturation vapor pressure, leading to the formation of vapor pockets (cavitation). This change of phase can be taken into account using mass transfer models like Schnerr-Sauer model [21], which considers phase transition from liquid to gas and vice versa based on a saturation pressure threshold. The evolution and interaction of two-phase fluid are described using the volume of fluid method (VoF).

Figure 4 shows the tridimensional model domain (5 times wider and 6 times longer than the fluidic nozzle width and height respectively) discretized through the *snappyHexMesh* utility employing different levels of refinement, where the maximum cell size is 0.0013 m and the minimum one is 0.00015 m. These refinement levels were chosen based on simulated oscillation frequency results, which determined the minimum

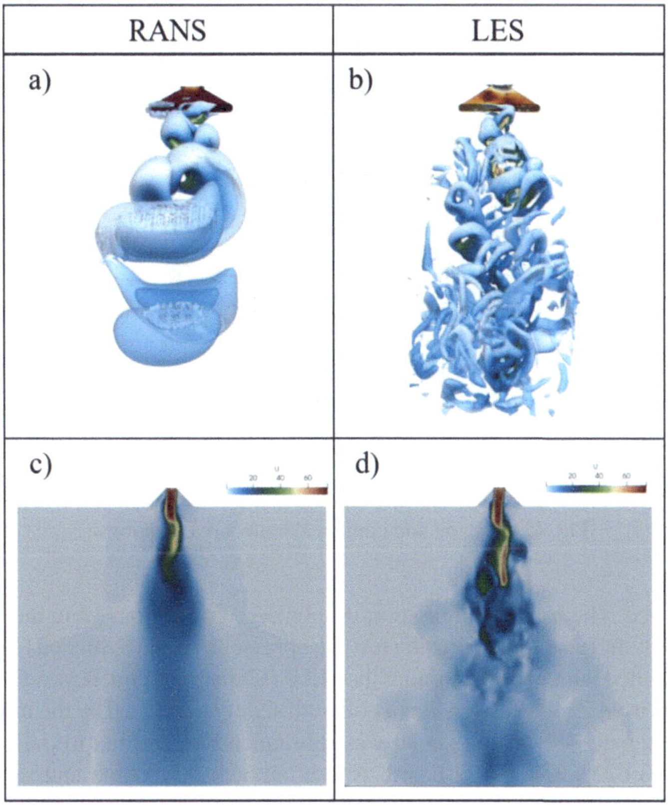

Fig. 3. CFD simulation results using RANS (left panel) and LES (right panel) models, where a) and b) show the coherent structures comparison using both methods; c) and d) show the sliced velocity contour of an oscillating fluid jet (expelled from the fluidic nozzle outlet) inside a fluid.

amount of cells needed to reach the mesh independency is 5 in the narrowest point of the nozzle and 5 in the chamber walls gap. The domain discretization is composed by 583005 hexahedral, 5594 prisms and 13044 polyhedral cells. The nozzle comprises a pressure inlet and wall boundary conditions, while the rest of the domain (cylindrical) uses a wall as upper boundary condition, and outlet pressure for the rest of the surfaces. Three layers of refinement (with size 0.00015 m) were added in all the fluidic nozzle walls.

2.3 Experimental Set up

The nozzle used for the generation of SEOJ was connected to a hydraulic line consisting in a Kärcher HD 7/18 pressure pump (max. flow rate of 11.7 l·min^{-1} at the operating pressure of 14.5 MPa). The tested nozzle was guided by an ABB 6 axis manipulator. The water pressure regulation was controlled with a bypass line considering a restriction valve for the reduction of the overflow. A standard USB pressure sensor was used for monitoring the online adjustment of the water pressure. One level of inlet pressure (25

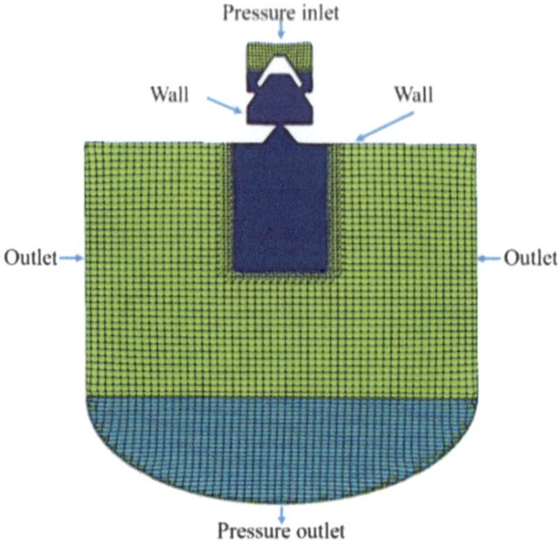

Fig. 4. Meshing and numerical boundary conditions.

bars) was tested. The basic configuration in submerged conditions were maintained for diagnostic tests and also for the erosion test. The pressure sensor Kistler 6211 was placed into the tank filled with water to capture the SEOJ oscillation frequency. A single charge amplifier was used for the conversion of coulomb signals generated by the impact sensor to a voltage signal. The analog signal was subsequently converted to a digital format using NI USB 6251 BNC DAQ board. A total amount of 20 thousand samples were acquired in a time interval of 1 s. The nozzle was placed at a standoff distance (SOD) of 20 mm above the sensor. Placing the robot manipulator to the measuring position and using an electronic trigger, the signal acquisition was activated. A schematic illustration of the experimental assembly is presented in Fig. 5.

For the visualization in submerged conditions, the following equipment was used: a high-speed camera LaVision v2011 CMOS (22.6 kHz, full resolution of 1680 × 1200 pixels); 2 x 120W spot high-energy pulsed LED lights driven by Veritas timing units; a LaVision, PTU 9 controller and DaVis 8.4.6 software for recording, image post-processing and analysis. The configuration was set according to the scheme presented in Fig. 6. The experimental images were acquired using time-resolve high-speed recording (91.2 kHz) with 51 × 95 mm of area of interest (389 × 744 pixels). The camera exposure time was set at 1 μs, the shortest value possible, as the LED lights were operated with pulse length of 3 μs.

Figure 7 shows the schematic representation of the nozzle and the soft plastic disposition for the erosion test, where the oscillation direction is orthogonal to the nozzle traverse motion. The SOD (standoff distance) used was 3 mm, with a traverse speed of 1 mm·sec-1. The experimental setup depicted in Fig. 5 is also employed to generate SEOJ. The workpiece material used was a PVC based plastic GelFlex with no important thermal sensitivity below 165 °C, selected mainly because of its lower mechanical resistance (Shore hardness A40), allowing the capture of some underwater hydrodynamic mechanisms effects at low hydraulic power levels.

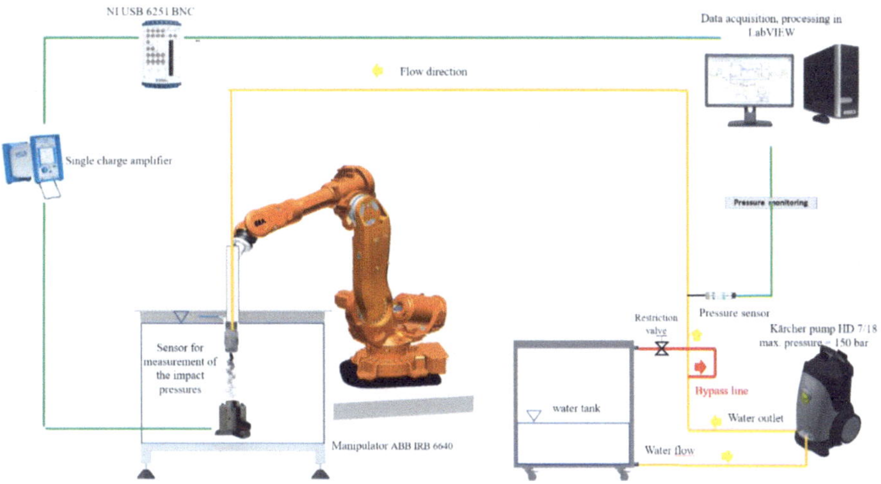

Fig. 5. Schematic representation of the experimental setup and frequency acquisition procedure.

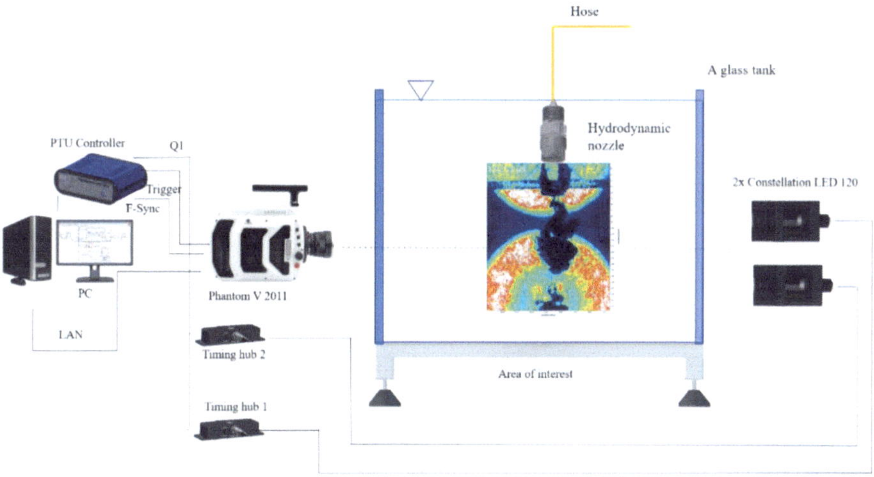

Fig. 6. Schematic illustration of the visualization technique.

As a first step, a digital microscope scanner Keyence VHX-5000 is used to get a clear idea of the groove shape generated during the test. The data collected by this method rely on the light access to the sample, which means in some cases a partial representation of the real volumetric shape generated by the erosion.

A complete representation of the erosion grooves can be given by X-Ray industry computed micro tomography (μ-XCT). This method provides 3D representation of the studied sample, allowing a complete and detailed description of the groove habitus created by the water jet and by extension a more accurate determination of the shape and removed material volume under water. The device XT H 225 ST - X-ray computed tomograph (Nikon Metrology) was used to scan 3D tomographic volumes of the eroded

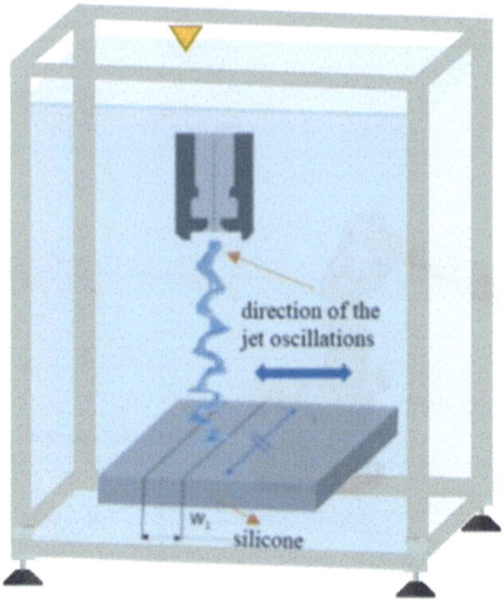

Fig. 7. Traverse water jet orientation.

plastic workpiece. The volume analysis and the visualization of the internal 3D structure of the computed tomograph (CT) scanned samples were carried out employing the "VGSTUDIO MAX 3.3.2" software from Volume Graphics GmbH (Table 1).

Table 1. X-ray setting parameters of the tomographic system and the resulting CT volumes (parameters for smaller test samples are listed in parentheses).

X-ray source setting (Reflection target)	Operating voltage (kV)	195
	Power outputs (W)	50
	Focal spot size (μm)	Cc 45
	Cu Filter, thickness (mm)	0.0
Scan parameters (scanning time cc. 1.5 h)	Radiograph image exposure (ms)	500
	Frame per one radiograph projections (–)	2
	16-bit Flat X-Ray detector, Number of pixels (–)	2000 x 2000
	Number of the radiographic projections (–)	3142
Origin 3D CT Volume	Cubic voxel size (μm)	40.1(16.0)
	CT volume data type	8 bit-256 gr. Level
	Number of Voxels in CT Volume (–)	Cc 3 000 000 000
	CT volume size (mm)	Cc 70 x 70 x 40 (23 x 23 x 20)

3 Results and Discussions

3.1 CFD Results: Velocity Field, Coherent Vortical Structures and Cavitation

Figure 8 shows the velocity field contour of the discharged water medium perturbed by an oscillating jet, by means of CFD simulations for an input pressure of 25 bars. The time-sequence images show three positions during a half period of the water jet oscillation. The maximum velocity of the jet core is stretched at the oscillation extremes, making it possible to reach 10 mm vertically away from the diffusor (panel "a" and "c"). On the other hand, the maximum velocity is not stretched at the oscillation middle zone and it can reach 8 mm vertically (panel "b"). Therefore, the freed energy is more concentrated at the middle oscillation zone than in the extreme ones. The maximum vertical velocity achieved is 76 m·s^{-1} and transversally the sweeps can reach up to 21 m·s^{-1}.

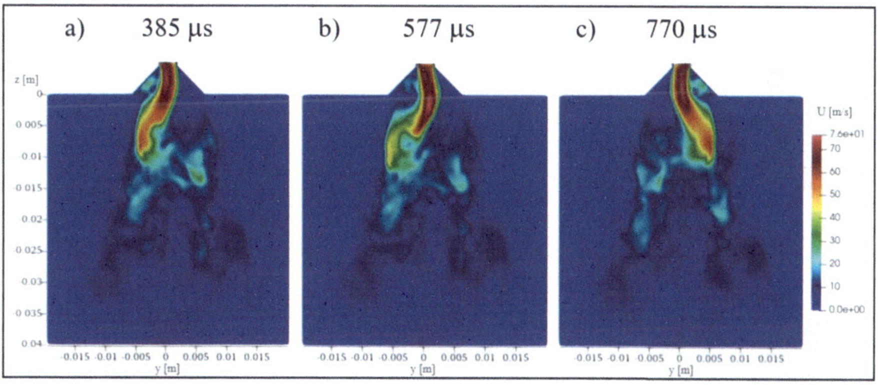

Fig. 8. Three contour plots of the velocity field of half period sweeping oscillation.

From the velocity information one can retrieve the fundamental oscillation frequency of the oscillating water jet by monitoring the velocity signals at a point close to the nozzle exit. Then the Fast Fourier Transform (FFT) is performed on the recorded signal. Figure 9 plots the frequency spectra of the signal, where the jet oscillation frequency is evident at 1280 Hz, for a nozzle input pressure of 25 bars in submerged conditions.

Another important analysis regards the vorticity dynamics presence during the oscillating jet injection in water. The coherent vortical structures generated during the injection of SEOJ in submerged conditions are obtained using the Q-criterion. Figure 10 shows the coherent structures detected during the fluid-fluid interaction at three different positions of one half of an oscillation period. At all times, the generation of a three-dimensional helical structure wrapped around the jet core is observed. This structure is deformed with the fast core sweeping motion and is cut at distances far enough from the nozzle exit, forming secondary sporadic ring, hair-pin, and horse-shoe vortical structures.

The permanent presence of vorticity and high-speed fluid motions induce low-pressure zones where cavitation is formed. Using a phase change model, one is able

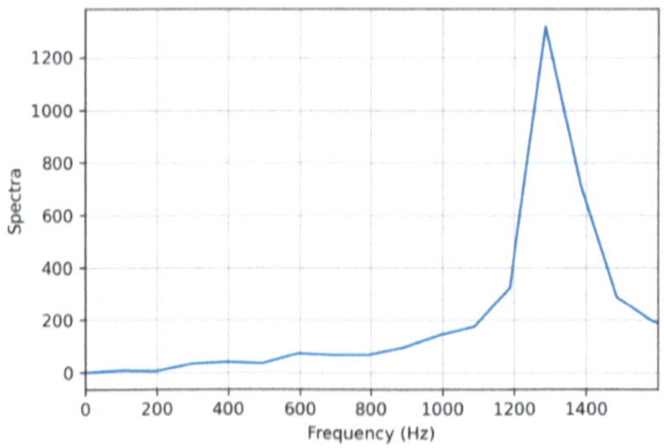

Fig. 9. Submerged SEOJ oscillation frequency calculated from CFD simulations.

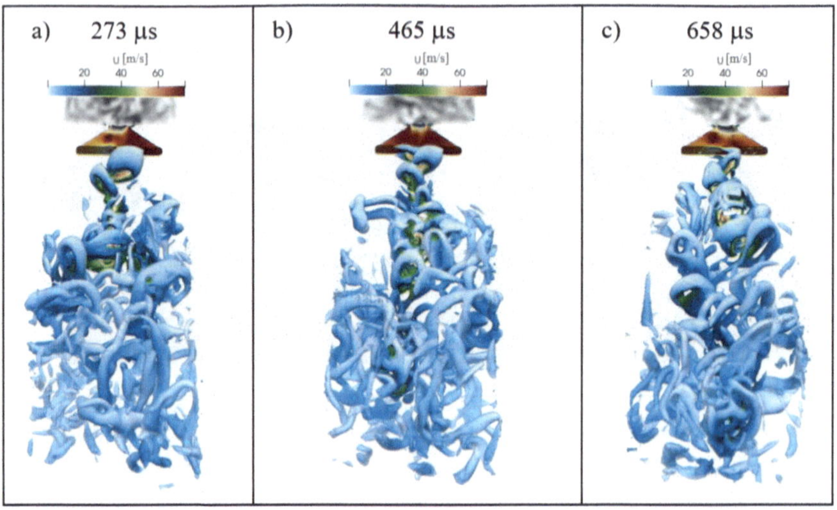

Fig. 10. Vorticity coherent structures using Q-criterion and colored by the velocity magnitude. Three panels based on the direction of the water jet: a) left directed WJ, b) vertical directed WJ and c) right directed WJ.

to numerically represent the cavitation intensity for an input pressure of 25 bars, atmospheric condition at the fluid bulk (assuming shallow water submergence) and temperature of 20 °C as observed in Fig. 11. The panels show sequentially half of an oscillation period, where the white colored phase represents the cavitation cloud evolution and the colored background shows the correlated velocity field contour. The amount of cavitation clouds during SEOJ operation is not negligible at short distances from the nozzle exit, as they might affect underwater processes, especially generate an extra damage to surfaces where the clouds periodically collide.

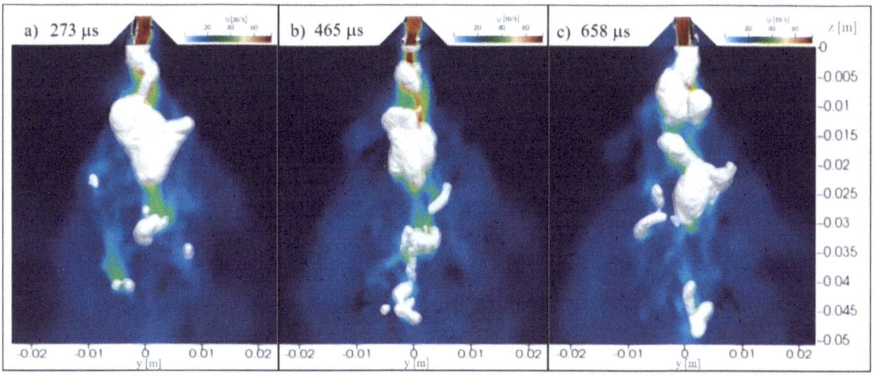

Fig. 11. Cavitation (vapor) phase represented as white with the velocity field contour as background at three SEOJ positions.

3.2 Experimental Results: Cavitation Visualization, Oscillation Frequency and Soft Erosion

As mentioned above, submerged jets injection in a fluid medium induces low-pressure zones. Cavitation-vortex interactions are therefore constantly generated during the studied process [22, 23]. The shadowgraph technique is applied to observe these sorts of dynamic interactions, and as a result (see Fig. 12), one can obtain in the footage constant ghosts-like blobs flowing out of the nozzle, resembling the cavitation clouds behavior presented in Fig. 11 (based on saturated vapor pressure models), which confirms qualitatively the results obtained using CFD. The experimental images in Fig. 12 allow the clear observation of the cavitation inception of the chosen nozzle at two places: i) at the outlet orifice (caused by high-speed jets) and ii) at the outlet diffusor walls, due to the rapid transversal motion of the jet. This footage also evidences the two-dimensional cavitation-vortex state inside of the 1mm diffusor orifice and the three-dimensional cavitation-vortex structure expansion after the jet exits this thin orifice. These vapor clouds are further advected by the oscillating jet towards the discharging medium where they eventually collapse.

The measured nozzle oscillation frequency with input pressure of 25 bars and a flowrate of 6.6 $\mathrm{l{\cdot}min^{-1}}$, was calculated from the impact pressure signal recorded for a short time. The signal was then taken to the spectral domain where the fundamental oscillation frequency of 1230 Hz was found (see Fig. 13).

The oscillating and the intermittent nature of water clusters impact on the workpiece surface leads to material fatigue, tearing and eroding it. Nevertheless, the erosion characteristics, e.g. shape, depth and width, depend more on the fluid jet structure and on the fluid behavior close to the material surface. Erosion thus helps to evaluate the oscillating jet and cavitation-vortical effects near a surface.

The erosion test consisted in the generation of grooves employing SEOJ injected underwater at constant pressure of 25 bars, where the nozzle was advanced along the plastic workpiece at constant SoD and at constant speed of 1 $\mathrm{mm{\cdot}s^{-1}}$. Figure 14 shows the groove shape and dimensions obtained using a digital scanner, where the analyzed scanned area of interest was 23x23 mm. The obtained average groove depth and width,

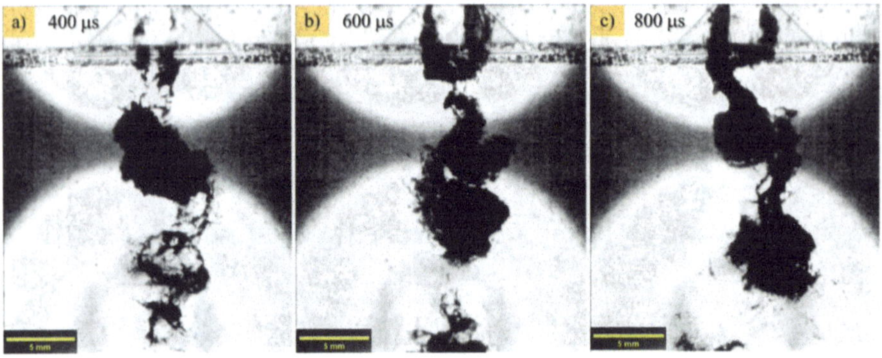

Fig. 12. Shadowgraph sequential visualization of submerged SEOJ during half period of oscillation: a) jet directed to the left extreme, b) jet directed towards the center and c) jet directed to the right extreme.

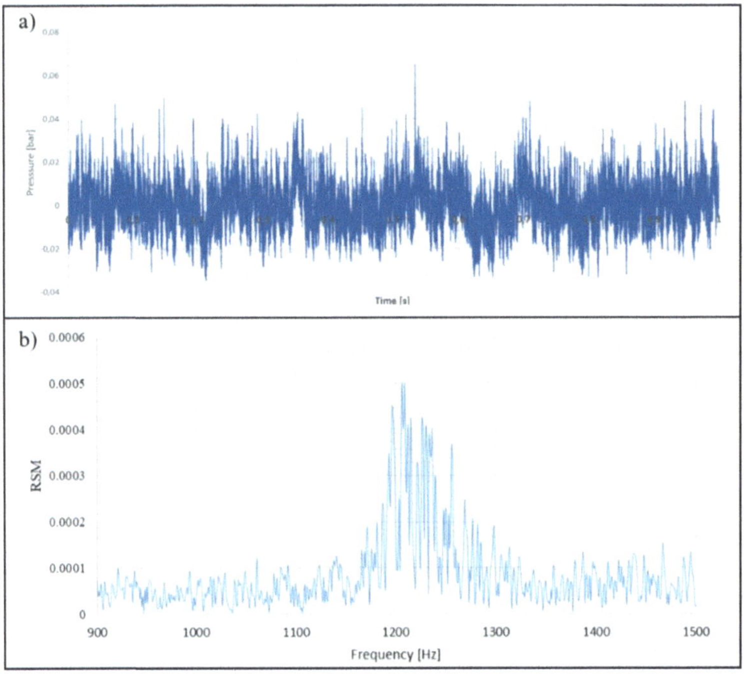

Fig. 13. a) Distribution of the base impact signal at time; b) FFT analysis of the impact signal under 25 bars (frequency of 1230 Hz).

from 10 measurements in a single groove, was 3.08 ± 0.41 mm and 5.20 ± 0.39 mm respectively.

Visual and physical inspection after the experiment showed the presence of intrinsic cuts along the middle zone of the groove, which were not captured by the microscope

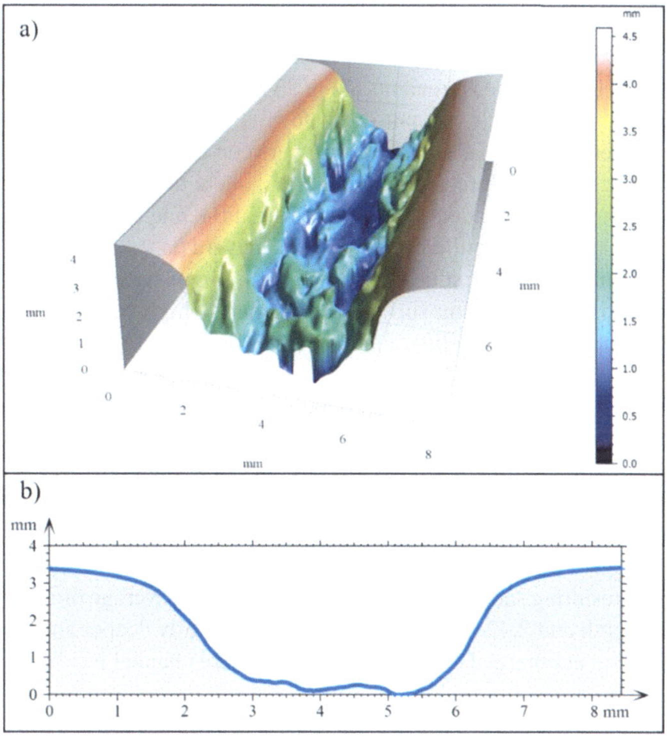

Fig. 14. Scan results using digital microscope. Panel "a" shows the 3D view of the eroded groove and panel "b" depicts the average groove shape obtained.

scan. In order to have better visual access to the cut using non-destructive methods, a CT scan was carried.

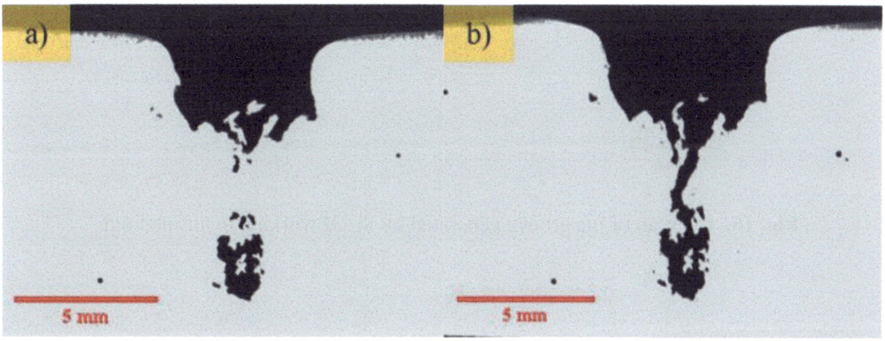

Fig. 15. Visualization of lateral CT slices of the a) submerged conditions, b) CT slice scan showing the connection with the lower channel.

In contrast to the digital microscope results, the CT images (Fig. 15) show that, parallel to the groove, a subsurface channel was created, having its lowest point at 8.83 ± 0.38 mm (average of 10 measurements along 23 mm in a single groove) from the non-damaged surface level, which was not detected by digital microscope analysis. Panel "a" shows that there is no apparent connection between the upper groove and the lower channel, since the cohesive forces of this material adhere the water-cut walls, however one can use external forces to detach them and reveal the fine cut connection existence (panel "b" shows the sample pulled transversally outwards).

As the cavitation bubbles, induced by jet injection in water medium, implode continuously close to the soft plastic surface, they generate high-impact force micro-jets enhancing further the erosion ability of the jet [24–26]. Figure 15 shows how the central region of the groove is a critical cavitation impact concentration zone, where the micro-jets are strong enough to cut through the material. Then the main jet squeezes several exhaust subflows through the cut, eroding it further and flowing backwards (with respect to the traversing nozzle direction), creating a subsurface channel parallel to the groove.

To confirm that this subsurface channel is a characteristic of a jet operating in submerged conditions, a soft erosion is carried-out at the same input pressure (25 bars) and using the same operational conditions but in air medium. Figure 16 shows the CT scan performed to its resulting sample. Although the groove, with average dimensions of 4.23 ± 0.32 mm of depth and 2.42 ± 0.27 mm of width, is slightly deeper and narrower than the one obtained in submerged conditions, no subsurface channel has been observed to form.

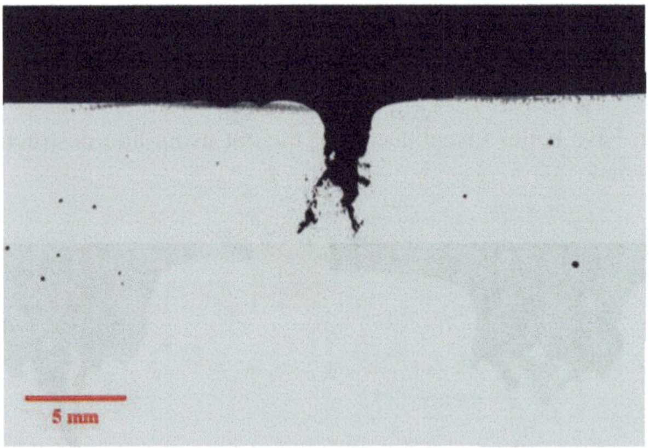

Fig. 16. CT scan of the groove generated by SEOJ working in air medium.

4 Conclusions

In this work we investigated the interactions produced by SEOJ operating in shallow-underwater conditions. Both numerical and experimental methodology was employed.

As a first step, CFD simulations were carried out in order to predict and understand the physical phenomenology involved in this process. The finite volume discretization method of the *OpenFOAM* platform was employed to solve the fluids momentum, continuity, multiphase approach and phase change model equations. Since the type of process studied in this work involves turbulent flows, RANS and LES turbulent models were compared prior their usage, determining that LES is more suitable for the accurate description of the fluid flow behavior caused by the analyzed process. Characteristic flow fields, coherent structures and cavitation clouds evolution of submerged SEOJ were qualitatively described for a nozzle with an input pressure of 25 bars. The maximum longitudinal velocity value reached during the oscillation was 76 $m \cdot s^{-1}$ and the maximum transversal velocity was 21 $m \cdot s^{-1}$. The velocity signal acquired from the numerical simulation was also analyzed at the nozzle output to determine the oscillation frequency of 1280 Hz.

The second part of the analysis relied on optical methods and direct measurement of the impact pressure. The stroboscopic shadowgraph technique was employed to observe the qualitative behavior of SEOJ. Although the jet could not be captured in the footage, it shows clearly how cavitation clouds evolve instead. The cavitation clouds' shapes, intensity and behavior correlate with the ones observed in the numerical simulations. The signal of the impact pressures was experimentally recorded with a pressure sensor and mapped to the spectral domain to obtain the dominant frequency of the signal of 1230 Hz, which is very close to the value calculated by the CFD simulations.

Finally, erosion tests were carried out in a soft material, in order to understand the cavitation-vortex effects during the oscillation process and the type of material decoupling created by these effects. The eroded sample was analyzed using digital microscope and computed tomography technology. CT gave better and more detailed visualization of the eroded grooves. Moreover, it allowed to discover a hidden channel formed several millimeters below the groove, because of cavitation induced micro-jets close to the soft plastic surface. A non-submerged erosion test, using the same input conditions, was performed and compared to the underwater results, revealing that the latter presents more removed volume.

To conclude, this work presented an initial study of the behavior and effects caused by the injection of SEOJ in submerged conditions. Further investigation is required to capture all the characteristics that this process implies. The suggested future work contemplates optical methods (e.g. PIV) usage to confirm the CFD calculated velocity vector fields and submerged jets morphology; it also contemplates more erosion tests for the comparison of in-air and underwater SEOJ performance.

Acknowledgment. This work was supported by the Czech Science Foundation under grant project 23-05235S.

References

1. Belsterling, C.: Fluidic systems design. Wiley-Interscience (1971)
2. Gregory, J.W., Sullivan, J.P., Raghu, S.: Visualization of jet mixing in a fluidic oscillator. J. Vis. **8**, 169–176 (2005). https://doi.org/10.1016/j.measurement.2023.113530
3. Seo, J.H., Zhu, C., Mittal, R.: Flow physics and frequency scaling of sweeping jet fluidic oscillators. AIAA J. **56**, 2208–2219 (2018). https://doi.org/10.2514/1.J056563
4. Woszidlo, R., Krüger, O.: Editorial for the special issue on "fluidic oscillators—devices and applications. Fluids **7**, 91 (2022). https://doi.org/10.3390/fluids7030091
5. Srinivas, T., Vasudevan, B., Prabhu, A.: Performance of fluidically controlled oscillating jet. In: Liepmann, H.W., Narasimha, R. (eds.) Turbulence Management and Relaminarisation, pp. 485–494. Springer Berlin Heidelberg, Berlin, Heidelberg (1988). https://doi.org/10.1007/978-3-642-83281-9_33
6. Gregory, J.W., Sullivan, J.P., Raman, G., Raghu, S.: Characterization of the microfluidic oscillator. AIAA J. **45**, 568–576 (2007). https://doi.org/10.2514/1.26127
7. Říha, Z., Zeleňák, M., Kruml, T., Poloprudský, J.: Comparison of the disintegration abilities of modulated and continuous water jets. Wear **478–479**, 203891 (2021). https://doi.org/10.1016/j.wear.2021.203891
8. Zelenak, M., et al.: Theoretical and practical research on micro hydrodynamic nozzles for modulated water jet generation as an innovation to replace continuous water jet in soft materials erosion. Measurement **221**, 113530 (2023). https://doi.org/10.1016/j.measurement.2023.113530
9. Tesař, V., Zhong, S., Rasheed, F.: New fluidic-oscillator concept for flow-separation control (2013)
10. Ostermann, F., Woszidlo, R., Nayeri, C.N., Paschereit, C.O.: The interaction between a spatially oscillating jet emitted by a fluidic oscillator and a cross-flow. J. Fluid Mech. **863**, 215–241 (2019). https://doi.org/10.1017/jfm.2018.981
11. Otto, C., Tewes, P., Little, J.C., Woszidlo, R.: Comparison of fluidic oscillators and steady jets for separation control on a wall-mounted hump (2018). https://doi.org/10.2514/6.2018-1281
12. Lin, J.C., Whalen, E.A., Eppink, J.L., Siochi, E.J., Alexander, M.G., Andino, M.Y.: Innovative flow control concepts for drag reduction. In: 54th AIAA Aerospace Sciences Meeting, American Institute of Aeronautics and Astronautics, San Diego (2016). https://doi.org/10.2514/6.2016-0864
13. Chen, Y., Fang, Z., Xiong, T., Hou, W., Zhang, Z., Shi, R.: An experimental study on the erosion of sandstone by self-excited oscillation cavitation waterjet in submerged environment. Ocean Eng. **279**, 114546 (2023). https://doi.org/10.1016/j.oceaneng.2023.114546
14. Yang, Y., Li, W., Shi, W., Wang, C., Zhang, W.: Experimental study on submerged high-pressure jet and parameter optimization for cavitation peening. Mech. **26**, 346–353 (2020). https://doi.org/10.5755/j01.mech.26.4.27560
15. Hutli, E., Nedeljkovic, M.S., Bonyár, A., Radovic, N.A., Llic, V., Debeljkovic, A.: The ability of using the cavitation phenomenon as a tool to modify the surface characteristics in micro- and in nano-level. Tribol. Int. **101**, 88–97 (2016). https://doi.org/10.1016/j.triboint.2016.04.006
16. Vidvans, A. N., Melkote, S. N., Sanders, D. G.: Enhancing cavitation intensity in co-flow water cavitation peening with organ pipe nozzles. J. Manufac. Sci. Eng. **143** (2021). https://doi.org/10.1115/1.4049649
17. Raghu, S.: Feedback-free fluidic oscillator and method. US6253782B1 (2001). https://patents.google.com/patent/US6253782B1/en. Accessed 2 Nov 2023
18. Tomac, M.N., Gregory, J.W.: Internal jet interactions in a fluidic oscillator at low flow rate. Experim. Fluids **55**, 1730 (2014). https://doi.org/10.1007/s00348-014-1730-8

19. Hunt, J.C.R., Wray, A.A., Moin, P.: Eddies, streams, and convergence zones in turbulent flows. In: Center for Turbulence Research Proceedings of the Summer Program 1988 (1988)
20. Chong, M.S., Perry, A.E., Cantwell, B.J.: A general classification of three-dimensional flow fields. Phys. Fluids A **2**, 765–777 (1990). https://doi.org/10.1063/1.857730
21. Schnerr, G.H., Sauer, J.: Physical and numerical modeling of unsteady cavitation dynamics (2001)
22. Ji, B., Luo, X., Arndt, R.E.A., Wu, Y.: Numerical simulation of three dimensional cavitation shedding dynamics with special emphasis on cavitation–vortex interaction. Ocean Eng. **87**, 64–77 (2014). https://doi.org/10.1016/j.oceaneng.2014.05.005
23. Kida, S., Goto, S., Makihara, T.: Low-pressure vortex analysis in turbulence: life, structure, and dynamical role of vortices. In: Bajer, K., Moffatt, H.K. (eds.) Tubes, Sheets and Singularities in Fluid Dynamics, pp 181–190, Springer Netherlands, Dordrecht (2002). https://doi.org/10.1007/0-306-48420-X_25
24. Dular, M., Požar, T., Zevnik, J., Petkovšek, R.: High speed observation of damage created by a collapse of a single cavitation bubble. Wear **418–419**, 13–23 (2019). https://doi.org/10.1016/j.wear.2018.11.004
25. Hutli, E., Nedeljkovic, M., Bonyár, A.: Dynamic behaviour of cavitation clouds: visualization and statistical analysis. J. Braz. Soc. Mech. Sci. Eng. **41**, 281 (2019). https://doi.org/10.1007/s40430-019-1777-9
26. Hutli, E., Nedeljkovic, M., Bonyár, A.: Controlled modification of the surface morphology and roughness of stainless steel 316 by a high speed submerged cavitating water jet. Appl. Surf. Sci. **458**, 293–304 (2018). https://doi.org/10.1016/j.apsusc.2018.07.007

Optimisation of Abrasive Waterjet Milling Technique for Pocket Milling Operations

Akash Nag[1(✉)], Dagmar Klichová[2], Petr Hlaváček[2], Jana Petrů[1], and Sergej Hloch[1]

[1] VŠB - Technical University of Ostrava, Faculty of Mechanical Engineering, Ostrava, Czechia
akash.nag@vsb.cz

[2] Institute of Geonics of the CAS, Ostrava, Czechia

Abstract. The study deals with statistical optimisation of the input parameters required for the implementation of mask-less milling of AISI 304 using abrasive waterjet technology. In this study, the effect of traverse speed (1000 and 3000 mm/min), abrasive flow rate (100 and 200 g/min) and standoff distance (4 mm and 15 mm) for a step-over distance of 0.6 mm were studied on the output responses of width, depth, surface roughness parameter (Rz). A total of 8 experiments using full factorial design were performed, and the experimental results were analyzed using the Minitab software. The results demonstrated that the maximum possible width is achieved by the execution of lower traverse speed, higher abrasive flow rate and higher standoff distance; the maximum possible depth is achieved by the execution of lower traverse speed, higher abrasive flow rate and lower standoff distance; and the execution of higher traverse speed, higher abrasive flow rate and lower standoff distance achieves the minimum possible Rz. By performing 'desirability approach' optimization, the optimum parameter configuration was found to be a traverse speed of 1622 mm/min, abrasive flow rate of 200 g/min, and standoff distance of 4 mm. The study concludes that the utilization of multiple input variables within the experimental domain to achieve more accurate results as per the desired output requirements.

Keywords: Abrasive Water Jet · Mask-Less Milling · Stainless Steel (AISI 304) · Optimization

1 Introduction

The rapid progression in material engineering has ushered in a new era of advanced materials characterized by exceptional mechanical properties, making them highly desirable for many applications necessitating high strength-to-weight ratios, robust corrosion resistance, remarkable fatigue endurance, and higher elastic moduli. Nevertheless, the fabrication of complicated engineering components from such materials presents a formidable challenge, particularly when employing conventional machining techniques [1]. Traditional machining methods encounter various limitations, including the generation of heat-affected zones and the requisite customization of cutting tool designs

L. Sitek et al. (Eds.): Advances in Water Jetting II, LNME, pp. 110–121, 2025.
https://doi.org/10.1007/978-3-031-72778-8_9

tailored to individual materials and applications. Furthermore, the effective production of complex item geometries having superior surface finishes and precise dimensions poses a significant hurdle with conventional machining methodologies [2]. The detrimental impacts ranging from heat generation at the cutting interface, tool longevity, residual stresses, and surface integrity associated with conventional machining practices have spurred the demand for the development of sustainable machining technologies, termed unconventional machining technologies, which aim to address environmental, social, and economic concerns [3]. Abrasive water jet (AWJ) is one of the unconventional machining technologies which offers versatile usage in terms of cutting complex parts made up of different materials ranging from plastics, composites [4], ceramics [5], and ferrous and non-ferrous metals [6]. The material removal rate (MRR) offered by AWJ is much higher compared to other unconventional machining methods. Also, due to the cold cutting process, the absence of a heat affected zone is favourable for machining thermal-sensitive materials. These advantages of the process make it very widely acceptable for machining hard-to-cut material used for automobile, aerospace, construction, marine, medical, etc. domains [7]. Initially, AWJ machining was focused only on cutting the material for the creation of specific geometries and shapes. However, with the increase in demand, the AWJ is now being used in various machining forms such as AWJ drilling, AWJ turning [8], AWJ milling, AWJ deburring, etc.

AWJ milling is a newer concept in which AWJ is used for milling complex shapes on hard-to-cut material, commonly using a sacrificial template or mask made of a material having higher hardness than the workpiece. The jet is passed over the mask with the desired speed, covering the entire area of the material and eroding it from the part of the workpiece uncovered by the mask, generating pockets of the desired shape and depth. Different authors have carried out various studies related to AWJ milling to optimize the output responses in terms of surface roughness of the generated pocket, MRR, and depth of cut achieved. It was observed that when creating pockets on AISI 304 stainless steel, faster traverse speed produced a better surface finish [9], and the material removal mechanism changed from cutting mode to deformation mode, lowering the MRR [15]. Also, when milling Ti6Al4V, a similar observation was made that traverse speed was a more significant factor in controlling the milling depth than stand-off distance and abrasive flow rate. Further, the kerf geometries are also influenced by the traverse speed when carrying out milling operations, making the kerf width narrower with increased traverse speed. However, the kerf width and angle can also be controlled and influenced by standoff distance, with increasing standoff distance, the kerf angle decreases, and the kerf width increases [10]. Standoff distance also affects the surface roughness of the generated milled surface. With the increase in the standoff distance, the surface roughness increased linearly, and it also increased the machining time [11]. Another technological parameter which affects the output responses is the step-over distance or the overlapping distance. Studies showed that the step over distance directly affects the depth of cut achieved in the final component, keeping all conditions the same [12]. Further, it was found that a lower step-over distance contributed to the increasing cutting depth and on the contrary, a larger step-over distance increased the surface roughness [13]. However, the step-over distance didn't affect the MRR largely. Researchers also investigated the effect of supply pressure on the generated pocket and found that an increase in the supply pressure elevated the MRR and the depth of the cut [14]. The effect of abrasive flow rate was also studied, and it was found that it mainly affects the surface roughness of the generated surface, with a higher abrasive flow rate showing lower surface roughness [15].

The above literature review shows that most of the work carried out relied on the investigation of individual input parameters on the output responses and used a masked milling approach. The masked milling is mainly used to protect against the undesired erosion of the material by sudden impingement starting and ending jet paths. Also, the sharp corners require a reduction in traverse speed, increasing erosion in those areas. However, in this approach, a sacrificial mask needs to be used, which increases the cost of the process. Also, the mask preparation is time-consuming, which further increases the manufacturing time and necessitates the investigation of the mask-less milling process. For mask-less milling, multi-variable optimization of the process is sparsely available in the open literature, which would provide an understanding of the relationship of various input parameters with the output responses. Therefore, in this study, a mask-less milling process on an AISI 304 stainless steel workpiece is carried out by taking three input parameters i.e., traverse speed, abrasive flow rate and standoff distance, investigating its effect on width, depth and surface roughness of the generated pocket. Two levels of each input parameter were taken such as traverse speed: 1000 mm/min and 3000 mm/min, abrasive flow rate: 100 g/min and 200 g/min and standoff distance: 4 mm and 15 mm. Multi response optimization was carried out using desirability optimization approach to select the optimal level of the factors to achieve the desired output.

2 Material and Method

2.1 Material

An AISI 304 plate with a thickness of 6 mm and dimensions 350 × 150 mm was used for the experiments. This workpiece material was used for its usage in various applications, majorly in aerospace applications, due to its structural integrity and inertness. The structural integrity demands strong withstanding capabilities for serviceable conditions. The material also possesses a high corrosion resistance to ensure component reliability and operation over wide temperature ranges. Austenitic stainless steels are commonly used in hydraulic and fuel pipelines, engine components and radiators in the aircraft industry. The chemical and mechanical properties of the stainless steel are shown in Tables 1 and 2, respectively.

Table 1. Chemical composition of stainless steel AISI 304.

Elements	C	Cr	Mn	Si	Ni	N	P	S
% vol	*<0.07*	*19.5*	2	*1*	10.5	0.11	0.05	0.02

Table 2. Mechanical properties of stainless steel AISI 304

Density	Melting point	Thermal Expansion	Modulus of Elasticity	Tensile strength	Elongation at break
8.00 g/cm^3	*1450* °C	*17.2 *10^{-6}/K*	*193 GPa*	*505 MPa*	70%

2.2 Experimental Method

The current milling operation was performed on PTV WJ2020-2Z-1xPJ abrasive waterjet machine. The abrasive water jet machine is equipped with – a high-pressure pump with a double-acting intensifier, allowing a pressure range from 40 to 400 MPa with a max flow of 7.8 L/min. Barton garnet was used as an abrasive grain, with MESH 80 having a hardness of 7.8 – 8 on the Mohr scale. The orifice diameter of 0.33 mm with a focusing tube of 76.2 mm and a nozzle diameter of 1.02 mm was used for the experiments. Rectangular pockets were selected for investigation and optimization of the input parameters. The total length of the milled pocket was kept at 1000 mm with a total of 10 adjacent passes for each experimental condition. The schematic diagram shows the pocket milling strategy used in Fig. 1. The starting point of the cutting head was placed on one corner of the rectangular groove to be created with the desired standoff distance over the material surface. The cutting head followed the linear trajectory shown in Fig. 1 with the set traverse speed. After covering the desired length of 100 mm, the cutting head travelled in the transverse direction with the selected step over distance and again followed the linear path back. Reaching the end, the cutting head again travelled in a transverse direction by the set amount, and the entire process was repeated 5 times to generate a rectangular milled pocket without using any mask. Similarly, using a mask-less approach, all the rectangular milled pockets were generated under different technological conditions to investigate their effects.

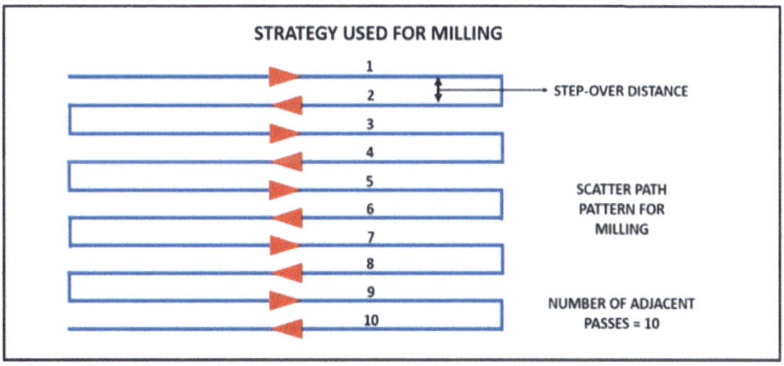

Fig. 1. Strategy used for pocket milling.

The experiments were carried out with 3 variable input parameters having two levels each, namely traverse speed (1000 and 3000 mm/min), abrasive mass flow rate (100 and 200 g/min) and standoff distance (4 and 15 mm). A step-over distance of 0.6 mm was used in the present study. In total, 8 experimental runs were designed using the full factorial method, with a step-over distance of 0.6 mm. The experimental conditions are shown in Table 3, followed by constant technological parameters used in the experiment, as shown in Table 4.

Table 3. Experimental run conditions.

Run No	Step-over distance (mm)	Traverse speed (mm/min)	Abrasive flow rate (g/min)	Standoff distance (mm)
1	0.6	1000	100	4
2	0.6	1000	100	15
3	0.6	1000	200	4
4	0.6	1000	200	15
5	0.6	3000	100	4
6	0.6	3000	100	15
7	0.6	3000	200	4
8	0.6	3000	200	15

Table 4 Constant process parameters and their levels.

Parameters	Value
Supply Pressure (Bar)	2000
Focusing nozzle diameter (mm)	1.02
Focusing tube length (mm)	76.2
Orifice diameter (mm)	0.33
Abrasive	Garnet
Mesh Size	80

2.3 Measurement and Analysis

The measurement of the surfaces of the milled pocket was done by the FRT MicroProf optical profilometer. The system makes use of an optical 3D surface sensor built on the white light interferometry technique. For each groove, the profile is measured at 10 evenly spaced linear lines across the region that extends 25 mm from the center on both sides (Fig. 2) of the milled area to avoid the deviation caused by the cutting head's deceleration at both the ends of the groove. The scanned data was then imported to the MountainsLab software to interpret the profile characteristics of the milled surface of the groove. The width of the milled region, depth of cut, and the roughness characteristics of the milled profile were obtained from the software. For each groove, 10 values were obtained for each output response of width and depth of cut (out of which 3 closely repetitive values are chosen), and one mean value for surface characteristics Rz is obtained and used in the further analysis.

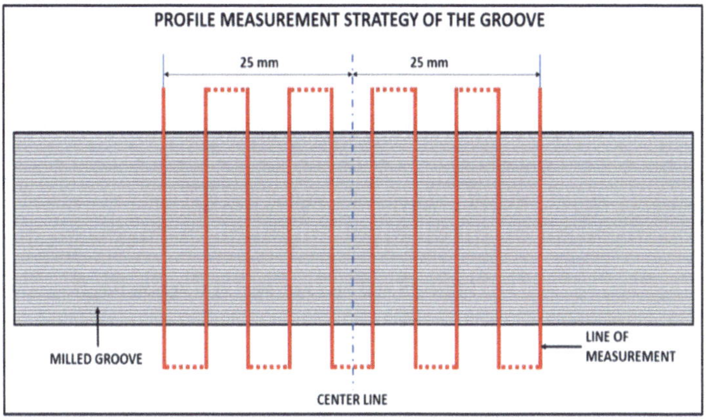

Fig. 2. Measurement strategy of the generated pockets.

3 Results and Discussion

A full factorial analysis was carried out with 3 factors 2 level model to determine the effect of each input parameters on the output responses such as width, depth and surface roughness. The statistical significance of the input parameters on the mean width, mean depth, and the surface roughness Rz of the generated pocket were calculated by the Analysis of variance (ANOVA) shown in Table 5. The ANOVA results clearly show that all the input parameters, i.e., traverse speed, abrasive flow rate and standoff distance, have statistically significant effects on the output response ($p < 0.05$). However, in comparison, the traverse speed and standoff distance are more statistically significant compared to the abrasive flow rate for all the responses. For the two-way interaction model, abrasive flow rate*standoff distance was significant for width ($p = 0.003$), traverse speed*standoff distance was significant for depth ($p = 0.008$) and traverse speed*abrasive flow rate ($p = 0.037$), and traverse speed*standoff distance ($p = 0.018$) were significant for surface roughness Rz. Further, the R-Sq values are significant for all the response models, which shows that the models can ensure respective relevance when used with some other parameter combination within the experimental domain. Also, the standard error value (S) is quite low for all the models, depicting the level of deviation in practical situations low. The regression equations formed by the statistical study for all the measured output responses are given by Eqs. (1), (2) and (3) for width, depth and surface roughness (Rz), respectively, as follows:

$$\begin{aligned}\text{Width} = {} & 6.6473 - 0.000112 * \text{Traverse speed} \\ & + 0.000128 * \text{Abrasive flow rate} + 0.00868 * \text{Standoff distance} \\ & + 0.000192 * \text{Abrasive flow rate} * \text{Standoff distance}\end{aligned} \quad (1)$$

$$\begin{aligned}\text{Depth} = {} & 1.858 - 0.000593 * \text{Traverse speed} \\ & + 0.001815 * \text{Abrasive flow rate} - 0.06941 * \text{Standoff distance} \\ & + 0.000022 * \text{Traverse speed} * \text{Standoff distance}\end{aligned} \quad (2)$$

$$\begin{aligned}\text{Surface roughness (Rz)} = {} & 55.82 - 0.012742 * \text{Traverse speed} \\ & - 0.1114 * \text{Abrasive flow rate} + 1.353 * \text{Standoff distance} \\ & + 0.000028 * \text{Traverse speed} * \text{Abrasive flow rate} \\ & - 0.000375 * \text{Traverse speed} * \text{Standoff distance}\end{aligned} \quad (3)$$

Table 5. ANOVA analysis of width, depth and surface roughness (Rz).

Source	DF	Adj SS	Adj MS	F-Value	P-Value
Width					
Model	4	0.539	0.135	436.12	0.000
Linear	3	0.516788	0.172263	557.48	0.000
Traverse speed	1	0.101250	0.101250	327.67	0.000
Abrasive Flow rate	1	0.076050	0.076050	246.12	0.001
Standoff distance	1	0.339488	0.339488	1098.67	0.000
2-Way interaction	1	0.022261	0.022261	72.04	0.003
Abrasive Flow rate*Standoff distance	1	0.022261	0.022261	72.04	0.003
Error	3	0.000927	0.000309		
Total	7	0.539975			
S = 0.01758, R-Sq = 99.83%, R-Sq (Adj) = 99.60% and R-Sq (Pred) = 98.78%					
Depth					
Model	4	1.49318	0.37329	120.98	0.001
Linear	3	1.37264	0.45755	148.29	0.001
Traverse speed	1	1.15824	1.15824	375.38	0.000
Abrasive Flow rate	1	0.06588	0.06588	21.35	0.019

(*continued*)

The factorial main effect plot shown in Fig. 3 shows the dependency of the input parameters and their levels on the output responses. The plot shows that for mean width, with

Table 5. (*continued*)

Source	DF	Adj SS	Adj MS	F-Value	P-Value
Standoff distance	1	0.14851	0.14851	48.13	0.006
2-Way interaction	1	0.12054	0.12054	39.07	0.008
Traverse speed*Standoff distance	1	0.12054	0.12054	39.07	0.008
Error	3	0.00926	0.00309		
Total	7	1.50244			
S = 0.05555, R-Sq = 99.38%, R-Sq (Adj) = 98.56% and R-Sq (Pred) = 95.62%					
Surface roughness (Rz)					
Model	5	1378.71	275.74	451.57	0.002
Linear	3	1329.36	443.12	725.68	0.001
Traverse speed	1	1178.55	1178.55	1930.07	0.001
Abrasive Flow rate	1	62.50	62.50	102.35	0.010
Standoff distance	1	88.31	88.31	144.63	0.007
2-Way interaction	2	49.35	24.68	40.41	0.024
Traverse speed*Abrasive flow rate	1	15.40	15.40	25.22	0.037
Traverse speed*Standoff distance	1	33.95	33.95	55.60	0.018
Error	2	1.22	0.61		
Total	7	1379.93			
S = 0.78142, R-Sq = 99.91%, R-Sq (Adj) = 99.69% and R-Sq (Pred) = 98.58%					

the increase in the traverse speed from v = 1000 mm/min to 3000 mm/min, the mean width decreased. This response can be attributed to the lesser time available for the abrasive particles to interact with the material surface due to the quicker movement of the nozzle, which results in a narrower cut compared to slower traverse speed. With respect to the abrasive flow rate, the mean width of the milled pocket increased with an increase in the flow rate (100 g/min to 200 g/min). The increase in the width is associated with a larger number of abrasive particles flowing through a specific area, leading to a collision between the particles and generating a wider interaction surface area, increasing the pocket width. Increasing the standoff distance, the mean width of the groove increased due to the flaring out of the jet with a higher standoff distance. Therefore, it can be summarized that the groove width increases with an increase in the abrasive flow rate and standoff distance and a decrease in traverse speed. The main effects result also complements the results from the ANOVA, which shows that the abrasive flow rate has lesser significance than other input parameters, such as traverse speed and standoff distance, for determining the mean width of the pocket.

The dependency of the milling depth on the input parameters and its level is shown in Fig. 4. It can be observed that with increase in the traverse speed from v = 1000 mm/min to 3000 mm/min, the generated groove depth decreased significantly. This can be due

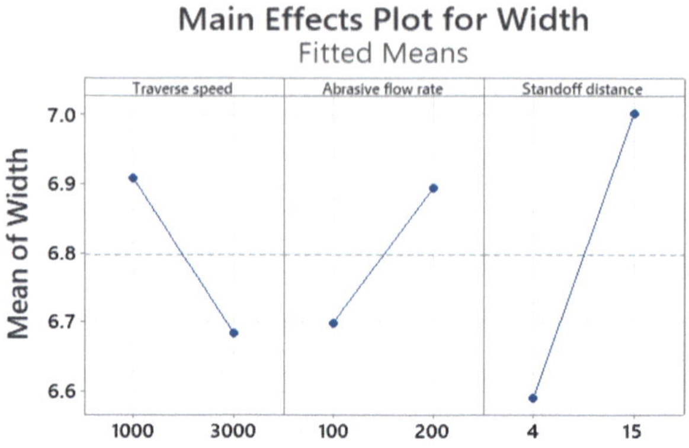

Fig. 3. Main effect plot for mean width.

to the lesser interaction time between the abrasive particle-laden jet and the material surface, diminishing the erosion effect. Further, with an increase in the abrasive flow rate from 100 g/min to 200 g/min, the pocket depth increases slightly due to a larger amount of abrasive eroding the surface. Moreover, the erosion depth shows a negative trend with increasing standoff distance. This is attributed to the flaring out of the jet, and the active core responsible for the erosion diminishes, declining the erosion depth. Correlating with the ANOVA results, the mean effect plots also show that changing the levels of the input parameters largely affects the depth, with traverse speed as the most significant parameter, followed by standoff distance and abrasive flow rate.

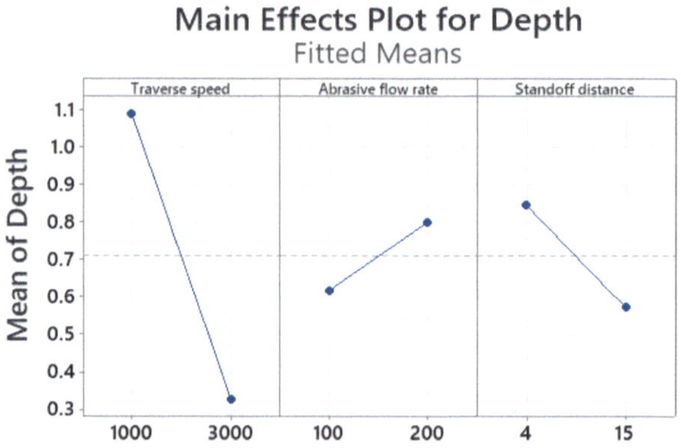

Fig. 4. Main effect plot for mean depth.

Another response studied is the surface roughness parameter, Rz (Fig. 5). Rz shows a similar trend with traverse speed as observed for depth and an opposite trend for abrasive flow rate and standoff distance. It is attributed to the interaction mechanism between the abrasive particles and the surface. At lower traverse speeds, the higher interaction of the abrasive grains with the surface increases, inducing the formation of higher peaks and valleys in the surface, increasing the roughness. Also, a lower concentration of the abrasive grains in the flow generates a uniform surface erosion, causing a smoother surface finish. Lastly, the increased standoff distance diminishes the concentrated effect of the jet on a particular area and also affects its surrounding, increasing the overall surface roughness. The results also confirm the ANOVA results showing traverse speed as the most significant parameter affecting the roughness values followed by standoff distance and abrasive flow rate.

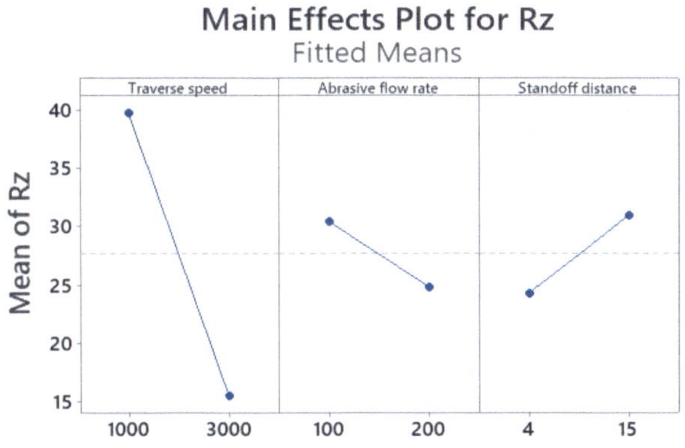

Fig. 5. Main effect plot for surface roughness parameter, Rz.

Multiple response optimization is carried out using the durability approach function available in Minitab Software. The goal of the output responses were selected according to our desired output. The surface roughness parameter goal was selected to be minimum, followed by depth to be maximum and the width to be target, whose value was set as 6.160 mm considering 10 consecutive lines with a step-over distance of 0.6 mm. The response optimization results are shown in Fig. 6. The optimal level of the input parameters was predicted to be 1622.5 mm/min, 200 g/min and 4 mm for traverse speed, abrasive flow rate and standoff distance, respectively. The overall composite desirability was calculated to be 0.6386. The software also predicted that an overall pocket width, depth and surface roughness achieved with the optimal level of parameters would be 6.6785 mm, 1.1265 mm and 24.848 μm, respectively.

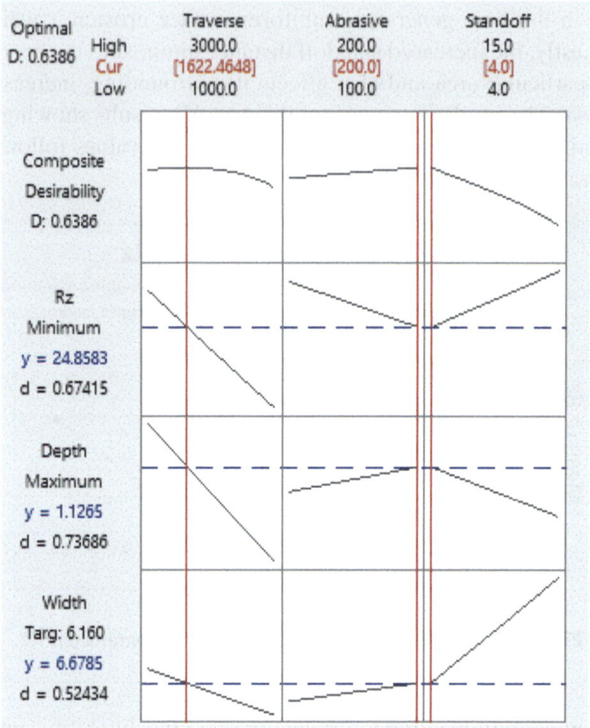

Fig. 6. Desirability plot for multiple response optimization.

4 Conclusion

The experimental study on process parameter optimization to implement mask-less milling using abrasive waterjet machining technology on stainless steel AISI 304 was carried out. The study aimed to establish the relationship between input parameters and the output responses for milling rectangular grooves of predetermined dimensions. The results attained are concluded as follows:

1. The maximum width of the groove is achieved by the incurrence of lower traverse speed, higher abrasive flow rate and higher stand-off distance.
2. The maximum depth of the groove is achieved by the incurrence of lower traverse speed, higher abrasive flow rate and lower stand-off distance.
3. The minimum surface roughness, Rz value is achieved by the incurrence of higher traverse speed, higher abrasive flow rate and lower stand-off distance.
4. Multiple response optimization showed that the optimal level of traverse speed, abrasive flow rate and standoff distance were 1000 mm/min, 200 g/min and 4 mm, respectively. A mean width of 6.75 mm, mean depth of 1.44 mm, and surface roughness Rz of 30.27 μm is achieved with a desirability level of 0.64.

Acknowledgment. This article was co-funded by the European Union under the REFRESH – Research Excellence For REgion Sustainability and High-tech Industries project number CZ.10.03.01/00/22_003/0000048 via the Operational Programme Just Transition.

References

1. Nyaboro, J., Ahmed, M., El-Hofy, H., El-Hofy, M.: Experimental and numerical investigation of the abrasive waterjet machining of aluminum-7075-T6 for aerospace applications. Adv. Manuf. **9**, 286–303 (2021)
2. Kibria, G., Bhattacharyya, B., Davim, J. P.: Non-traditional Micromachining Processes. Springer (2017)
3. Jagadish, Gupta, K.: Abrasive Water Jet Machining of Engineering Materials. Springer (2020)
4. Nag, A., et al.: Hybrid aluminium matrix composite AWJ turning using olivine and Barton garnet. Int. J. Adv. Manuf. Technol. **94**, 2293–2300 (2018)
5. Saurabh, S., et al.: Processing of alumina ceramics by abrasive waterjet-an experimental study. Mater. Today Proc. **5**, 18061–18069 (2018)
6. Cárach, J., Hloch, S., Petrů, J., Nag, A., Gombár, M., Hromasová, M.: Hydroabrasive disintegration of rotating Monel K-500 workpiece. Int. J. Adv. Manuf. Technol. **96**, 981–1001 (2018)
7. Nag, A., et al.: Influence of abrasive water jet turning parameters on variation of diameter of hybrid metal matrix composite. In: Lecture Notes in Mechanical Engineering, pp. 495–504 (2018). https://doi.org/10.1007/978-981-10-5329-0_36
8. Srivastava, A.K., et al.: Hardness measurement of surfaces on hybrid metal matrix composite created by turning using an abrasive water jet and WED. Measurement **131**, 628–639 (2019). https://doi.org/10.1016/j.measurement.2018.09.026
9. Gupta, T.V.K., Ramkumar, J., Tandon, P., Vyas, N.S.: Role of process parameters on pocket milling with Abrasive Water Jet Machining technique. Int. J. Mech. Mechatron. Eng. **7**, 2021–2026 (2013)
10. Yuan, Y., Chen, J., Gao, H., Wang, X.: An investigation into the abrasive waterjet milling circular pocket on titanium alloy. Int. J. Adv. Manuf. Technol. **107**, 4503–4515 (2020)
11. Pal, V.K., Tandon, P.: Identification of the role of machinability and milling depth on machining time in controlled depth milling using abrasive water jet. Int. J. Adv. Manuf. Technol. **66**, 877–881 (2013)
12. Kanthababu, M., Ram, R.M., Emannuel, N.P., Gokul, R., Rammohan, R.: Experimental investigations on pocket milling of titanium alloy using abrasive water jet machining. FME Trans. **44**, 133–138 (2016)
13. Goutham, U., Hasu, B.S., Chakraverti, G., Kanthababu, M.: Experimental investigation of pocket milling on Inconel 825 using abrasive water jet machining. Int. J. Curr. Eng. Technol. **6**, 295–302 (2016)
14. Pal, V.K., Choudhury, S.K.: Surface characterization and machining of blind pockets on Ti6Al4V by abrasive water jet machining. Proc. Mater. Sci. **5**, 1584–1592 (2014)
15. Badgujar, P.P., Rathi, M.G.: Analysis of surface roughness in abrasive water jet cutting of stainless steel. Int. J. Eng. Res. Technol. (IJERT). **3** (2014)

Multiple-Criteria Optimization of the Water Jet Cutting Process Using WISP Methodology

Andrzej Perec[1(✉)], Elżbieta Kawecka[1], Aleksandra Radomska-Zalas[2], Frank Pude[3,4], and Greg Galecki[5]

[1] Faculty of Technology, Jacob of Paradies University, Gorzów Wielkopolski, Poland
aperec@ajp.edu.pl
[2] Gorzów Technology Center, Science and Industrial Park, Gorzów Wielkopolski, Poland
[3] Inspire AG (ETH Zurich), Zurich, Switzerland
[4] Steinbeis Consulting Center High-Pressure Waterjet Technology, Horgau, Germany
[5] Missouri University of Science & Technology, Rolla, MO, USA

Abstract. The article presented optimization outcomes of alloy steel machining by abrasive water jet (AWJ). The Weighted Sum Product Method (WISP) is introduced for the optimization of machining parameters like working pressure, abrasive flow rate (AFR), and feed rate on the cutting depth of cut, the inclination of the cut kerf, and its surface roughness. The WISP method is grounded on the joining of four measures that denote the influence of beneficial and non-beneficial criteria grounded on the use of Weighted Sum (WS) and Weighted Product (WP) methods. Based on the optimization calculation it was initiated that the best possible set of control parameters and presented an SEM view of the cutting surface at optimal process conditions.

Keywords: Abrasive Water Jet · AWJ · Surface Roughness · Alloy Steel · Cutting Parameter Optimization

1 Introduction

Optimization methods for advanced manufacturing methods include several techniques aimed at improving the efficiency, effectiveness, and quality of production processes. One of the key methods is the optimization of process parameters, i.e. tuning parameters such as speed, feed, and cutting depth to minimize energy consumption improve the quality of the processed material, and tool life, and increase the quality of the final product [1].

For this purpose, advanced statistical and mathematical methods [2], data analysis [3], and machine learning methods, including AI [4, 5] and metaheuristic algorithms such as Antlion [6], and others [7–9] as well as decision support methods [10–12] are used.

Optimization of Abrasive Water Jet machining is indispensable for the reason of more control parameter numbers in collated to traditional cutting technologies [13, 14].

L. Sitek et al. (Eds.): Advances in Water Jetting II, LNME, pp. 122–133, 2025.
https://doi.org/10.1007/978-3-031-72778-8_10

The objective of the article is to fit optimal parameters for alloy steel cutting by AWJ technology.

There are various methods using Multi-Criteria Decision Making (MCDM) is used in industrial activities. Simple Additive Weighting Method (SAW)—is the uncomplicated and frequently used MCDM method [15]. The SAW method utilizes the aggregation method proposed in the Weighted Sum (WS) method. The SAW method is utilized with several normalization procedures [16].

The multi-objective optimization based on the simple ratio analysis (MOOSRA) approach [17] utilizes the proportion of utility received on the grounds of beneficial and non-beneficial criteria, properly, for establishing the output score of alternatives indicating the set of beneficial and set of non-beneficial criteria, respectively. The MOOSRA approach exploits vector standardization mode, and the normalization mode is the identic for the beneficial and non-beneficial criteria.

The weighted aggregated sum product assessment (WASPAS) approach [18, 19] joins two methods for establishing the output score of alternatives and shows the relative importance of alternatives grounded on the Weighted Sum (WS) and Exponentially Weighed Product (EWP) method, properly. The WASPAS approach utilizes maximum normalization mode, where beneficial and nonbeneficial criteria are handled differently and beneficial and non-beneficial criteria are standardized in other ways, using other equations.

The operational competitiveness rating (OCRA) approach [20] may be supposed as the application of the SAW approach with max-min normalization mode. Stanujkic et al., [21] used the modified OCRA approach to the global efficiency ratings concerning the beneficial and non-beneficial criteria to establish the linear efficiency ratings.

The multi-objective optimization based on ratio analysis with the full multiplicative form (MULTIMOORA) approach [22] integrates three methods: the ratio system method, the reference point method, and the full multiplicative form. The use of alternatives is grounded on the ratio system and on the reference point way. As in the case of the ratio system method, a full multiplicative form may be investigated of the WS model involving beneficial and nonbeneficial criteria. Dissimilar to the earlier-considered methods, MULTIMOORA does not exploit a general variable to regard the output score of alternatives. The ranking of alternatives, and the choice of the most acceptable, in the MULTIMOORA approach, is grounded on the dominance theory.

The machining of D2 heat-treated steel by Abrasive Water Jet (AWJ) was presented Doreswamy [23]. The research was conducted to discover the impact of process parameters such as stand-off distance and the traverse speed on the kerf width and the surface roughness (Ra) level of the kerf caused by AWJ.

Also Kumar et al. [24] presented experimental tests into AWJ cutting tests of D2 steel by various stream attack angles and abrasive grain sizes. In addition, the results of influencing cutting parameters, jet attack angle, and abrasive grain dimensions on the output factors as the stream penetration, material removal rate (MRR), taper ratio, and cut kerf surface roughness, were studied.

Perec et al. [25] present a study of the cutting performance of hard-to-machine tool steels by abrasive waterjet (AWJ) using recycled abrasives. Reusing such abrasives has great potential to reduce processing costs while remaining environmentally friendly. In this case, the RSM method was used for modeling and optimization.

Pude et al. [26] present an innovative method of using the water jet with a hammer effect caused by forced multiple impacts of drops to generate a structural surface with the desired values of titanium surface profile parameters. It was found that changing the input parameters causes significant changes in the generated surface. This means that roughening the surface increases the surface area.

Kawecka [27] introduced the opportunity of Whale Optimization Algorithm (WOA) usage in AWJ cutting of tool steel. Grounded on the process parameters of tool steel machining by AWJ, the cost function was assigned through the RSM approach. The achieved on the basis WOA maximum was benchmarked with the best effect set out experimentally.

Also, Armagan [28] presented the AWJ cutting of St37 steel plates in stacked form. Experimental factors were stand-off distance, traverse speed, and abrasive flow rate. Surface roughness and kerf widths were tested to determine the machining efficiency. The most productive factors for surface roughness and kerf width in almost all cases were abrasive mass flow rate and stand-off distance, respectively.

The choice of optimal parameters was performed with the interactive MCDM methodology [29], specifically the Weighted Sum Product Method (WISP) example [30]. For analysis, the abrasive flow rate, pressure, and traverse speed were selected as the independent variables and were appraised. Whereas depth of cut (H_{max}), kerf inclination angle (δ), and kerf surface roughness (Sq), were chosen as dependent variables for each experimental run (see Fig. 1).

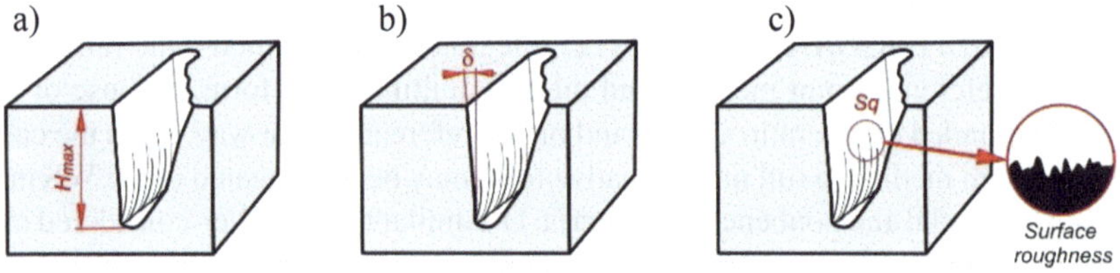

Fig. 1. Schematic view of the cut kerf properties: a) cutting depth, b) inclination angle, c) cut kerf roughness.

In summary, methods for optimizing advanced manufacturing processes focus on comprehensive improvement of performance, energy efficiency, and production quality using analytical and computational tools, especially advanced ones.

The WISP method uses a simpler standardization procedure, includes utility measures to determine the overall usefulness of alternatives, and allows for a much easier ranking of alternatives. The essential idea is partly based on multi-objective optimization through ratio analysis and on the full multiplicative form and product evaluation feature method of weighted sum and combined trade-off solution methods.

This paper fills a research gap because no reports noted of used to optimize an advanced manufacturing process such as AWJ machining grounded on the integration of weighted sum and weighted product approaches (WISP).

2 Materials and Methods

2.1 Target Material

18CrNiMo7-6 steel was selected for cutting tests as the target. This material uses chromium, nickel, and molybdenum as alloy-strengthening additives. 18CrNiMo7-6 steel is used to produce parts and components that are subjected to heavy loads and high pressures. It is also used for special bearings, pins, shafts, or gears. It is appropriate for applications that require high wear resistance.

The chemical composition is presented in Table 1. It is also known by its other designations such as 1.6587 and 821M22. The contains of chromium, nickel, and molybdenum (Table 1) makes this steel suitable for applications that require a good balance of toughness, wear resistance and high strength.

Table 1. Chemical composition of 18CrNiMo7-6 steel.

Element	Carbon	Chromium	Nickel	Molybdenum
Contain [%]	0.15–0.21	1.5–1.8	1.4–1.7	0.25–0.35

The case-hardening assets mean that the surface of the steel can be hardened while maintaining a tough core, meeting it ideal for applications such as gears, shafts, and other components subject to high stress and wear. The machinability of 18CrNiMo6-7 can be defined as moderate to difficult due to its alloying elements, which can influence its cutting and machining properties.

2.2 Abrasive Material

The abrasive material used in the research was the crushed rock garnet type J80A comes from Jinhong Mining deposit, located in Jiangsu, China. Its mineral content, chemical composition, and physical characteristics are presented in Table 2.

Table 2. Chosen properties of garnet abrasive.

Mineral Content [%]						
Amandine	Ilmenite	Omphacite	Rutile	Quartz	Hornblende	Free Silica
90–96	1.0	1.5	0.6	<0.1	<0.5	<0.5%
Physical Characteristics						
Density [kg/dm^3]	Bulk Gravity [kg/dm^3]	Mohs Hardness	Color	Grain shape	Conductivity	HCl solubility
3.8–4.1	2.3–2.4	7.5–8.0	Dark red	Sub angular	<25 S/m	<1.0%

2.3 Machining Methods

Research tests were carried out on a CNC Water Jet machining center OMAX 60120 (see Fig. 2).

Fig. 2. Example view of the OMAX 60120 JetMachining Center.

The material, 18CrNiMo7-6 alloy steel, was cut perpendicular to the Abrasive Water Jet axis, by a straight in-line shift with the given traverse speed. The thickness of the target material was chosen relatively as big to avoid total separation of material to exactly determination of the cutting depth at the following variable parameters:

- abrasive flow rate: 250; 350 and 450 g/min,
- pressure: 360; 380; 400 MPa,
- traverse speed: 50; 150 and 250 mm/min.

The other parameters were constant and amounted to abrasive grains: almandine garnet #80, water nozzle internal diameter: 0.33 mm, focusing tube internal diameter: 0.76 mm, stand-off distance: 2 mm. The experiments were conducted as stated in the L27 research plan. Process control parameters, such as pressure, abrasive flow, and feed, have been optimized from the point of view of three criteria: cutting depth, kerf inclination angle, and cut kerf surface roughness.

Side surface roughness measurements were carried out in the middle of the cutting zone in a 0.95 mm $\times$ 0.95 mm elemental area on Olympus DSX1000 3D microscope. The surface structure was observed on a Thermofisher Scientific Axia ChemiSEM operating at 20 keV accelerating voltage in the low-vacuum mode. Root means square height (Sq) was chosen as the roughness parameter. It expresses the root mean square for area $Z(x,y)$ within the evaluation space. Is also defined as the RMS value. The parameter enables stable measurement results since the parameter is slightly influenced by scratches, contamination, and other measurement noise.

2.4 Optimization Method

Recently, a lot of different multi-criteria decision-making methods were proposed to facilitate the solution of various problems. Using the WISP method, the best control parameters combination was found under maximal cutting depth, and minimal kerf inclination angle and surface roughness were attained. Weighted Sum Product Method-Simple WISP approach is in the following steps [31]:

- *Step 1*: Create a standardized decision-making array. The standardized grades, without transforming the influence of nonbeneficial into the influence of beneficial criteria, are determined as follows:

$$r_{ij} = \frac{x_{ij}}{\max x_{ij}} \tag{1}$$

- *Step 2*: Determine the values of utility measures for example:

$$u_i^{wsd} = \sum_{j\epsilon\Omega_{max}} r_{ij}w_j - \sum_{j\epsilon\Omega_{min}} r_{ij}w_j \tag{2}$$

$$u_i^{wpd} = \prod\nolimits_{j\epsilon\Omega_{max}} r_{ij}w_j - \prod\nolimits_{j\epsilon\Omega_{min}} r_{ij}w_j \tag{3}$$

$$u_i^{wsr} = \frac{\sum_{j\epsilon\Omega_{max}} r_{ij}w_j}{\sum_{j\epsilon\Omega_{min}} r_{ij}w_j} \tag{4}$$

$$u_i^{wpr} = \frac{\prod_{j\epsilon\Omega_{max}} r_{ij}w_j}{\prod_{j\epsilon\Omega_{min}} r_{ij}} \tag{5}$$

where u_i^{wsd} and u_i^{wpd} indicate the difference among the impact of beneficial and non-beneficial criterium on the ultimate utility of alternative achieved on the grounds of WS and WP models then u_i^{wsr} and u_i^{wpr} indicate the relation among the impact of beneficial and nonbeneficial criteria on the final utility of alternative achieved on the ground of WS and WP models.

- *Step 3*: Recalculate utility measures worth, for example:

$$\overline{u}_i^{wsd} = \frac{u_i^{wsd}}{\left(1 + u_{max_i}^{wsd}\right)} \tag{6}$$

$$\overline{u}_i^{wpd} = \frac{u_i^{wpd}}{\left(1 + u_{max_i}^{wpd}\right)} \tag{7}$$

$$\overline{u}_i^{wsr} = \frac{u_i^{wsr}}{\left(1 + u_{max_i}^{wsr}\right)} \tag{8}$$

$$\overline{u}_i^{wpr} = \frac{u_i^{wpr}}{\left(1 + u_{max_i}^{wpr}\right)} \tag{9}$$

where $\overline{u}_i^{wsd}$, $\overline{u}_i^{wpd}$, $\overline{u}_i^{wsr}$ and $\overline{u}_i^{wpr}$ are normalized values of u_i^{wsd}, u_i^{wpd}, u_i^{wsr} and u_i^{wpr}. The u_i^{wsd} and u_i^{wsr} factors take values from the range into interval [0, 1].

- *Step 4*: Determine the overall utility factor u_i for every one alternative.

$$u_i = \frac{\overline{u}_i^{wsd} + \overline{u}_i^{wpd} + \overline{u}_i^{wsr} + \overline{u}_i^{wpr}}{4} \tag{10}$$

- *Step 5*: Rate the alternatives and choose the finest one.

The alternatives are classified in descending order, and the alternative with the maximum worth is the best one.

3 Test Results and Discussion

In the research cutting control parameters were chosen on the full L27 matrix basis. This matrix involves three independent variables (AFR, working pressure, and feed rate) with three levels each, as presented in Table 3.

The cutting depth (beneficial parameter), the inclination of the cut kerf, and its surface roughness (non-beneficial parameters) were taken as response parameters.

Table 4 shows the calculated values of the measures according to the WISP approach. Overall utility factors u_i for alternative sets of control parameters are also included here.

The set of parameters with the highest value of this factor is the most preferred and takes first place in the ranking. In this case, it is the control parameter set number 13, as indicated in bold italics. The optimum of effects in the form of cutting depth equal 7.93 mm, inclination of the cut kerf angle $\delta = 2.84°$ and surface roughness $Sq = 3.81$ μm, were obtained for the following control parameters:

- AFR = 350 g/s,
- Pressure = 380 MPa,
- Traverse speed = 50 mm/min.

An example view of the cut groove surface is shown in Fig. 3.

In the upper part of the cut kerf, we can observe parallel traces of shallow erosion. In the central part, we can observe a deviation of the erosion traces and an increase of erosion in depth. In the lower part, there is an increase in the deviation of erosion traces and a further increase of erosion.

4 Conclusion

Research was led according to L27 orthogonal matrix with 3 independent variables, each taking 3 levels. The principal purpose of the research was to reach such a level of control parameters affecting cutting depth, inclination angle, and surface roughness that the optimization process would be successful. The carried-out tests were fixed to the virtue of this approach realization in optimizing the AWJ cutting process.

The WISP method is a good accessibility appropriate when cutting new materials. It significantly reduces the amount of test process by reducing the time to obtain reliable results and reducing the cost of obtaining them. In the case of cutting the 18CrNiMo7-6 alloy steel, the optimal machining parameters were determined.

Grounded on the WISP measures and factors calculation, it was established that the best possible combination of control parameters to reach optimal output factors from

Table 3. Control and output parameters of AWJ cutting process.

No	*AFR* [g/min]	*P* [MPa]	*Vp* [mm/min]	*Hmax* [mm]	*d* [deg]	*Sq* [mm]
1	250	360	50	7.48	3.48	3.677
2	250	360	150	4.86	6.76	4.337
3	250	360	250	2.36	9.46	5.250
4	250	380	50	7.71	3.97	3.070
5	250	380	150	5.09	5.45	4.059
6	250	380	250	2.60	7.87	4.730
7	250	400	50	8.16	4.23	3.264
8	250	400	150	5.14	6.57	4.347
9	250	400	250	2.93	7.48	5.300
10	350	360	50	7.84	3.29	3.664
11	350	360	150	4.99	6.66	4.899
12	350	360	250	3.00	8.30	5.634
13	350	380	50	7.93	2.84	3.805
14	350	380	150	5.12	7.42	5.996
15	350	380	250	3.06	8.48	8.832
16	350	400	50	8.70	3.47	4.791
17	350	400	150	5.22	6.96	5.707
18	350	400	250	3.36	8.80	7.126
19	450	360	50	7.18	3.78	2.869
20	450	360	150	4.60	5.60	4.012
21	450	360	250	3.31	8.08	8.096
22	450	380	50	7.59	3.77	3.818
23	450	380	150	4.86	5.82	5.152
24	450	380	250	3.07	8.01	5.974
25	450	400	50	7.64	3.42	3.922
26	450	400	150	4.89	6.48	4.326
27	450	400	250	3.17	8.06	4.954

all considered points of view: cutting depth, cut kerf inclination angle, and side surface roughness *Sq*, are the following:

- Abrasive flow rate: 350 g/min,
- Pressure: 380 MPa,
- Traverse speed 50 mm/min.

These control parameters achieved the following effects:

Table 4. WISP measures and calculation effects.

No	u_i^{wsd}	u_i^{wpd}	u_i^{wsr}	u_i^{wpr}	$\overline{u}_i^{wsd}$	$\overline{u}_i^{wpd}$	$\overline{u}_i^{wsr}$	$\overline{u}_i^{wpr}$	u_i	Rank
1	0.03	–0.52	1.10	0.65	0.97	0.90	0.93	0.98	0.9446	6
2	–0.21	–0.86	0.46	0.49	0.74	0.26	0.65	0.89	0.6348	14
3	–0.44	–1.19	0.17	0.35	0.53	–0.36	0.52	0.81	0.3756	26
4	0.04	–0.50	1.16	0.66	0.98	0.93	0.96	0.99	0.9660	3
5	–0.15	–0.77	0.57	0.52	0.80	0.43	0.70	0.91	0.7084	10
6	–0.35	–1.08	0.22	0.38	0.61	–0.15	0.54	0.82	0.4555	21
7	0.04	–0.51	1.15	0.66	0.98	0.91	0.96	0.99	0.9597	5
8	–0.20	–0.84	0.50	0.50	0.76	0.30	0.67	0.89	0.6552	12
9	–0.35	–1.07	0.24	0.39	0.62	–0.13	0.55	0.83	0.4665	20
10	0.05	–0.49	1.18	0.66	0.99	0.95	0.97	0.99	0.9746	2
11	–0.23	–0.88	0.46	0.49	0.73	0.22	0.65	0.89	0.6208	16
12	–0.39	–1.12	0.23	0.39	0.58	–0.21	0.55	0.83	0.4344	23
13	***0.06***	***–0.46***	***1.25***	***0.68***	***1.00***	***1.00***	***1.00***	***1.00***	***1.0000***	***1***
14	–0.29	–0.96	0.40	0.47	0.67	0.07	0.62	0.87	0.5591	18
15	–0.51	–1.26	0.19	0.36	0.46	–0.47	0.53	0.81	0.3314	27
16	0.03	–0.54	1.10	0.65	0.97	0.86	0.93	0.98	0.9373	8
17	–0.26	–0.92	0.43	0.48	0.70	0.14	0.64	0.88	0.5896	17
18	–0.45	–1.18	0.22	0.38	0.52	–0.33	0.54	0.82	0.3904	24
19	0.03	–0.49	1.14	0.66	0.98	0.94	0.95	0.99	0.9645	4
20	–0.17	–0.80	0.51	0.50	0.78	0.37	0.67	0.90	0.6787	11
21	–0.46	–1.19	0.21	0.38	0.51	–0.36	0.54	0.82	0.3784	25
22	0.01	–0.54	1.05	0.64	0.96	0.85	0.91	0.98	0.9242	9
23	–0.21	–0.86	0.47	0.49	0.74	0.25	0.65	0.89	0.6339	15
24	–0.39	–1.12	0.23	0.39	0.58	–0.22	0.55	0.83	0.4346	22
25	0.02	–0.52	1.09	0.65	0.97	0.89	0.93	0.98	0.9408	7
26	–0.20	–0.85	0.48	0.49	0.75	0.29	0.66	0.89	0.6467	13
27	–0.35	–1.06	0.26	0.40	0.62	–0.11	0.56	0.84	0.4766	19

- cutting depth $H_{max} = 7.93$ mm,
- inclination angle of the cut kerf $\delta = 2.84°$,
- surface roughness $Sq = 3.81$ μm.

WISP, like other decision-making methods, has its strengths and limitations. WISP handles uncertainty and vagueness well to produce meaningful results. Additionally, the quality of the decision heavily relies on the accuracy of criteria weights and dominance degree calculations, which can be subjective and may require expert opinion.

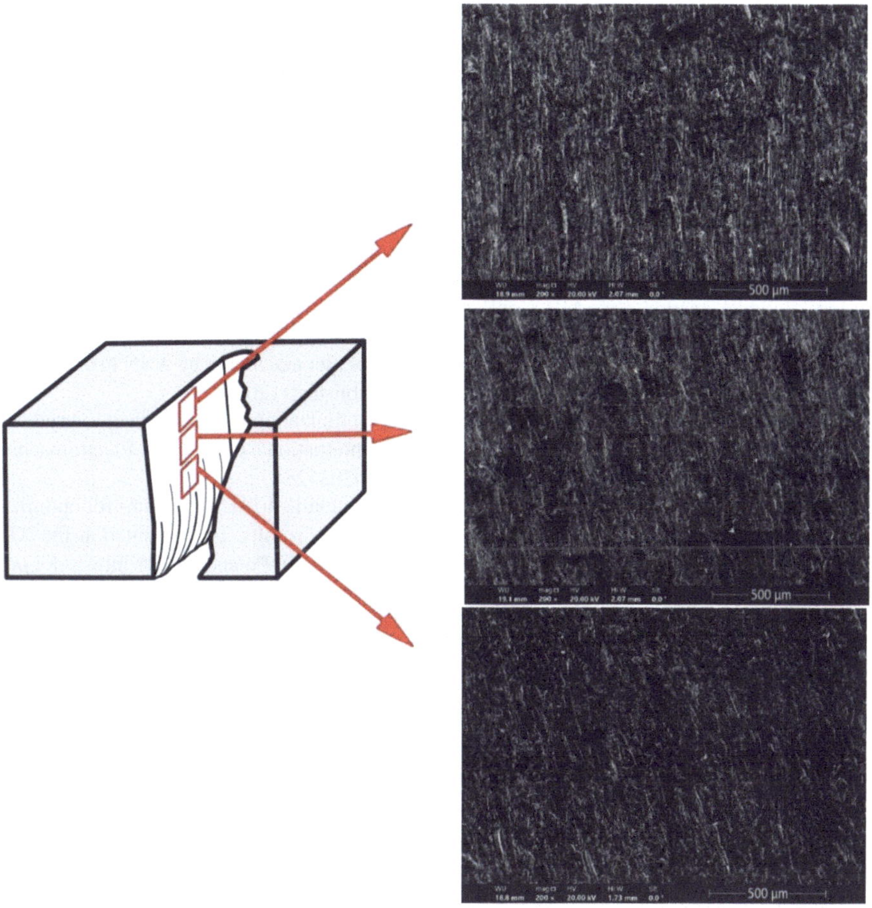

Fig. 3. Example view of the cut surface and its location on the kerf.

The simple WISP approach is first effortless to use and encourages increasing the reliability of the decision chosen. Above all, WISP's simple approach is easy to use and encourages greater reliability in decision-making.

Further research will be carried out on the impact of other control parameters of the AWJ cutting process.

References

1. Jerman, M., Orbanić, H., Valentinčič, J.: CFD analysis of thermal fields for ice abrasive water jet. Int. J. Mech. Sci. **220**, 107154 (2022). https://doi.org/10.1016/j.ijmecsci.2022.107154
2. Perec, A., Radomska-Zalas, A., Fajdek-Bieda, A., Pude, F.: Process optimization by applying the response surface methodology (RSM) to the Abrasive suspension water jet cutting of phenolic composites. Facta Univ. Ser. Mech. Eng. **21**, 575–589 (2023). https://doi.org/10.22190/FUME211123010P

3. Szada-Borzyszkowska, M., Kacalak, W., Banaszek, K., Pude, F., Perec, A., Wegener, K., Królczyk, G.: Assessment of the effectiveness of high-pressure water jet machining generated using self-excited pulsating heads (to be published) (2024). https://doi.org/10.21203/rs.3.rs-4083190/v1
4. Szatkiewicz, T., Perec, A., Radomska-Zalas, A., Banaszek, K., Balasz, B.: Preliminary studies into cutting of a novel two component 3D-printed stainless steel-polymer composite material by abrasive water jet. Materials **16**, 1170 (2023). https://doi.org/10.3390/ma16031170
5. Perec, A.: The artificial neural networks in modeling the high pressure, suspension waterjet cutting. Arch. Mech. Technol. Autom. **33**, 21–29 (2014)
6. Kawecka, E.: The use of metaheuristic optimization algorithm in abrasive water jet machining of white marble. In: AIP Conference Proceedings. XIV International Conference Electromachining 2023, 020015 (2024). https://doi.org/10.1063/5.0203448
7. Perec, A., Kawecka, E.: Modeling of the abrasive water jet machining by ANN in uncertainty conditions. In: Procedia Computer Science (to be published) (2024)
8. Radomska-Zalas, A.: Multi-criteria methods in the optimization of the abrasive waterjet cutting process. In: AIP Conference Proceedings. XIV International Conference Electromachining 2023, 020030 (2024). https://doi.org/10.1063/5.0203525
9. Rawicki, S., Podhajecki, J.: Investigation on selected meta-heuristic algorithms for optimization the abrasive water jet cutting process—Preliminary results. In: Presented at the XIV International Conference Electromachining 2023, Bydgoszcz, Poland (2024). https://doi.org/10.1063/5.0207626
10. Perec, A., Musial, W.: Multiple criteria optimization of abrasive water jet cutting using entropy-VIKOR approach. In: Hloch, S., Klichová, D., Pude, F., Krolczyk, G. M., Chattopadhyaya, S. (eds.) Advances in manufacturing engineering and materials II, pp. 50–62. Springer International Publishing, Cham (2021). https://doi.org/10.1007/978-3-030-71956-2_5
11. Perec, A., Kawecka, E., Radomska-Zalas, A., Pude, F.: Optimization of abrasive waterjet cutting by using the CODAS method with regard to interdependent processing parameters. SV-JME **69**, 367–375 (2023). https://doi.org/10.5545/sv-jme.2023.647
12. Radomska-Zalas, A.: Experimental research on the use of a selected multi-criteria method for the cutting of titanium alloy with an abrasive water jet. Materials **16**, 5405 (2023). https://doi.org/10.3390/ma16155405
13. Jandačka, P., Ščučka, J., Martinec, P., Lupták, M., Janeček, I., Mahdi Niktabar, S.M., Zeleňák, M., Hlaváček, P.: Optimal abrasive mass flow rate for rock erosion in AWJ machining. In: Klichová, D., Sitek, L., Hloch, S., Valentinčič, J. (eds.) Advances in water jetting, pp. 81–90. Springer International Publishing, Cham (2021). https://doi.org/10.1007/978-3-030-53491-2_9
14. Tripathi, D.R., Vachhani, K.H., Kumari, S., Dinbandhu, Abhishek, K.: Experimental investigation on material removal rate during abrasive water jet machining of GFRP composites. Mater. Today Proc. **26**: 1389–1392 (2020). https://doi.org/10.1016/j.matpr.2020.02.280
15. Pavan, M., Todeschini, R.: Multicriteria decision-making methods. In: Comprehensive chemometrics, pp. 591–629. Elsevier (2009). https://doi.org/10.1016/B978-044452701-1.00038-7
16. Zavadskas, E.K., Turskis, Z.: A new logarithmic normalization method in games theory. Informatica **19**, 303–314 (2008). https://doi.org/10.15388/Informatica.2008.215
17. Bhowmik, C., Dhar, S., Ray, A.: Comparative analysis of MCDM methods for the evaluation of optimum green energy sources: A case study. Int. J. Decis. Supp. Syst. Technol. **11**, 1–28 (2019). https://doi.org/10.4018/IJDSST.2019100101
18. Radomska-Zalas, A.: Application of the WASPAS method in a selected technological process. Proc. Comput. Sci. **225**, 177–187 (2023). https://doi.org/10.1016/j.procs.2023.10.002
19. Perec, A., Radomska-Zalas, A.: WASPAS optimization in advanced manufacturing. Proc. Comput. Sci. **207**, 1193–1200 (2022). https://doi.org/10.1016/j.procs.2022.09.175

20. Parkan, C., Wu, M.-L.: Decision-making and performance measurement models with applications to robot selection. Comput. Ind. Eng. **36**, 503–523 (1999). https://doi.org/10.1016/S0360-8352(99)00146-1
21. Stanujkic, D., Zavadskas, E.K., Liu, S., Karabasevic, D., Popovic, G.: Improved OCRA method based on the use of interval grey numbers. J. Grey Syst. **29** (2017)
22. Brauers, W.K.M., Zavadskas, E.K.: Project management by multimoora as an instrument for transition economies. Technol. Econ. Dev. Econ. **16**, 5–24 (2010). https://doi.org/10.3846/tede.2010.01
23. Doreswamy, D.: Machining of D2 heat treated steel using abrasive water jet: The effect of Standoff distance and feed rate on Kerf width and surface roughness. Int. J. Res. Eng. Technol. **3**, 417–421 (2014). https://doi.org/10.15623/ijret.2014.0308065
24. Yuvaraj, N., Kumar, M.P.: Investigation of process parameters influence in abrasive water jet cutting of D2 steel. Mater. Manuf. Process. **32**, 151–161 (2017). https://doi.org/10.1080/10426914.2016.1176183
25. Perec, A., Radomska-Zalas, A., Fajdek-Bieda, A., Kawecka, E.: Efficiency of tool steel cutting by water jet with recycled abrasive materials. Materials **15**, 3978 (2022). https://doi.org/10.3390/ma15113978
26. Klichova, D., et al.: Utilising of water hammer effect for surface roughening of Ti6Al4V. Int. J. Adv. Manuf. Technol. **126**, 5633–5647 (2023). https://doi.org/10.1007/s00170-023-11521-y
27. Kawecka, E.: The whale optimization algorithm in abrasive water jet machining of tool steel. Proc. Comput. Sci. **225**, 1037–1044 (2023). https://doi.org/10.1016/j.procs.2023.10.091
28. Armağan, M.: Cutting of St37 steel plates in stacked form with abrasive water jet. Mater. Manuf. Process. **36**, 1305–1313 (2021). https://doi.org/10.1080/10426914.2021.1906895
29. Radomska-Zalas, A., Perec, A., Fajdek-Bieda, A.: IT support for optimisation of abrasive water cutting process using the TOPSIS method. IOP Conf. Ser. Mater. Sci. Eng. **710**, 012008 (2019). https://doi.org/10.1088/1757-899X/710/1/012008
30. Stanujkic, D., Popovic, G., Karabasevic, D., Meidute-Kavaliauskiene, I., Ulutaş, A.: An integrated simple weighted sum product method—WISP. IEEE Trans. Eng. Manage. **70**, 1933–1944 (2023). https://doi.org/10.1109/TEM.2021.3075783
31. Ulutaş, A., Stanujkić, D., Karabašević, D., Popović, G., Novaković, S.: Pallet truck selection with MEREC and WISP-S methods. Strategic Manage. **27**, 23–29 (2022). https://doi.org/10.5937/StraMan2200013U

Relation Between Applied Traverse Speed and Quality of Metal Surfaces Created by Abrasive Water Jet Cutting

Petr Pětroš(✉), Libor M. Hlaváč, and Jakub Gřunděl

VSB—Technical University of Ostrava, 17. Listopadu 2172/15, Ostrava, Czechia
petr.petros.st@vsb.cz

Abstract. The abrasive waterjet is nowadays widely used especially for cutting hard and difficult to machine materials. However, achieving the desired surface quality of the material to be cut, which can be described by roughness (*Ra*) and waviness *(Wa)* values, is challenging and is related to the setting of the jet parameters. The traverse speed of the cutting head was selected as the most suitable parameter that can influence the roughness and waviness, as it has a direct economic impact in addition to influencing the surface quality. Therefore, the aim of this paper is to determine the effect of the traverse speed on roughness and waviness values. However, these also depend on the depth of cut and the effect of speed may be different for different depths. For this reason, the dependence of roughness and waviness on the depth of cut has also been considered. To investigate these effects, experiments were carried out on different materials. The results show that roughness and waviness are essentially constant for small depth of cut, then increasing non-linearly with increasing of the depth of cut. There is an increase in roughness and waviness with increasing traverse speed, but this also becomes apparent only from a certain depth. These conclusions apply to all materials, but material properties, such as hardness, have an effect on specific roughness values. Further research work will be carried out to find the specific dependence of roughness and waviness on the depth of cut and the traverse speed of the cutting head.

Keywords: Abrasive Water Jet · Roughness · Waviness

1 Introduction

Abrasive waterjet cutting is a commonly used technology today, especially for cutting and machining hard and tough materials where traditional methods would cause excessive damage to the conventional machine tool. In addition, abrasive waterjet technology causes essentially no thermal effect on the material to be cut and is, therefore, more suitable than laser cutting for certain materials. Abrasive water jet can also cut materials of greater thickness than laser. In addition, abrasive waterjet cutting is very precise and is, therefore, used in the manufacturing of huge parts with high demands on precision [1].

L. Sitek et al. (Eds.): Advances in Water Jetting II, LNME, pp. 134–141, 2025.
https://doi.org/10.1007/978-3-031-72778-8_11

However, it is very difficult to achieve the required surface quality when cutting such resistant materials, e.g. stainless steel. This quality can be described by roughness (*Ra*) and waviness (*Wa*), whose values depend on many parameters of the abrasive water jet. These parameters are primarily those that are unchangeable or only very difficult to change during the cutting process, such as the diameter of the nozzle (orifice) and focusing tube, the type of abrasive used or the distance of the focusing tube from the workpiece [2]. In industry, these parameters are normally kept the same and not changed even between cutting processes, as this would cause excessive complications and time loss. Therefore, although it would be theoretically possible to influence the roughness and waviness values using these parameters, it is not advisable.

Secondly, roughness and waviness depend on jet parameters that can be changed very easily during the cutting process, especially pressure, abrasive mass flow rate and cutting head traverse speed [3]. In industrial applications, the pressure close to the maximum pressure of the pump is used most often, as this is how the highest quality cut can be achieved. The abrasive mass flow rate is then selected appropriately to the specific pressure, with higher abrasive mass flow rates the quality of the cut increases, but if the abrasive mass flow rate was set too high the abrasive feed could clog or mix poorly with the water jet in the mixing chamber, which would then have negative effects on the quality of the cut.

The most suitable parameter for achieving the desired surface quality is, therefore, the traverse speed of the cutting head, which affects not only the quality of the cut but it is also directly related to the length of the cutting process. Reducing the traverse speed theoretically reduces the roughness and waviness values, but at the same time increases the cutting time (with a direct economic impact). Therefore, it is necessary to choose an appropriate traverse speed to achieve the desired quality with the least economic loss [4]. For this purpose, it is necessary to analyse in detail the effect of the traverse speed on the roughness and waviness values, which is the subject of this paper.

2 Theoretical Assumptions

To study roughness and waviness and how it is affected by the traverse speed of the cutting head, the cutting process itself needs to be studied in more detail. During the cutting process, a jet of water containing accelerated abrasive particles hits the surface of the material. The abrasion of these particles against the workpiece then leads to the actual breaking of the material. The higher the kinetic energy of these particles is the higher the quality of the cut. The deeper the jet passes through the material, the lower the kinetic energy of the particles and the more irregularly it is distributed among the particles [5]. Because of this, the surface of the cut is striated and the waviness becomes more evident.

In addition to the kinetic energy of the incident particles the number of particles is also important. When the traverse speed increases, the number of incident particles decreases, which should inevitably lead to an increase in roughness and waviness. These dependencies should be the same in basic form for most machined materials, but the specific values of roughness and waviness will be, of course, influenced by the material properties.

This theoretical description can be summarised in the following three assumptions, which are tested in the course of the work.

1. Roughness values should increase with depth of cut with waviness significantly predominating for greater depths.
2. Roughness values will increase with increasing traverse speed of the cutting head; this increase may be different for different depths of cut.
3. The qualitative shape of the dependence of roughness and waviness on depth of cut and traverse speed should be the same for all materials. However, the specific roughness and waviness values are influenced by the properties of the materials, mainly their hardness.

The roughness and waviness values are of course also influenced by the previously mentioned parameters that cannot be changed during the cutting process, mainly the size of the abrasive grain, which directly interacts with the cutting surface and influences its quality.

Thus, based on these theoretical assumptions, all of the previously mentioned parameters can be expected to influence the roughness values. One of the possible expressions including the influence of abrasive particle size together with the material resistance and either the depth dependence or the traverse speed impact on roughness can be expresses through these equations:

$$Ra = \frac{a_n}{i} + Hr^j\left(\frac{h}{h_{lim}}\right)^k Ra = \frac{a_n}{i} + Hr^j\left(\frac{v_P}{v_{Plim}}\right)^k \tag{1}$$

The respective variables are: Ra—roughness, a_{n}—mean size of the abrasive particles after the mixing process, Hr—hardness of cut material, h – depth of cut (h_{lim}—the limit depth of cut), v_{P}—traverse speed (v_{Plim}—the limit traverse speed). The equation was derived similarly to the equation for the declination angle derived by Hlaváč [6]. Determining of factors i, j and k is the task of the future research. Nevertheless, the first approximation of the values is presented in section Results.

3 Design of Experiments

To verify the theoretical assumptions, a series of experiments were carried out. To achieve greater versatility, experiments were conducted on four different materials. The blocks of each material were firstly prepared with dimensions of approximately 90×90×20 mm. The five cuts were made into each of these blocks by an abrasive water jet, each of the five cuts being made using a different traverse speed. Experiments were performed at the VSB—Technical University of Ostrava using high pressure pump PTV 19/60 on the Flow HSQ 5× basis and the cutting table X-Y CNC WJ1020-1Z-EKO. The specific traverse speeds selected for each one of studied materials and some materials properties can be seen in Table 1. The unchanged abrasive waterjet parameters identical for all cuts are given in Table 2.

The velocity values were selected for each material so that the exit angle of the jet from the material sample is approximately identical for all materials, i.e. the ratios of

Table 1. Applied traverse speeds for selected materials and material properties.

Material number	Material type	Hardness HB Max	Yield Strength (MPa)	Traverse speeds (mm/min)
1.3343	High speed tool steel	255	850	20, 40, 60, 80, 100
1.4828	Heat resisting steel	223	650	20, 40, 60, 80, 100
2.1176	Bronze alloy	65	180	30, 60, 90, 120, 150
3.2315	Aluminium alloy	83	295	80, 120, 160, 200, 240

Table 2. Factors and settings applied in all experiments.

Variable	Value
Pump pressure	380 MPa
Water nozzle diameter	0.25 mm
Focusing tube diameter	0.76 mm
Focusing tube length	76 mm
Abrasive mass flow rate	250 g/min
Abrasive material mean grain size (input)	0.250 mm
Abrasive material type	Australian garnet (GMA80)
Stand-off distance	2 mm

Fig. 1. Material with completed cuts.

the used velocities relative to the limit velocities for the selected material are identical for all of them [7]. The block with the cuts made can be seen in Fig. 1.

When all five cuts were made, the cut perpendicular to them was performed to produce the plates, the surfaces of which were the walls of the testing cuts. Since surfaces were created using different traverse speeds of the cutting head, it is possible to measure several points of the surface roughness and waviness dependence on the traverse speed.

The roughness and waviness of the individual surfaces were then measured using an optical instrument. This was a laser profilometer that used the principle of triangulation. Thus, a laser line was incident on the surface to be measured and reflected from the surface to the digital camera, and from the recorded position on the digital camera, the height profile of the line was then determined, from which the roughness and waviness values for the specific line were obtained. After measuring one line, the next line was then measured until the entire surface of the sample was scanned. The distance of lines from each other was chosen as 1 mm. The measurements therefore resulted in roughness and waviness values for each millimeter of created surface.

Since the length of one measurement line was only 4 mm, which is less than the length of the plate, multiple lines were measured on each surface and the results were averaged to give the results for the entire surface.

4 Results

Matlab was used to process the measured data. Due to the large number of values, the measured roughness and waviness values for individual cuts were first evaluated from the created graphs of roughness/waviness dependence on the depth of cut. An example of such a plot is shown in Fig. 2.

In this way, a graphical dependence for the roughness and waviness of each surface was created and these dependencies were similar in nature. Along with the measured values, the calculated average roughness/waviness values for the individual depths were also entered into the graphs, indicated by red dots in the graphs. The range of the measured values can be clearly visible from these plots demonstrating that the range is getting greater with increasing depth of cut. The measurement errors as the standard deviations from the arithmetic mean were determined for each depth of section. These errors rarely exceeded ten percent which is a very good result considering the variability of jet parameters, materials properties and other factors, all of them fluctuating often in the range of the 5–10% around their average values.

The measured roughness and waviness values obtained for the different traverse speeds were then compared with each other, separately for each material. For this comparison, the average values calculated for a given surface belonging to a given speed were always used. It was concluded from the previously generated graphs that the average values adequately represent the surface. This conclusion is supported by the relatively low measurement error. The values belonging to one material were then compared with each other. An example of such a comparison is shown in Fig. 3.

Finally, the roughness and waviness values for each of the tested materials were compared with each other. In this way, graphs containing the values measured on surfaces made in different materials but created at corresponding speeds were prepared. This comparison can be seen in Fig. 4.

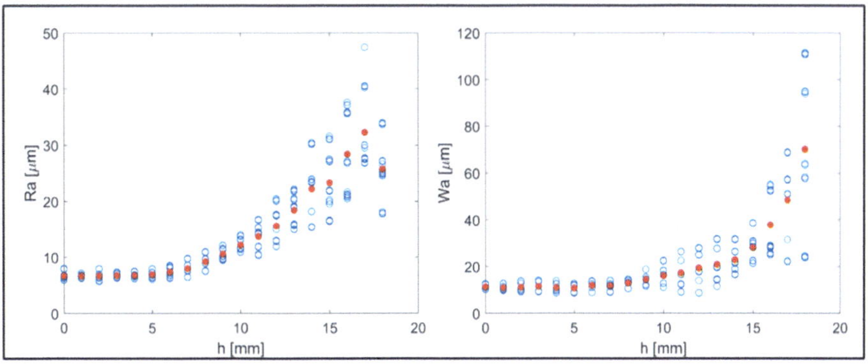

Fig. 2. Roughness and waviness dependencies on the depth of cut.

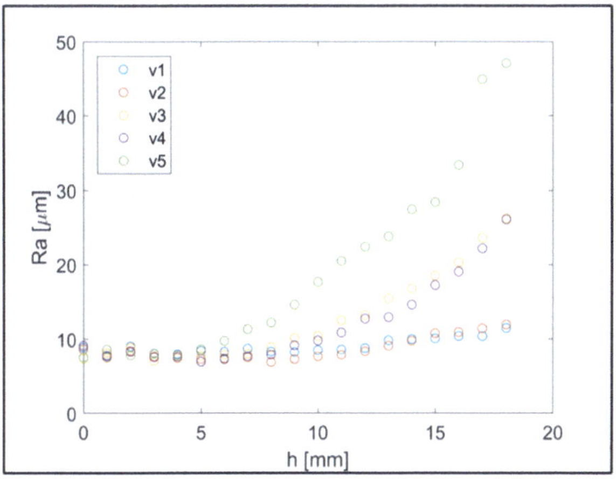

Fig. 3. Dependency of roughness on depth of the cut for different traverse speeds for material 1.4828.

It is evident that the more ductile materials have lower increase of roughness with the depth of jet penetration. Simultaneously, all dependencies determined for roughness seem to demonstrate the identical qualitative relationship as the one determined for the declination angle [6]. Assuming the relationship determined for declination angle the factors i, j, and k can be determined to have these values: $I = 6$, $j = 3/4$, $k = 3/2$. Then the curves calculated from the proposed equations for dependence on the depth in cut or the traverse speed approximately follow the measured dependencies. However, this result is just a preliminary one and needs further very extensive research to prove it.

5 Discussion of Results

Several conclusions can be drawn from the measurements. The roughness values are indeed dependent on the depth of cut. It is an increasing non-linear dependence and qualitatively similar for all applied traverse speeds and material types. This relationship

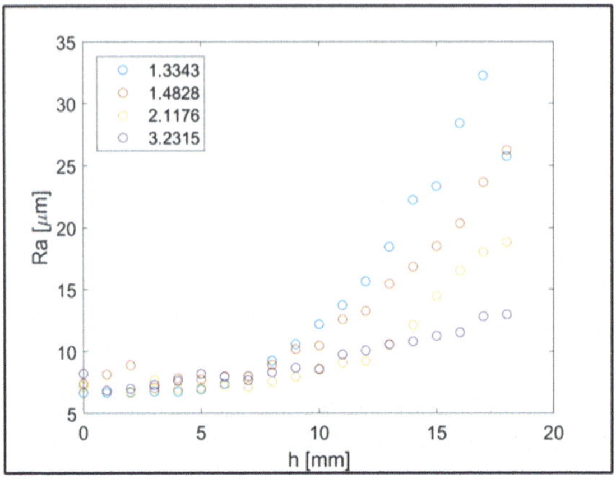

Fig. 4. Dependency of roughness on depth of the cut for different materials.

is also similar to the depth dependence of the declination angle. Surfaces created at higher relative speeds (at least 60% of the limit one) show significant waviness, in these cases the dependence of waviness on depth of cut is a non-linear increasing function. Both roughness and waviness initially increase only minimally with increasing depth and remain almost constant, but after a certain depth, there is a more noticeable increase in values with further increasing depth. Beyond this depth, the waviness values tend to be higher than the roughness values.

From the measured values, the assumption that there will be an increase in roughness and waviness with increasing traverse speed of the cutting head has been confirmed, but this increase is very irregular and not uniform throughout the depth of cut. Moreover, in this case too, the increase in roughness and waviness is only apparent from a certain point in the cut.

The previously described conclusions are valid for all the materials examined. However, the specific values of roughness and waviness are noticeably dependent on the material type. The highest values are achieved at the tool steel 1.3343 (with the highest hardness), followed by the stainless steel 1.4828. Materials 2.1176 (bronze) and 3.2315 (aluminium alloy) have noticeably lower hardness than steels and higher ductility, so the values of waviness and roughness are lower. The surface roughness and waviness of the individual materials only started to differ from each other after a certain depth of cut. Thus, all the investigated dependencies became apparent only from a certain depth of cut below which the roughness and waviness values were almost constant, as it was indicated during results evaluation. This depth is likely to be related to the change from the cutting removal mode to the deformation one [8] and it differs according to the specific cutting conditions determined by the jet parameters and material properties.

Further research work will be devoted to finding of specific dependencies of roughness and waviness on the traverse speed and thickness of used materials.

Acknowledgements. Presented research was supported by project SP 2023/036 of the Ministry of Education, Youth and Sports of the Czech Republic.

References

1. Wang, H., Yuan, R., Zhang, X., Zai, P., Deng, J.: Research progress in abrasive water jet processing technology. Micromachines **14**(8), 1526 (2023). https://doi.org/10.3390/mi14081526
2. Płodzień, M., Żyłka, Ł, Żak, K., Wojciechowski, S.: Modelling the kerf angle, roughness and waviness of the surface of inconel 718 in an abrasive water jet cutting process. Materials **16**(15), 5288 (2023). https://doi.org/10.3390/ma16155288
3. Qasem, I., Hussien, A.A., Alrawashdeh, K.A., Kataraki, P.S., Janvekar, A.A.: Performance analysis of cutting parameters on 304 stainless steels using abrasive water jet technique. Annales de Chimie—Science des Matériaux **47**(1), 9–16 (2023). https://doi.org/10.18280/acsm.470102
4. Begic-Hajdarevic, D., Cekic, A., Mehmedovic, M., Djelmic, A.: Experimental study on surface roughness in abrasive water jet cutting. Procedia Eng. **100**, 394–399 (2015). https://doi.org/10.1016/j.proeng.2015.01.383
5. Abdullah, R., Mahrous, A., Barakat, A., Zhou, Z.: Surface quality of marble machined by abrasive water jet. Cogent Eng. **3**(1), (2016). https://doi.org/10.1080/23311916.2016.1178626
6. Hlaváč, L.M.: Investigation of the abrasive water jet trajectory curvature inside the kerf. J. Mater. Process. Technol. **209**(8), 4154–4161 (2009). https://linkinghub.elsevier.com/retrieve/pii/S0924013608007310
7. Hlaváč, L.M.: Revised model of abrasive water jet cutting for industrial use. Materials **14**(14), 4032 (2021). https://doi.org/10.3390/ma14144032
8. Hashish, M.: A modeling study of metal-cutting with abrasive waterjet. J. Eng. Mater. Technol. Trans. ASME **106**(1), 88–100 (1984). https://doi.org/10.1115/1.3225682

Influence of Erosion Time on HVOF-Sprayed Cermet Coating: Preliminary Study

Jakub Poloprudský[1,2(✉)], Tomáš Vražina[1,2], Akash Nag[3,4], Gabriel Stolárik[3,4], Alice Chlupová[1], Karel Zábranský[5], Pavel Gejdoš[5], Ladislav Čelko[5], and Sergej Hloch[3,4]

[1] Institute of Physics of Materials CAS, Brno, Czechia
poloprudsky@ipm.cz
[2] Faculty of Mechanical Engineering, Brno University of Technology, Brno, Czechia
[3] Institute of Geonics of the CAS, Ostrava, Czechia
[4] VSB—Technical University of Ostrava, Ostrava, Czechia
[5] CEITEC—Central European Institute of Technology, Brno University of Technology, Brno, Czechia

Abstract. The vast majority of hard coatings are known for their abrasion resistance. They provide excellent protection against both water erosion and solid particle abrasion. These coatings represent the option on how to protect the substrate in many areas of industrial applications. However, the possible resistance of the coating to the adverse conditions of high-frequency water droplets is still not well investigated. This study deals with the assessment of erosion resistance of WC-CrC-Ni cermet coatings against multiple water droplet impingements. Water droplets were generated using an ultrasonic pulsating water jet with a nominal frequency of 40 kHz. The cermet coating was subjected to different exposure times to assess erosion development and evaluate the integrity of the coating with the substrate. To provide a comparison, the samples were also subjected to continuous water jet utilizing the same conditions. The erosion pits were examined using optical microscopy and digital image analysis for general observations, while electron microscopy was utilized to identify the mechanisms of erosion. The shape, depth, and volume of the pits were measured using a confocal laser scanning microscope to evaluate the intensity of the damage caused by both continuous and discontinuous water jets. The effect of increasing time exposure of the pulsating water jet was evaluated and recalculated to the number of droplet impingements.

Keywords: Pulsating Water Jet (PWJ) · Droplet Impingement (DI) · High-Velocity Oxygen Fuel (HVOF) · Cermets · Erosion

1 Introduction

During the service life, the surface of components is exposed to various conditions determined by the application. Surfaces separate components from their surroundings and ensure their performance and reliability in the given environment. One of the important characteristics is the erosion resistance created by a coincidence of corrosion, abrasion,

L. Sitek et al. (Eds.): Advances in Water Jetting II, LNME, pp. 142–155, 2025.
https://doi.org/10.1007/978-3-031-72778-8_12

cavitation erosion and water droplet erosion resistance. To increase the material surface resistance against erosion, many techniques have been developed, including the application of various coatings [1]. These techniques include, for example, high-velocity oxygen fuel (HVOF) spraying, which is an effective technique that does not significantly thermally affect the substrate. Generally, the HVOF technique is considered to have better wear resistance than many other thermal processes [2]. The HVOF is used mainly in the industry for surface treatment of components that are exposed to external forces of water or gas such as compressor blades, turbine blade roots or helicopter rotor joints to prevent surface degradation of functional parts [3], and as well in many other applications in energy sector [1–3].

In real-world applications, hard and erosion-resistant materials like ceramics and cermets are often used to protect hydrodynamic pump impeller blades, whether as coating or bulk [4]. However, material selection is not the sole pivotal factor affecting erosion pit formation. Other important factors to consider are the presence of surface defects or an increased level of surface roughness, which can noticeably amplify the erosive effects induced by pulsating water jet (PWJ) [5]. The surface roughness combined with the presence of surface and subsurface defects may have a detrimental effect on the erosion resistance of hard and brittle coatings. Cermet coatings studied in the present work were applied onto a substrate by the HVOF method. The structural quality of the coating is strongly affected by a high variety of spraying parameters, including liquid/gaseous medium pressure, nozzle diameter, stand-off distance, the shape and size distribution of the powder particles, etc. Consequently, due to the lower HVOF plume temperature (compared to the atmospheric plasma spray technique), the coating may exhibit non-uniform density and contain several gaps between the splats and internal pores. These characteristics further significantly contribute to erosion damage [6, 7].

In recent years, there has been growing interest in the continuous water jet [8] (CWJ) and abrasive water jet (AWJ) techniques as an alternative process for cutting and drilling through a wide range of materials such as metals [9–11], ceramics [12], and composites [13]. These techniques also proved enormous effectiveness in the removal of damaged material layers and coatings [14].

Continuous water jet can be split into individual droplets by several methods [15], resulting in repetitive impact pressure acting on the surface of the material. The impact pressure is significantly higher than the stagnation Bernoulli pressure [16].

One modification of the water jet, known as the PWJ method achieves jet disintegration by the formation of pressure pulses directly in the acoustic chamber of a device located at the Institute of Geonics in the Czech Republic [17]. The pressure pulses are created by oscillations of the sonotrode, which is connected to a pulsating piezoelectric actuator and then transformed into the velocity pulses upon reaching the nozzle. The velocity pulses then lead to the shattering of the jet into the individual droplets. The droplet impact causes the impact pressure followed by lateral outflow at the material surface. The lateral jet is several times faster than the velocity of impacting droplets [18]. Repetition of this can have a severe effect on the defects present in the coating. CWJ causes the impact pressure and lateral outflow jetting only during the first contact with the surface. PWJ causes impact pressure and lateral outflow at the frequency of the sonotrode.

While extensive research has been conducted to assess erosion behaviour during the exposition of various materials to waterjet, most of these studies were primarily focused on CWJ and AWJ methods [19, 20]. PWJ testing has been focused predominantly on metallic materials such as aluminium, titanium, and stainless steel [21, 22] and also on minerals and rocks [23].

The selection of the HVOF technique through the other alternatives such as atmospheric plasma spray (APS), and detonation gas spraying (DGS) was driven by its ability to maintain the almost original chemical composition of the powder particles used. In the case of Amperit 551.074, its chemical composition is notably pure, containing primarily the main elements of WC-CrC-Ni [24]. This level of purity distinguishes it from other commercially produced WC-CrC-Ni powders, where the presence of $(W,Cr)_2C$ [25] is more common. The disadvantage of the DGS method is the formation of an extremely brittle cubic WC1-x phase [26]. In the case of APS, nearly all of the initial Cr_3C_2 content transforms into the $(W,Cr)_2C$ phase [27], nevertheless, the initial cermet structure of the powder feedstock was found irreversibly disrupted mainly due to extremely high plasma plume temperature of APS. Also, when using the same powder for APS, the coating itself reveals significantly lower hardness, of approximately 200 HV0.3, compared to those coatings produced by the HVOF technique. In the case of vacuum plasma spray (VPS), the widest range of hardness values was observed, indicating a noticeable heterogeneity within the coating [25].

In the literature, the experiments focused on the evaluation of HVOF droplet erosion resistance and comparing them with other erosion-resistant solutions can be found. Mann et al. [2] conducted a comparative study involving HVOF-coated and laser-treated steel and titanium alloy. Later, Mann et al. [28] studied the HVOF erosion utilizing advanced fuel injectors leading to a substantial improvement in fracture toughness and the erosion resistance of the HVOF coatings. Mahdipoor et al. [29] observed significant erosion rate differences for coatings with similar chemical compositions. Zhang et al. [30] compared HVOF coating with atmospheric plasma-sprayed Ni coating. Gujba [31] compared HVOF coatings with high-velocity air fuel (HVAF) coatings. This comparison revealed that HVAF delays erosion at lower impact speeds 250 m/s and 300 m/s about 2–5 times compared to HVOF coatings [31]. Tieu et al. [32] used a forced pulsed water jet (FPWJ) for the removal of HVOF-coated coupons and coated landing pins. The removal was done at a pressure of 104 MPa with a 1.37 mm orifice and a flow rate of 36.3 l/min. Removing the coating from pins took only 16 min compared to several hours in the case of the conventional method [32]. The parameters of the HVOF coating and erosion parameters from the discussed works are summarized in Table 1. These works used WC-Co or WC-Co-Cr systems. The reason for widespread testing of this system is that in applications where abrasive or erosive wear is assumed WC-Co powders are used with the possible addition of nickel or chrome. Chrome addition (WC-Co-Cr) is used when there is also a demand for corrosion resistance [2].

Despite the widespread use of cermet coatings with cobalt matrices, there are emerging concerns related to this metal matrix due to the high cost of cobalt, its price instability, geopolitical risks with its mining as well as a range of health-related issues, including carcinogenicity [33]. Furthermore, environmental challenges are connected to its extraction, processing, and utilisation. Therefore cobalt-free binders are rising in popularity. This includes cermets like WC-CrC-Ni, with only 7% nickel in the matrix, as well as other variants containing chromium carbide (CrC) and nickel (Ni). These alternatives exhibit almost similar corrosion resistance and wear properties in comparison with WC-Co-Cr cermets [34].

Oke et al. [35] used a waterjet apparatus at a high standoff distance of 200 mm to observe the erosion rate of WC-20 CrC -7Ni at 7 different pressure levels. The same material was selected for this study. The difference between this system and a widely used WC-Co system is the presence of nickel as well as 2 types of carbides based on wolfram and chromium. Korobov et al. [36] compared the cavitation erosion resistance of WC-Co-Cr to WC-CrC-Ni. The result proved lower volume loss of WC-CrC-Ni compared to WC-Co-Cr [36], and the suspected reasons were finer structure and lower porosity of WC-CrC-Ni coating which reduces cavitation damage, lowers sprayed roughness, and increased plasticity of the metal matrix.

Table 1. Erosion testing of HVOF coatings.

Coating	Technology	Impact velocity	Coating thickness	Coating hardness	Porosity %	Source
WC-10Co-4Cr	Rotating disc	147.4 m/s 147.7 m/s	250 ± 30 μm	1090–1226	–	[2]
WC10Co4Cr	Rotating disc	147.6 m/s	250 ± 30 μm	1450–1500	–	[28]
WC-12Co	Rotating disc	250 m/s 300 m/s 350 m/s	360 μm	874 ± 30	2.9	[29]
WC-11/13Co	Rotating disc	250 m/s 300 m/s 350 m/s	330 μm	1232 ± 14	1.4	[29]
WC-17Co	Rotating disc	638 m/s	387 μm 251 μm	1270 1103	1.04 0.98	[30]
WC-10Co-4Cr	Rotating disc	250 m/s 300 m/s 350 m/s	170 μm	910	<1	[31]
WC-Co-Cr	FPWJ	410 m/s*	180 μm	–	–	[32]
WC-20CrC-7Ni	WJ apparatus	191 m/s for 50 MPa	300 μm	1672*	–	[35]
WC-20CrC-7Ni	PWJ	285 m/s	194 ± 16 μm	1074 ± 140	1%	This work

* Calculated from the parameters in the article

Table 1 presents and compares all documented studies related to water-jetting on cermet coatings. Notably, very few of these studies delineate the distinct differences in damage caused by impact pressure and stagnation pressure. This research fills that gap by comparing PWJ and CWJ techniques and focusing extensively on the qualitative and quantitative analysis of the mechanisms of damage and their development over time when interacting with brittle coatings possessing an initial value of surface roughness.

2 Materials and Methods

2.1 Material Formation and Composition

The specimens used in this study were cylindrical having 30 mm in diameter and 10 mm in height. The specimens were split up into 2 parts. One was subjected to PWJ and CWJ and the second one was used for the microstructural analysis and assessment of surface quality before the waterjet erosion. These specimens contained WC-CrC-Ni coating firmly adhered to an austenitic stainless steel substrate (AISI 304) through the Ni20Cr bond coat as shown in Fig. 1(a).

The WC-CrC-Ni cermet coating, with a composition consisting of 73% WC, 20% CrC, and 7% Ni, was formed during the deposition process by employing commercially available Amperit 551.074 powder. This powder, manufactured by Höganäs, has particle sizes ranging from 15 to 45 μm. The resulting cermet coatings were of approximate thickness of 194 ± 16 μm, hardness of 1074 ± 140 HV0.3, the roughness of Ra 7.47 ± 0.44 μm and porosity roughly around 1% as shown in Fig. 1(b). The top coat and bond coat system altogether was found 322 ± 11 μm thick. The optimisation of the coating deposition process is described in more detail in the literature [24].

For the coating deposition, an ethane/oxygen gas-fueled HVOF technique was employed, specifically the GLC model from GTV with a GLC5 gun. This tool was mounted on a six-axis IRB 2600 Robot for precise gun motion. Before deposition, a preparatory treatment was applied to the austenitic steel coupons, involving grit blasting using Al_2O_3 abrasive particles. The deposition parameters were set as follows: a spray distance of 200 mm, an oxygen flow rate of 200 slpm (standard litres per minute), and an ethylene flow rate of 80 slpm. No post-processing techniques were employed to modify the surface quality.

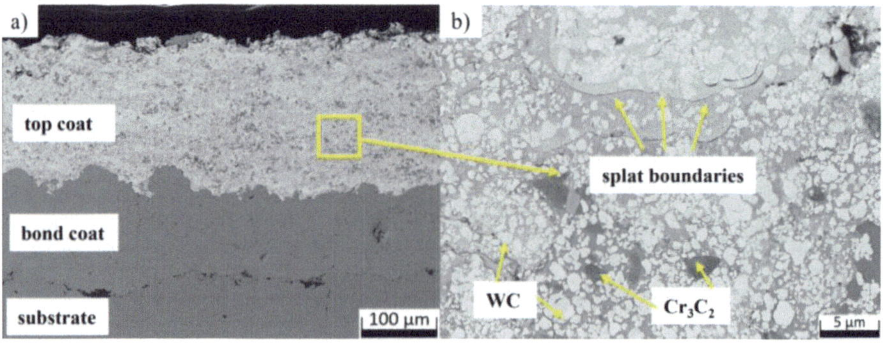

Fig. 1. Cross section of WC-CrC-Ni coating: a) general overview, b) detail of microstructure.

2.2 Experiments

The surface treatment of cermet coatings was performed using an ultrasonically generated PWJ. It acted periodically with increasing time exposure on defined points of the HVOF coating. The control points were eroded by CWJ. The experimental schematic is shown in Fig. 2. The input technological hydraulic parameters were as follows: pressure $p = 50$ MPa, nozzle diameter $d = 0.4$ mm and frequency of sonotrode $f = 40.3$ kHz (for PWJ). The pressure was chosen to achieve a flow rate of 285 m/s as most of the research conducted on this topic oscillates between impact velocities of 250 m/s and 300 m/s. The technology currently allows two frequencies 20 and 40 kHz. The frequency of 40 kHz was chosen to shorten the period between the impingement of the individual water clusters.

The optimal length of acoustic chamber l_c was evaluated before the experiment according to [37]. The optimal standoff distance z was evaluated by stairs trajectory according to [38] to achieve the culmination point. The selected PWJ and CWJ parameters are shown in Table 2. The semicircular sample with a specific thickness of the coating was subsequently treated by the PWJ with a time exposure t varying between 10–60 s with an increment of 10 s, leading to 6 exposure areas. This process was repeated twice (L4, L5). The same time increment was also used in the CWJ control tests, where the time exposure was set to $t = 20$–60 s, leading to 5 exposure areas (L1). Due to the necessity of a more detailed observation at the start of the coating penetration, the test with a time exposure of $t = 2$–12 s was performed for PWJ with a time increment of 2 s, leading to 6 exposure areas. This process was repeated twice (L2, L3). The total number of evaluated pits was 24 for PWJ and 5 for CWJ. Each PWJ line exhibited six pits, with the procedure being repeated 2 times.

Table 2. CWJ and PWJ parameters.

Line	p [MPa]	f [kHz]	Nozzle diameter d [mm]	Standoff z [mm]	Chamber length [mm]	Flow speed [m/s]	Exposure time t [s]	Time increment [s]
L1	50	-	0.4	45	10	285	20–60	10
L2, L3	50	40.3	0.4	45	10	285	2–12	2
L4, L5	50	40.3	0.4	45	10	285	10–60	10

2.3 Methods of Characterization

Characterization of coating and observation of the eroded sample surface was performed using a scanning electron microscope LYRA TESCAN 3 (SEM) both before and after the water jet treatment, specifically for the assessment of erosion pit characteristics. Additionally, optical microscopy was employed on the DVM6 Digital Microscope (Fig. 3a). To determine the coating thickness, analysis was conducted based on a minimum of five

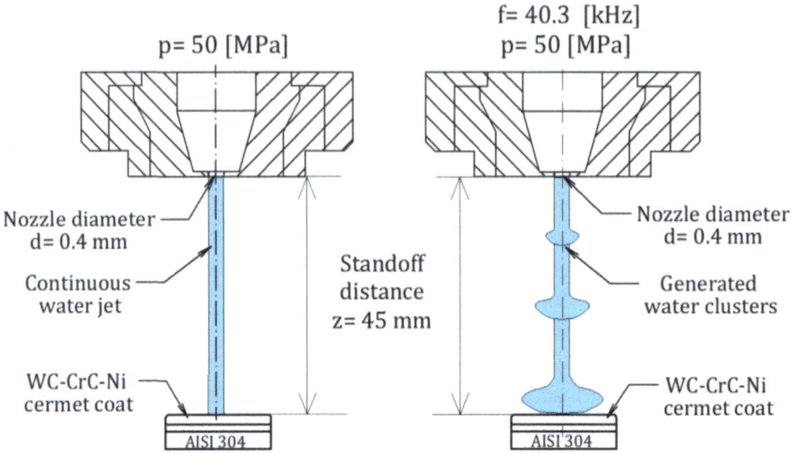

Fig. 2. Experimental set-up for CWJ and PWJ erosion points of the HVOF layer.

SEM images. For a more precise examination of pit morphology, comprising aspects such as shape, height, width, and volume, as well as the surface roughness parameters of the deposited cermet coatings, measurements were carried out using a confocal laser scanning microscope, the LEXT 5100 (CLSM).

3 Results and Discussion

Figure 1(a) and (b) show the initial microstructure from the cross-section. Figure 1(b) highlights features such as intersplat boundaries, porosity, and a distinct phase composition region. Bright white particles correspond to WC, and dark regions to CrC-carbides as documented also in [25, 26]. Porosity primarily develops at the particle-matrix interface or the splat boundaries. Intersplat boundaries are created during the spraying process, where a new layer is added onto the previous one. This may cause variations in thermal gradients, resulting in microsegregation and the formation of metastable phases.

3.1 Qualitative Observation

As described in Sect. 2.2, there are five rows of erosion pits at the surface of the coated specimen, created by either PWJ (L2-L4) or CWJ (L1), as shown in Fig. 3. The erosion pits produced by the CWJ method on the WC-CrC-Ni coating (Fig. 3b) are relatively shallow and can only be detected because of variations in reflectivity caused by surface topography.

The first penetration using the PWJ method occurred within 2 s using the PWJ method. As illustrated in Fig. 3(d) in the vicinity of the pit, there is a noticeable horizontal crack. This is presumably the result of the combined effects of water hammering and lateral flow, which causes the rapid growth of preexisting internal cracks as mentioned in [29]. The influence of lateral flow is also evident in Fig. 3(e) and 3(f), where it significantly contributes to crack development behaviour. In Fig. 3(e), water droplets initiated erosion at two positions independently, subsequently coalescing into a larger one. Prolonged exposure to PWJ resulted in the formation of circular pits.

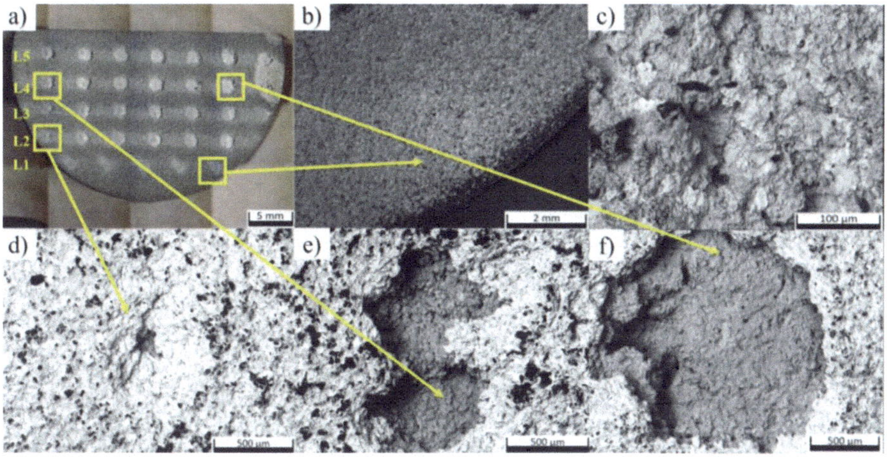

Fig. 3. Macroscopic observation of eroded surface: a) general overview, b) CWJ exposure of 60 s, c) detail of CWJ erosion, d) detail of PWJ erosion after 2 s, e) detail of PWJ erosion after 10 s, f) detail of erosion after 60 s.

Figure 4 shows HVOF coating exposed to CWJ for 60 s compared to HVOF exposed to PWJ for 3 levels of exposure time. When comparing the initial state to the state after CWJ exposure, there is a minimal difference, even when considering pits caused by CWJ exposure over 60 s. In contrast, just 4 s of PWJ exposure managed to penetrate the coating and reach the material substrate. Several erosion pits even exhibit slight asymmetry, as depicted in Fig. 3(e) or Fig. 4(c). This asymmetry is most likely due to initiation at two specific locations where surface defects were present.

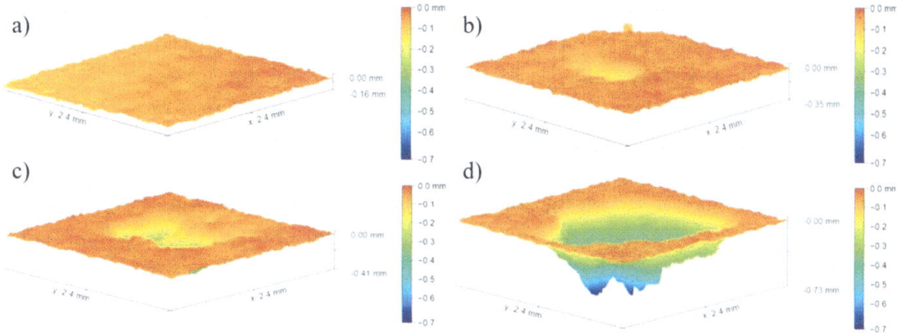

Fig. 4. A 3D visualization by CLSM depicting the erosion evolution after a) CWJ with time exposure of 60 s, b) PWJ exposure of 2 s, c) PWJ exposure of 10 s, d) PWJ exposure of 60 s.

3.2 Quantitative Observation

Three parameters were selected to evaluate the exposure to water jets: width, volume, and depth. The width of the pit was further divided into horizontal width (along the x-axis) and vertical width (along the y-axis) to describe the asymmetry observed during the water jetting process. The width parameter represents the width of the affected region as depicted in Fig. 5.

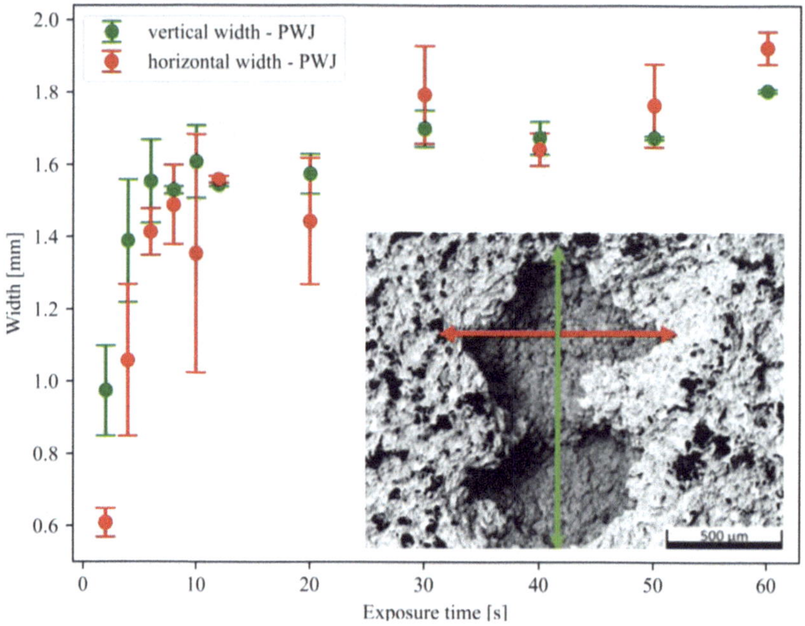

Fig. 5. Plot depicting the relationship between eroded width and exposure time of PWJ.

In Fig. 5, the asymmetry in the impacted area of the erosion pit resulting from PWJ exposure becomes apparent. Larger vertical width values are noticeable for low exposures up to 20 s. Nevertheless, after the 20 s exposure, the pits gradually adopt a more uniform appearance. This pattern was also documented in a prior study [30] where the width increased faster than the depth. After a certain period, it stabilized at a saturated value, representing the effective area of the water jet. The elevated error values in Fig. 5 indicate that even with the same exposure time, the formation of pits can vary significantly. Lower error rates are observed, particularly in the case of vertical width after 4 s. This behaviour can be attributed to the fact that the coating is penetrated after the 4 s exposure, and the asymmetry in the vertical direction is presumably a result of the asymmetry in the water jet. Data from CWJ were not considered in Fig. 5 because no macroscopic material removal was observed and therefore it would be problematic to distinguish the area boundaries.

Figure 6 reveals that the depth of the erosion pit reaches a saturation point after 4 s exposure, which corresponds to the time required to penetrate both the top-coat and bond-coat. This short duration can be compared to results from publications using an ASTM G73 rotating disc to generate water clusters [2, 28, 29]. This may be attributed to the high frequency of impacts (40 kHz) used in this study. Research that closely matches the findings reported in this study is [35], where the material was similar. When considering only the speed of clusters, the depth of penetration would be comparable. Further deepening occurs at the 50 s exposure, indicating that the PWJ possesses sufficient power to penetrate the substrate. In contrast, the CWJ results in negligible deepening, as no significant material removal was observed on the surface, and the presence of small holes can be attributed to pre-existing surface defects before the CWJ exposure.

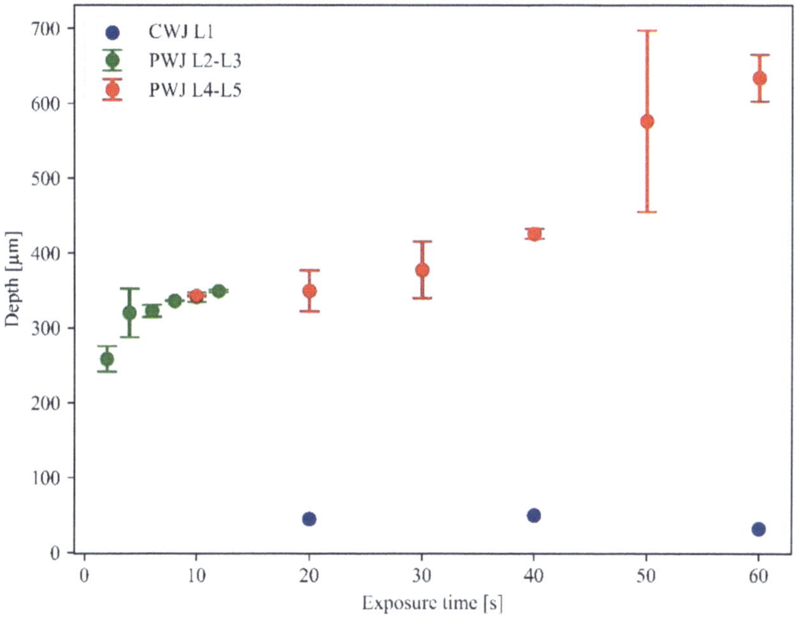

Fig. 6. Plot of the relationship between eroded depth and exposure time for both CWJ and PWJ.

In Fig. 7, the extent of damage caused by PWJ and CWJ is illustrated through the measurement of removed material volume. This figure demonstrates that a more than 2 s exposure to PWJ results in significantly higher damage compared to a 60 s exposure to CWJ. The volume removal by CWJ is expected to range from 0.001 to 0.002 mm3, consistent with surface evaluations before CWJ treatment. Distinguishing between pits formed before and after CWJ proves to be challenging due to the subtle effects of the CWJ process on the coatings. The Fig. 7 also illustrates that after the coating is penetrated the values of removed material volume start to saturate. The biggest error was recorded at about 10 s within lines 2–3 subjected to PWJ.

The significant difference in behaviour between CWJ and PWJ suggests that, although cermet material can withstand the flow of abrasive particles and wear caused by grinding machines and other factors [39], hardness is not the sole factor influencing the erosion behaviour of the material during PWJ. Other factors to consider include

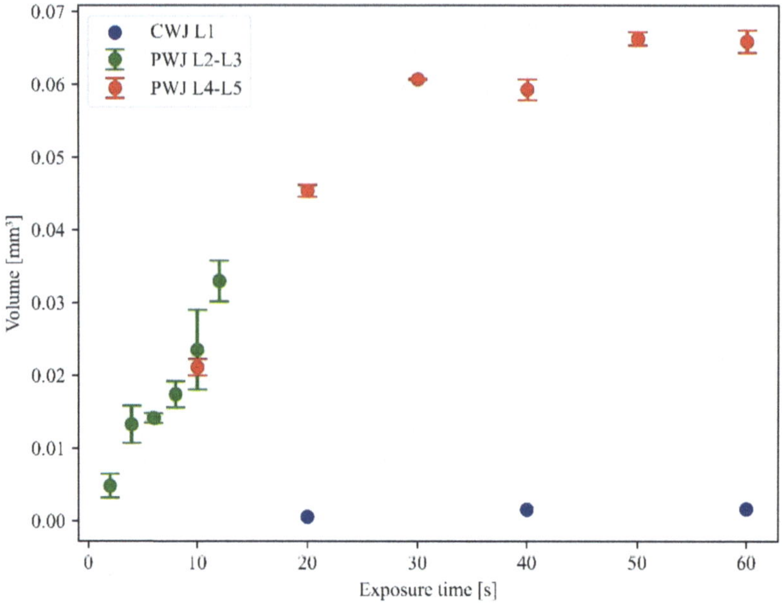

Fig. 7. Scatter plot depicting the relationship between eroded volume and exposure time for both CWJ and PWJ.

surface quality (such as roughness and the presence of surface porosity) and internal factors: splat boundaries and other defects within the coating, which create pathways for the water jet to follow. In this case, the more detrimental effect of PWJ on the coatings in this study, compared to the research highlighted in Table 1, can also be attributed to surface roughness. This is especially apparent when comparing it to other studies [29] where surface roughness reached values approximately $Ra = 3$ μm, and in some cases even lower, down to $Ra = 1$ μm [36]. Significant differences in erosion behaviour due to varying surface roughness were also documented among metallic materials [40]. Polishing the samples led to a doubling of the pit's incubation period in the treated material.

High erosion effectivity and low time required for coating penetration reached by PWJ at 50 MPa were achieved.

4 Conclusions

An experimental study of the effect of PWJ and CWJ water on cermet coating prepared by the HVOF technique resulted in the following conclusions:

- PWJ exposure resulted in erosion of HVOF deposited WC-CrC-Ni coating after only 2 s at given conditions and another 2 s results in penetration through the Ni20Cr bond coat.
- CWJ exposure at 60 s did not lead to significant damage to the HVOF coating, as confirmed through visual assessment and analysis of eroded volume data under the given hydrodynamic conditions.

- PWJ erosion depth data reveals substantial asymmetry within the pit formation. This asymmetry can be attributed to the initiation process of erosion pit formation, as well as the penetration of both the coating and the base material.

Further research should elucidate the difference in water erosion results acquired using spinning disc erosion testing and PWJ droplet erosion. Furthermore, the results obtained in this article can be compared further with widely used industrial WC-Co-Cr coatings and the possibility of polishing the coating can also be taken into consideration.

Acknowledgements. The financial support by the Czech Science Foundation, project GACR 23-05372S, is gratefully acknowledged. The authors would like to acknowledge the technical support provided by Pragolab and CEITEC (Central European Institute of Technology), as well as the Institute of Geonics CAS.

References

1. Krella, A.: Degradation and protection of materials from cavitation erosion: a review. Materials **16**(5), 2058 (2023)
2. Mann, B., Arya, V.: HVOF coating and surface treatment for enhancing droplet erosion resistance of steam turbine blades. Wear **254**(7–8), 652–667 (2003)
3. Pradeep, D., et al.: Review on tribological and mechanical behavior in HVOF thermal-sprayed composite coatings. J. Bio. Tribo. Corros. **8**, 30 (2022)
4. Zhang, S., et al.: Investigation on microstructure, surface properties and anti-wear performance of HVOF sprayed WC–CrC–Ni coatings modified by laser heat treatment. Mater. Sci. Eng. B **162**(2), 127–134 (2009)
5. Di, J., et al.: Experimental investigation on effect of surface strengthening process and roughness on water droplet erosion behavior in turbomachinery. Tribol. Int. **153**, 106647 (2021)
6. Murugan, K., et al.: Optimizing HVOF spray process parameters to attain minimum porosity and maximum hardness in WC–10Co–4Cr coatings. Surf. Coat. Technol. **247**, 90–102 (2014)
7. Ribu, D., et al.: Investigating the water jet erosion performance of HVOF-sprayed WC-10Co coatings on 35CrMo steel utilizing design of experiments. Coatings **12**, 482 (2022)
8. Raj, P., et al.: A review on continuous and pulsed water jet machining. Mater. Today Proc. **27**, 2596–2604 (2020)
9. Poloprudský, J., et al.: Effects of liquid droplet volume and impact frequency on the integrity of Al alloy AW2014 exposed to subsonic speeds of pulsating water jets. Wear **488–489**, 204136 (2022)
10. Arola, D., et al.: Waterjet and abrasive waterjet surface treatment of titanium: a comparison of surface texture and residual stress. Wear **249**, 943–950 (2001)
11. Azhari, A., et al.: Improving surface hardness of austenitic stainless steel using waterjet peening process. Int. J. Adv. Manuf. Technol. **63**, 1035–1046 (2012)
12. Raj, P., et al.: Investigation of sandstone erosion by continuous and pulsed water jets. J. Manuf. Process. **42**, 121–130 (2019)
13. Foldyna, V., et al.: Effects of continuous and pulsating water jet on CNT/concrete composite. Stroj. Vestn. J. Mech. Eng. **63**, 583–589 (2017)
14. Kong, M., et al.: Aspects of material removal mechanism in plain waterjet milling on gamma titanium aluminide. J. Mater. Process. Technol. **210**, 573–584 (2010)

15. Říha, Z., et al.: Comparison of the disintegration abilities of modulated and continuous water jets. Wear **478–479**, 203891 (2021)
16. Burson-Thomas, C., et al.: Water droplet erosion of aeroengine fan blades: the importance of form. Wear **426–427**, 507–517 (2019)
17. Foldyna, J.: Use of acoustic waves for pulsating water jet generation. In: Beghi M. (ed.) Acoustic Waves—From Microdevices to Helioseismology, pp. 323–342. Intech, Rijeka (2011)
18. Field, J., et al.: The effects of target compliance on liquid drop impact. J. Appl. Phys. **65**, 533–540 (1989)
19. Gugliemi, G., et al.: Life cycle environmental and economic comparison of water droplet machining and traditional abrasive waterjet cutting. Sustainability **13**, 12275 (2021)
20. Natarajan, Y., et al.: Abrasive water jet machining process: a state of art of review. J. Manuf. Process. **49**, 271–322 (2020)
21. Srivastava, M., et al.: Surface treatment of AISI 304 using pulsating water jet peening. In: Singh, M., Kushvah, B., Seth, G., Prakash, J. (eds.) Applications of Fluid Dynamics, pp. 535–548. Springer, Singapore (2018)
22. Chlupová, A., et al.: Effect of pulsating water jet processing on erosion grooves and microstructure in the subsurface layer of 25CrMo4 (EA4T) steel. Wear **524–525**, 204774 (2023)
23. Tripathi, R., et al.: Influence of frequency change during sandstone erosion by pulsed waterjet. Mater. Manuf. Process. **35**, 187–194 (2020)
24. Vražina, T., et al.: Influence of the deposition parameters on microstructure and properties of HVOF sprayed WC-Crc-Ni coating. In: Proceedings 31st International Conference on Metallurgy and Materials, pp. 487–492 (2022)
25. Berger, L., et al.: Influence of feedstock powder characteristics and spray processes on microstructure and properties of WC–(W, Cr)2C–Ni hardmetal coatings. Surf. Coat. Technol. **205**, 1080–1087 (2010)
26. Kirner, K.: Metallurgy of tungsten carbides in thermal spraying. In: Metallurgie der Wolframcarbide beim thermischen Spritzen. Schweissen und Schneiden, vol. 41(11), pp. 583–586. Germany F.R (1989)
27. Sarin, V.: Comprehensive hard materials, 1st edn. Elsevier, Amsterdam (2014)
28. Man, B., et al.: Advanced high-velocity oxygen-fuel coating and candidate materials for protecting LP steam turbine blades against droplet erosion. J. Mater. Eng. Perform. **14**, 487–494 (2005)
29. Mahdipoor, M., et al.: HVOF sprayed coatings of nano-agglomerated tungsten-carbide/cobalt powders for water droplet erosion application. Wear **330–331**, 338–347 (2015)
30. Zhang, Z., et al.: Experimental study on water droplet erosion resistance of coatings (Ni60 and WC-17Co) sprayed by APS and HVOF. Wear **432–433**, 202950 (2019)
31. Gujba, A., et al.: Water droplet impingement erosion performance of WC-based coating sprayed by HVAF and HVOF. Wear **484–485**, 203904 (2021)
32. Tieu, A., et al: Considerations in the use of pulsed water jet techniques for the removal of HVOF coatings. In: Proceeding of 2007 American WJTA Conference and Expo. Waterjet Technology Association, St. Louis, MO (2007)
33. Van Den Bink, S., et al.: Identifying supply risks by mapping the cobalt supply chain. Resour. Conserv. Recycl. **156**, 104743 (2020)
34. Picas, J., et al.: Cobalt free metallic binders for HVOF thermal sprayed wear resistant coatings. Surf. Coat. Technol. **456**, 129243 (2023)
35. Oka, Y., Miyata, H.: Erosion behaviour of ceramic bulk and coating materials caused by water droplet impingement. Wear **267**, 1804–1810 (2009)
36. Korobov, Y., et al.: Cavitation resistance of WC-10Co4Cr and WC-20CrC-7Ni HVAF coatings. J. Therm. Spray Technol. **31**, 234–246 (2022)

37. Nag, A., et al.: Acoustic chamber length performance analysis in ultrasonic pulsating water jet erosion of ductile material. J. Manuf. Process. **47**, 347–356 (2019)
38. Hloch, S., et al.: Effect of pressure of pulsating water jet moving along stair trajectory on erosion depth, surface morphology and microhardness. Wear **452–453**, 203278 (2020)
39. Thakur, L., Arora, N.: Solid particle erosion behavior of WC-CoCr nanostructured coating. Tribol. Trans. **56**, 781–788 (2013)
40. Kirols, H., et al.: The effect of initial surface roughness on water droplet erosion behaviour. Wear **342–343**, 198–209 (2015)

Influence of Abrasive Water Jet Cutting Parameters on the Surface Properties of Modern Plain Bearing Materials

Jarosław Praźmo[1], Elżbieta Kawecka[2], Monika Forysiewicz[1], Aleksandra Radomska-Zalas[3], and Andrzej Perec[2](✉)

[1] Institute of Engineering and Technology, University of Applied Sciences, Walcz, Poland
[2] Faculty of Technology, Jacob of Paradies University, Gorzów Wielkopolski, Poland
aperec@ajp.edu.pl
[3] Gorzów Scientific and Industrial Technology Center, Gorzów Wielkopolski, Poland

Abstract. The article presents an analysis of the possibility of using hydro-abrasive machining to shape Permaglide P10 plain bearing elements. The standard material for Permaglide plain bearings with the P10 designation used in the tests is characterized by a very low tendency to stick-slip vibrations, high wear resistance, good chemical resistance, low coefficient of friction, no tendency to weld with metal, resistance to swelling and a very important feature, from the point of view of the treatment used in the tests, the material does not absorb water. To the identify of optimal combination of control parameters that ensure a high quality cut the Taguchi approach was utilized. The signal-to-noise ratio η (S/N) was used to determine the current scatter of values. Signal (S) is gained from factors are set or under the control of the user, but the noise (N) shows the factors affecting the signal beyond user control. In the case of cutting the KS Permaglide P10 material, the validity of using the AWJ technology was confirmed, and the optimal machining parameters were determined.

Keywords: Plain bearing materials · Abrasive water jet · AWJ · Cutting performance · Surface roughness

1 Introduction

The use of composite materials is becoming increasingly important and common in industry. A composite material most often consists of two or more materials with different physical properties. The composite obtained in this way has properties that differ significantly in relation to the properties of each of the components separately. The connection matrix and reinforcement make it possible to produce materials with special properties in various aspects of strength, hardness, stiffness, corrosion resistance or better sliding properties. The parameters of composites largely depend on the properties of the composite components, their quantity in the volume of individual layers, the structure of the particle distribution, as well as their geometric features. The dispersed phase contained in individual composite layers may have larger or smaller particles and

L. Sitek et al. (Eds.): Advances in Water Jetting II, LNME, pp. 156–168, 2025.
https://doi.org/10.1007/978-3-031-72778-8_13

a different share in its volume. Composites in which the particles are approximately 1 μm in size and the filling level in the matrix ranges from 20% to even 90%, most often it is the surface layer and the matrix that transfer the loads that affect them. In this layer, elastic stresses occur between the matrix and the particles, which strengthen the layer and at the same time limit the occurrence of plastic stress, protecting against mechanical damage [1].

Composites reinforced with smaller particles (dispersively) are usually made of a metal matrix, reinforced with very fine ceramic or metallic inclusions. The share of these particles usually ranges from 1% to 15% and their size ranges from 0.01–0.1 μm. With this reinforcement, the mechanical strength of the composite does not improve at lower temperatures. The improvement in mechanical properties is noticeable at higher temperatures, up to 80% of the melting point. In this type of composites, the creep phenomenon is reduced compared to the matrix material [2].

An important group of composite materials are PTFE composites, having a polymer matrix with fillers in the form of metal powders. The dispersion filler used in these composites increases wear resistance while reducing the friction coefficient. The aspect of thermal stability is also important here, because only at temperatures above 425 °C the matrix used undergoes rapid thermal decomposition [3]. PTFE maintains stable mechanical properties in the temperature range −215 to +260 °C, is non-flammable and practically completely chemically resistant. Unfortunately, the mechanical properties of PTFE, such as tensile and compressive strength and flow even under small loads, are quite low. The introduction of an appropriate filler into PTFE is therefore a very important aspect in the production of composites based on it.

Shaping composites, due to the multi-layer nature and different properties of the components, is quite a significant problem. Machining tools are usually adapted to one type of material being processed. In contact with composites, they often cause the effects of their work to be unsatisfactory. Mechanical processing of a material with a relatively soft surface layer causes its fraying, local detachment, or sticking of the tool with the matrix material. To eliminate these unfavorable phenomena, attempts are made to use other cutting techniques to process this type of materials, e.g., laser [4], electro-erosion [5, 6], pulsating water jet [7, 8] or abrasive water jet [9–11].

Conventional cutting techniques often do not give the expected results, the tools used to shape composites wear out quite quickly, causing discontinuities and delamination of the composite [12, 13]. Problems also arise when machining more complex shapes. Laser thermal processing, however, due to its specificity, causes problems related to overheating of the cut details, which directly affects their quality [14, 15].

Abrasive Water Jet is an advanced manufacturing technology. It is characterized by more control parameters than common cutting technologies [16, 17]. This makes it necessary to optimize the cutting process using this separation method [18–20].

In research and optimization of AWJ processing parameters on a glass–epoxy reinforced aluminum composite of the GLARE type [21] was conducted to an AWJ drilling process with varying input parameters: the nozzle–material distance, feed, abrasive flow rate, pressure. Output parameters were used delamination and surface roughness.

Investigation on cutting phenolic resin composites reinforced with cotton fabric using an abrasive water suspension jet (AWSJ) was shown by Perec et al. [22]. A process model was created in the form of a second-degree polynomial equation using the response surface method (RSM). The research indicates the optimal factors controlling the process to obtain the best cutting depth parameters.

The possibility of the titanium and carbon fiber multilayer composite machining was tested by the abrasive water jet machining presented Pahuja et al. [23]. The authors noticed that the surface roughness and slot width instability was strong at poor jet power level. Mathematical regression models were proposed to slot width prediction and the energy based semi-analytical model was proposed to predict the kerf properties.

The preliminary research of machining the carbon fiber composite with abrasive water jet, which can reduce delamination published by Popan et al. [24]. The authors designed a special device and set up based on the proposed machining method and it is ready to use for common cutting head.

Maneiah et al. [25] published results of AWJ cutting of a modern type of material—aluminum metal matrix composite. Tests were concentrated on increasing mechanical properties and the machinability of materials. Research was carried out by selecting the common control parameters as a stand of distance, traverse speed, and abrasive flow. The significant impact of process effects showed the most significant control parameters was traverse speed and abrasive flow.

Abrasive water jet can use also for composite turning. Nag et al. [26] presented the research of the water jet turning control process parameters as abrasive type and abrasive mass flow rate of the special $A359/Al_2O_3/B_4C$ composite material on output geometry. The authors analyzed the variation of real diameter with the target diameter of metal matrix composite. The results revealed that the abrasive type had a greater influence on the deviation of diameter from the target diameter as compared to mass flow rate. Carach et al. [27] published similar studies for Inconel 925 material.

The Chenrayan et al. [28] present to minimize the machining damages of the composite made glass and carbon fibers as reinforcements in the epoxy matrix, by applying a hybrid grey relational analysis (GRA)-principal component analysis (PCA) mathematical model. The research effects show that the abrasive flow and pressure was parameters to remarkable minimize the kerf damage, whereas the mass flow of abrasive and stand-off distance are the parameters to minimize the delamination damage. The adequacy and accuracy of the proposed mathematical model is being performed with the value of indicators R square.

In new research [29] was introduced a progressive composite material made with 3D-printing technology using the SLM method, and the successful research on its cutting using AWJ.

The comparison of conventional cutting process with AWJ treatment of TiGr [30] showed the superiority and higher effectiveness of AWJ with less cut surface damage.

Whereas the tests of the machining effectiveness of the Ti6Al4V and CFRP composite by AWJ [31] recognized deep linear dependency between the power-to-velocity ratio and the cut depth for CFRP and titanium singly, and then to cutting depth prediction of the Ti-CFRP composite.

Recently, research has been undergoing around natural fiber reinforced composite as an alternative for metals and plastics in automobile application, due to its good strength and lighter weight.

Karthikeyan et al. [32] published research of machinability of pineapple fiber epoxy composite by the abrasive water jet machining at a high traverse speed. The optimal parameter for maximizing the material removal rate is done using the signal-to-noise ratio analysis. Also, an artificial neural network (ANN) algorithm is developed to predict the responses. The results of the ANN model show that the model predicts the effects with an accuracy of almost 95%. Similar research results, but for three-ply layer perceptron type network with an error Broyden—Fletcher—Goldfarb—Shanno (BFGS) learning algorithm [33], showed even better model compliance.

Balaji et al. [34] presented the research of the fly ash filler affects the machinability of hybrid Banana/Coir/Bagasse fiber reinforcing composites. Surface roughness and Material Removal Rate are negatively affected by the high hardness of 6% Bagasse Fly ash and 6% Coir Fly ash in composites. Cracks and matrix breakages were observed to minimize by using these filler powders.

AWJ shows useful machining effects due to its minimal mechanical and thermal impact during treatments compared to ordinary cutting processes [35, 36]. On the one hand, the treatment of composites made of varied materials could be the reason for the divergences in roughness and kerf geometry on the cut surface of each constituent item [37, 38]. On the other hand, impact of water and abrasives may deteriorate the touch zone between both components. These phenomena could have an influence on the deterioration in the target quality and increase the cutting time.

Based on state-of-issue research, the AWJ occurs applicable for cutting composites made of polymers and metals, despite their properties can be acutely different. AWJ technology can be applied for the machining of almost all multi-layer composites, though, the aleatory purposed nature of the abrasive water jet (AWJ) treatment on a multi-layered structure (MLS) can cause some complications in form of non-regular kerf properties.

Additionally, after state-of-issue analysis, the research gap was established in the existing literature, as no research on cutting by AWJ the composite for sliding bearings was reported so far.

Due to the need to maintain the mechanical properties of the Permaglide composite material after machining was resolved to carry-out of cutting research by AWJ technology. In addition, the machinability by the AWJ can be ensure a good property in the further operation of components, made of such composite material.

The purpose of this study was to test the impact of the main control parameters of the machining process on the status of the kerf surface of the composite. The concentration was placed on defining the behavior of the material at the composite layers contact area.

2 Materials and Method

Machining of the plain bearing material used for is a demanding process. The specificity of this material forces the use of an appropriate processing technique that will guarantee that the sliding properties will not change or will change to an exceedingly small extent. Processing by plasma cutting or cutting with a concentrated stream of a laser beam causes significant overheating of the material, which in practice excludes the possibility of using these techniques to shape plain bearing components. The article presents an analysis of the possibility of using hydro-abrasive machining to shape Permaglide P10 plain bearing elements.

2.1 Cut Material

KS Permaglide P10© material [39] is a leaded, robust sliding composite with maximal tribological implementation. It is aimed for maintenance-free, dry-running applications, however, can too be working in systems with liquid lubrication.

The standard material for Permaglide plain bearings with the P10 designation used in the tests is characterized by a very low tendency to stick-slip vibrations, high wear resistance, good chemical resistance, low coefficient of friction, no tendency to weld with metal, resistance to swelling and a very important feature, from the point of view of the treatment used in the tests, the material does not absorb water. Figure 1 illustrates the Permaglide P10 layering properties.

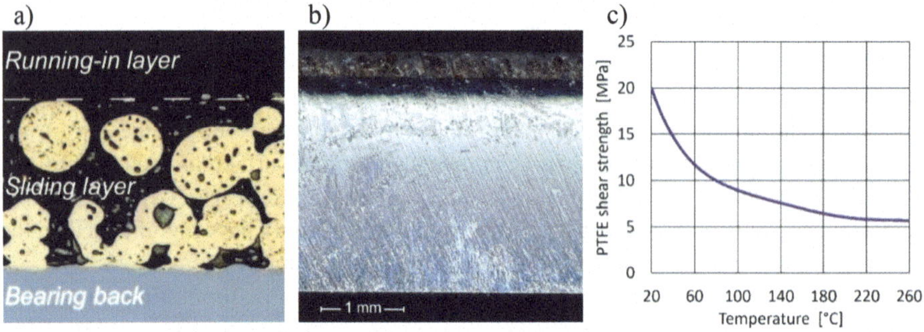

Fig. 1. P10 material properties: **a** layer system, **b** microscopic view, **c** influence of temperature on P10 shear strength

During the processing of this type of materials, temperature is an extremely important external factor, which affects the reliability and life of the bearing system and has a particular effect on the mechanical properties of the solid lubricant. In the temperature range up to about 100 °C, the coefficient of friction is slightly lower than at room temperature. When the operating temperature significantly exceeds 100 °C, this effect is reversed. The PTFE polymer is very with temperature change. The use of an abrasive water jet to shape bearings minimizes the occurrence of high temperatures, which in turn makes it possible to produce bearing systems that are not loaded with a temperature factor. The tests carried out are aimed at checking the impact of such treatment on the condition and quality of the sliding layer.

2.2 Abrasive Material

As abrasive the crushed rock garnet type J80A from Jinhong Mining located in Jiangsu, China was used. This garnet is a pure natural abrasive, consisting mainly of almandine. The chemical composition (Table 1) and grain distribution (Fig. 2c) can vary slightly within a narrow range but are characterized by stable cutting performance.

The grains are characterized by a deep red color inclining to purple and have dodecahedral and trapezohedron shape and often are modified combinations of the two [40]. The near to isometric form of the almandine grain, its high hardness and density doing it utilized in abrasive jet machining.

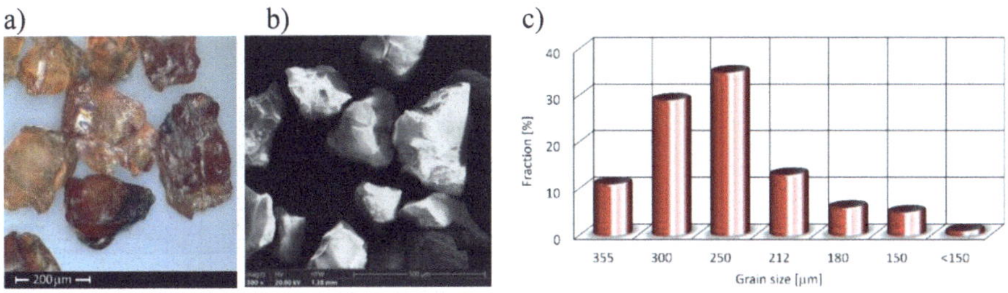

Fig. 2. J80A properties: **a** optical microscope view, **b** SEM view, **c** grain distribution

Table 1. Garnet J80A properties.

Mineral Content [%]							
Amandine	Ilmenite	Omphacite		Rutile	Quartz	Hornblende	Free Silica
90–96	1.0	1.5		0.6	< 0.1	< 0.5	< 0.5%
Chemical Composition [%]							
Fe_2O_3	SiO_2	TiO_2	Al_2O_3	FeO	CaO	MgO	MnO
17	39	0.05	21	8	9.5	5	0.4
Physical Characteristics							
Density [kg/dm^3]	Bulk Gravity [kg/dm^3]	Mohs Hardness		Color	Grain shape	Conductivity	Acid solubility (HCL)
3.8–4.1	2.3–2.4	7.5–8.0		Dark red	Sub angular	< 25 S/m	< 1.0%

2.3 Test Method

The cutting tests were carried out on the WaterJet CNC OMAX 60120 machining center. The materials were cut perpendicular to top surface, directing Abrasive Water Jet to the workpiece, and linear moving it with a specific traverse speed [18]. Surface inspection, cut kerf inclination angle and it surface roughness measurements (Fig. 3) were conducted on the Olympus DSX1000 high-resolution digital microscope. Additionally, detailed observations of the cut groove were performed on a Phenom ProX SEM microscope.

2.4 Design of Experiment

To the identify of optimal combination of control parameters that ensure a high quality cut the Taguchi approach was utilized [41]. The signal-to-noise ratio η (S/N) was used to determine the current scatter of values [42, 43]. Signal (S) is gained from factors are set or under the control of the user, but the noise (N) shows the factors affecting the signal beyond user control. The focus of the design is the choice of experimental parameters of process to maximize the ratio η (S/N). In the research the case "smaller is better" was

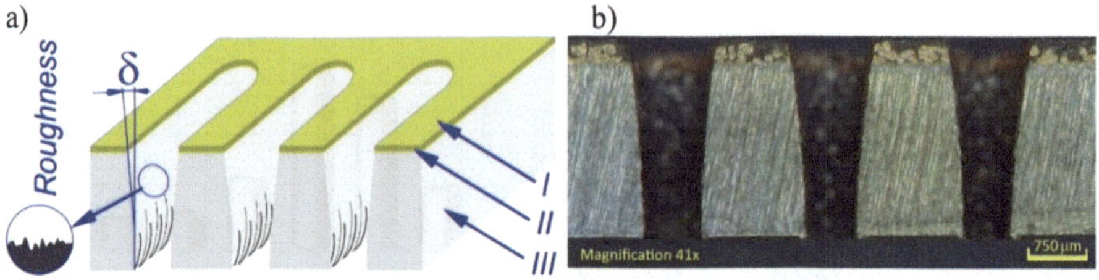

Fig. 3. Detail of cut kerf: **a** schematic view, **b** kerfs intersection view. I—running-in layer, II—sliding layer, III—bearing back.

chosen, described by the following equation:

$$\eta = -10log\left(\frac{1}{n}\sum_{i=1}^{n} y_i^2\right) \tag{1}$$

where:

n is the number of repetitions using the same measurement conditions,
y is analyzed the result of the measurement,
i is the number of variables in orthogonal array.

3 Tests Results and Discussion

The AWJ treatment process was carried out with the following variable control parameters:

- Pressure: 360; 380; 400 MPa,
- Feed: 100; 200 and 300 mm/min,
- Abrasive flow rate: 250; 350 and 450 g/min

as shown in Table 2 and following constants parameters:

- Water nozzle diameter: 0.33 mm,
- Water-abrasive nozzle diameter: 0.76 mm,
- Nozzle-material distance: 2 mm.

The relatively small difference between the levels of pressure implemented in the tests resulted from need maintaining the highest possible productivity, because pressure is a parameter that strongly influences on machining efficiency [44, 45].

The highest S/N values were obtained for the following control parameters: AFR = 350 g/min, pressure = 400 MPa and traverse speed = 100 mm/min (Figs. 4 and 5). The view of the surface cut under such conditions is shown in Figs. 6 and 7, and Table 3 presents both the roughness coefficient Sz obtained during the tests and the prediction result. The difference is equal 1.16%.

Table 2. Control and output parameter of cutting process

No	AFR [g/min]	Pressure [MPa]	Traverse speed [mm/min]	Sz [μm]	Delta [°]
1	250	360	100	33.73	5.52
2	250	380	200	36.03	6.83
3	250	400	300	39.06	7.13
4	350	360	200	34.47	6.29
5	350	380	300	38.85	7.20
6	350	400	100	30.22	4.90
7	450	360	300	46.40	7.49
8	450	380	100	38.62	5.00
9	450	400	200	42.47	6.79

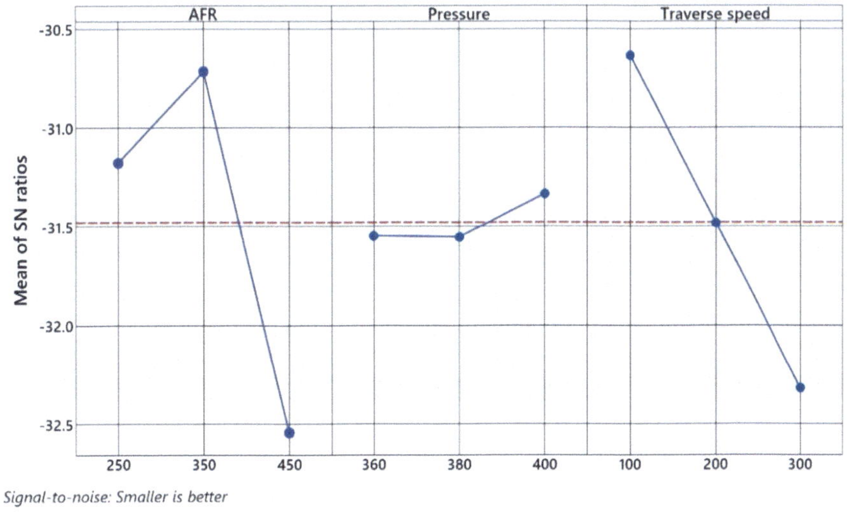

Fig. 4. S/N ratio for surface roughness Sz

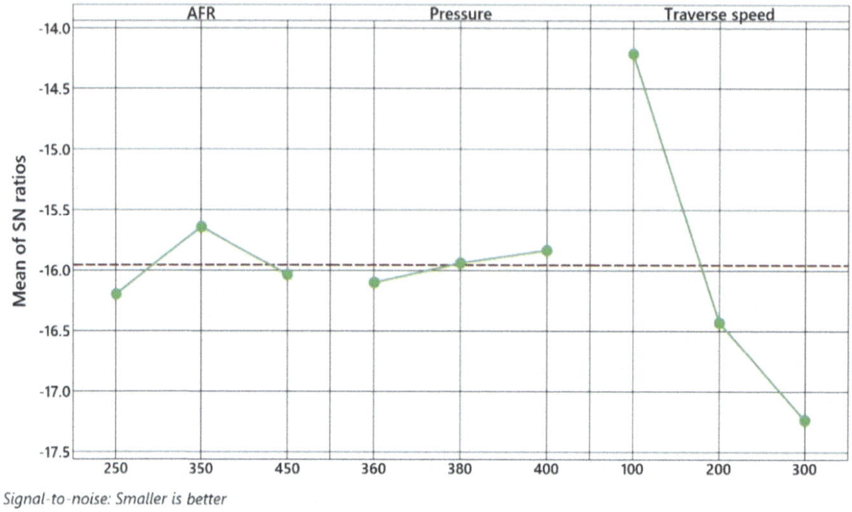

Fig. 5. S/N ratio for δ average cut kerf inclination angle.

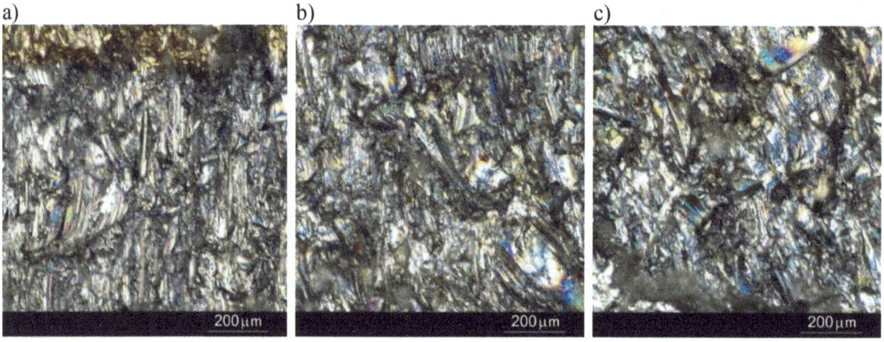

Fig. 6. Exemplary optical microscope view of cut surface: **a** top, **b** middle, **c** bottom areas

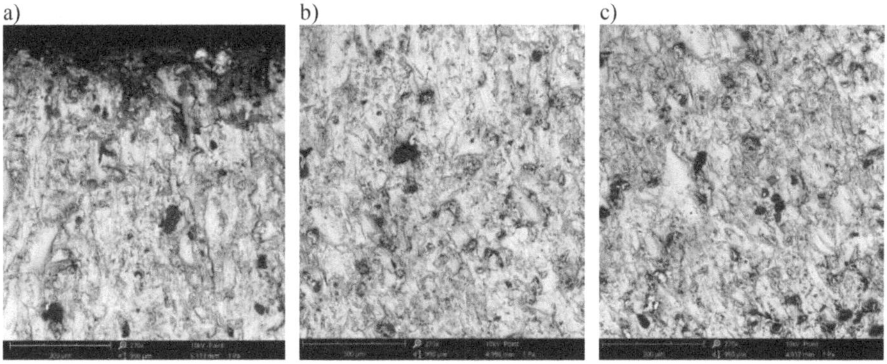

Fig. 7. Exemplary SEM microscope view of cut surface: **a** top, **b** middle, **c** bottom areas

Table 3. Effect of surface roughness and inclination angle comparison

Predicted Sz roughness value [μm]	Achieved Sz roughness value [μm]	Difference [%]
30.57	30.22	1.16
Predicted δ value [°]	Achieved δ value [°]	Difference [%]
4.84	4.9	1.23

4 Conclusion

Taguchi approach is a great convenience appropriate when cutting new materials. It significantly reduces the amount of test process by reducing the time to obtain reliable results and reducing the cost of obtaining them. In the case of cutting the Permaglide material, the validity of using the AWJ technology was confirmed, and the optimal machining parameters were determined.

Based on the calculation of the coefficients, it was found that the best possible combination of process parameters to achieve optimal output parameters from both side surface roughness Sz and the cut kerf inclination angle points of view, the optimal input parameters are following:

- Abrasive flow rate: 350 g/min,
- Pressure: 400 MPa,
- Traverse speed: 100 mm/min.

The method also allows for the prediction of output parameters with high confidence. In the case of both cut surface roughness and inclination angle, the deviation of the predicted values to the actual values turned out to be less than 1.5%.

Additionally, no delamination was observed between the surface layer and the matrix, which confirms the validity of using AWJ to process such materials.

References

1. Kurtyka, P., Wierzbinski, S., Faryna, M.: Selected mechanical properties of composites based on aluminum alloys reinforced Al2O3 particles. Composites **2**, 185–190 (2002)
2. Dobrzanski, L., Włodarczyk, A., Adamiak, M.: Composite materials based on the aluminum alloy AlCu4Mg1 reinforced with ceramic particles. In: Proceedings of 12th International Scientific Conference on Achievements in Mechanical and Material Engineering, pp. 297–300 (2003)
3. Rymuza, Z.: Trybologia polimerów ślizgowych. Wydawnictwa Naukowo-Techniczne, Warszawa (1986)
4. Calabrese, L., et al.: Micro-milling process of metals: a comparison between Femtosecond laser and EDM techniques. JMMP **5**, 125 (2021). https://doi.org/10.3390/jmmp5040125
5. Nowicki, B., Pierzynowski, R., Spadło, S.: New possibilities of machining and electrodischarge alloying of free-form surfaces. J. Mater. Process. Technol. **109**, 371–376 (2001). https://doi.org/10.1016/S0924-0136(00)00828-1
6. Valentincic, J., Kuser, D., Smrkolj, S., Blatnik, O., Junkar, M.: Machining parameters selection for varying surface in EDM. IJMPT **29**, 344 (2007). https://doi.org/10.1504/IJMPT.2007.013128
7. Nag, A., Hvizdos, P., Dixit, A.R., Petrů, J., Hloch, S.: Influence of the frequency and flow rate of a pulsating water jet on the wear damage of tantalum. Wear **477**, 203893 (2021). https://doi.org/10.1016/j.wear.2021.203893
8. Lehocka, D., et al.: Effect of pulsating water jet disintegration on hardness and elasticity modulus of austenitic stainless steel AISI 304L. Int. J. Adv. Manuf. Technol. **107**, 2719–2730 (2020). https://doi.org/10.1007/s00170-020-05191-3
9. Bankowski, D., Spadlo, S.: The use of Abrasive Waterjet cutting to remove flash from castings. Arch. Foundry Eng. **19**, 94–98 (2019). https://doi.org/10.24425/AFE.2019.129617
10. Jandačka, P., Ščučka, J., Martinec, P., Lupták, M., Janeček, I., Mahdi Niktabar, S.M., Zeleňák, M., Hlaváček, P.: Optimal abrasive mass flow rate for rock erosion in AWJ machining. In: Klichová, D., Sitek, L., Hloch, S., Valentinčič, J. (eds.) Advances in Water Jetting, Lecture Notes in Mechanical Engineering, pp. 81–90. Springer International Publishing, Cham (2021). https://doi.org/10.1007/978-3-030-53491-2_9
11. Kinik, D., Gánovská, B., Hloch, S., Monka, P., Monková, K., Hutyrová, Z.: On-line monitoring of technological process of material abrasive water jet cutting—On-line praćenje tehnološkog postupka rezanja materijala abrazivnim vodenim mlazom. Techničky vjesnik **22**, 351–357 (2015). https://doi.org/10.17559/TV-20130904111939
12. Oczoś, K.E.: The problems encountered in shaping parts from polymer or ceramic matrix fibrous composites by means of the material dislodging operations. Part 1. Mechanik R. **84**(1), 5–10 (2011)
13. Oczoś, K.E., Bazan, A.: Fiber reinforced polymer matrix composites forming methods. Part 2. Mechanik R. **85**(4), 272–277 (2012)
14. Oczoś, K.E.: Fibre reinforced composites—properties, application, machining. Mechanik R. **81**(7), 579–592 (2008)
15. Zaborski, S., Poroś, D., Sradomski, W.: Obróbka stumieniowo-ścierna powierzchni blach ocynkowanych. Mechanik R. 85(8–9CD) (NSOŚ) (2012)
16. Forysiewicz, M., Kukielka, L., Gotowala, K.: Finite element simulation of physical phenomena in real conditions of a single grain cutting process. MSF **862**, 288–297 (2016). https://doi.org/10.4028/www.scientific.net/MSF.862.288
17. Chomka, G., Kasperowicz, M., Chodór, J., Chudy, J., Kukiełka, L.: Possibilities of rock processing with a high-pressure abrasive Waterjet with an aspect terms to minimizing energy consumption. Materials **16**, 647 (2023). https://doi.org/10.3390/ma16020647

18. Perec, A.: Multiple response optimization of abrasive water jet cutting process using response surface methodology (RSM). Procedia Comput. Sci. **192**, 931–940 (2021). https://doi.org/10.1016/j.procs.2021.08.096
19. Kawecka, E.: The whale optimization algorithm in abrasive water jet machining of tool steel. Procedia Comput. Sci. **225**, 1037–1044 (2023). https://doi.org/10.1016/j.procs.2023.10.091
20. Kawecka, E., Puzio, P.: Moth flame metaheuristic algorithm in optimization of the abrasive water jet cutting process parameters. In: Procedia Computer Science, 28th International Conference on Knowledge-Based and Intelligent Information & Engineering Systems (KES 2024), to be published in 2024
21. Gnanavelbabu, A., Rajkumar, K., Saravanan, P.: Investigation on the cutting quality characteristics of abrasive water jet machining of AA6061-B4C-hBN hybrid metal matrix composites. Mater. Manuf. Process. **33**, 1313–1323 (2018). https://doi.org/10.1080/10426914.2018.1453146
22. Perec, A., Radomska-Zalas, A., Fajdek-Bieda, A., Pude, F.: Process optimization by applying the response surface methodology (RSM) to the abrasive suspension water jet cutting of phenolic composites. Facta Universitatis, Series: Mechanical Engineering **21**, 575–589 (2023). https://doi.org/10.22190/FUME211123010P
23. Pahuja, R., Ramulu, M., Hashish, M.: Surface quality and kerf width prediction in abrasive water jet machining of metal-composite stacks. Compos. B Eng. **175**, 107134 (2019). https://doi.org/10.1016/j.compositesb.2019.107134
24. Popan, I., Balc, N., Popan, A.: Avoiding carbon fibre reinforced polymer delamination during abrasive water jet piercing: a new piercing method. Int. J. Adv. Manuf. Technol. (2021). https://doi.org/10.1007/s00170-021-08294-7
25. Maneiah, D., Shunmugasundaram, M., Raji Reddy, A., Begum, Z.: Optimization of machining parameters for surface roughness during abrasive water jet machining of aluminium/magnesium hybrid metal matrix composites. Materials Today: Proceedings **27**, 1293–1298 (2020). https://doi.org/10.1016/j.matpr.2020.02.264
26. Nag, A., Srivastava, A.K., Dixit, A.R., Chattopadhyaya, S., Mandal, A., Klichová, D., Hlaváček, P., Zeleňák, M., Hloch, S.: Influence of abrasive water jet turning parameters on variation of diameter of hybrid metal matrix composite. In: Singh, M.K., Kushvah, B.S., Seth, G.S., Prakash, J. (eds.) Applications of Fluid Dynamics, Lecture Notes in Mechanical Engineering, pp. 495–504. Springer Singapore, Singapore (2018). https://doi.org/10.1007/978-981-10-5329-0_36
27. Cárach, J., et al.: Tangential turning of Incoloy alloy 925 using abrasive water jet technology. Int. J. Adv. Manuf. Technol. **82**, 1747–1752 (2016). https://doi.org/10.1007/s00170-015-7489-0
28. Chenrayan, V., et al.: Algahtani, A: an experimental and empirical assessment of machining damage of hybrid glass-carbon FRP composite during abrasive water jet machining. J. Mater. Res. Technol. **19**, 1148–1161 (2022). https://doi.org/10.1016/j.jmrt.2022.05.042
29. Szatkiewicz, T., Perec, A., Radomska-Zalas, A., Banaszek, K., Balasz, B.: Preliminary studies into cutting of a novel two component 3d-printed stainless steel-polymer composite material by abrasive water jet. Materials 16 (2023). https://doi.org/10.3390/ma16031170
30. Pahuja, R., Ramulu, M., Hashish, M.: Abrasive water jet machining (AWJ) of hybrid Titanium/Graphite composite laminate: preliminary results. BHR Group—22nd International Conference on Water Jetting 2014, pp. 83–95 (2014)
31. Pahuja, R., Ramulu, M.: Abrasive water jet machining of Titanium (Ti6Al4V)–CFRP stacks—a semi-analytical modeling approach in the prediction of kerf geometry. J. Manuf. Process. **39**, 327–337 (2019). https://doi.org/10.1016/j.jmapro.2019.01.041
32. Karthikeyan, R., Leela Prasad, G., Saraswathy, R., Pratap Singh, D., Rajkamal, M.D., Madhu, S.: Abrasive water jet machining of pineapple fibre epoxy composite at high transverse speed.

Materials Today: Proceedings S2214785323008854 (2023). https://doi.org/10.1016/j.matpr.2023.02.321
33. Perec, A., Kawecka, E.: Modeling of the abrasive water jet machining by ANN in uncertainty conditions. In: Procedia Computer Science, 28th International Conference on Knowledge-Based and Intelligent Information & Engineering Systems (KES 2024), to be published in 2024
34. Balaji, R., Raja, S., Jeevanandam, P., Kailasavalli, S., Kaarthik, M., Pitchandi, P.: Study of abrasive water jet machining (AWJM) of coir/banana epoxy composites by adding of fly ash fillers. Materials Today: Proceedings S2214785323034533 (2023). https://doi.org/10.1016/j.matpr.2023.06.074
35. Perec, A., Radomska-Zalas, A., Fajdek-Bieda, A., Kawecka, E.: Efficiency of tool steel cutting by water jet with recycled abrasive materials. Materials **15**, 3978 (2022). https://doi.org/10.3390/ma15113978
36. Radomska-Zalas, A.: Multi-criteria methods in the optimization of the abrasive waterjet cutting process. AIP Conference Proceedings XIV International Conference Electromachining **2023**, 020030 (2024). https://doi.org/10.1063/5.0203525
37. Radomska-Zalas, A.: Experimental research on the use of a selected multi-criteria method for the cutting of Titanium Alloy with an abrasive water jet. Materials **16**, 5405 (2023). https://doi.org/10.3390/ma16155405
38. Perec, A., Kawecka, E., Radomska-Zalas, A., Pude, F.: Optimization of abrasive waterjet cutting by using the CODAS method with regard to interdependent processing parameters. SV-JME **69**, 367–375 (2023). https://doi.org/10.5545/sv-jme.2023.647
39. KS Permaglide, Plain bearing [WWW Document]. URL https://www.permaglide.com/en/catalogues-leaflets/downloads/, last accessed 2023, August 29
40. Minerals Net, Almandine garnet: The mineral Almandine information and pictures, https://www.minerals.net/mineral/almandine.aspx, last accessed 2022, November 26
41. Taguchi, G., Chowdhury, S., Wu, Y.: Taguchi's Quality Engineering Handbook. 1st edn. Wiley (2004). https://doi.org/10.1002/9780470258354
42. Perec, A.: Abrasive suspension water jet cutting optimization using orthogonal array design. Procedia Eng. **149**, 366–373 (2016). https://doi.org/10.1016/j.proeng.2016.06.680
43. Radomska-Zalas, A., Puzio, P.: The application of the Taguchi method in the optimization of AWJ machining process. In: Procedia Computer Science, 28th International Conference on Knowledge-Based and Intelligent Information & Engineering Systems (KES 2024), to be published in 2024
44. Perec, A.: Multiple-criteria optimization of abrasive water jet (AWJ) cutting using TODIM approach. In: AIP Conference Proceedings, XIV International Conference—Electromachining 2023. AIP Publishing, New York, United States (in print)
45. Lebar, A., Jerman, M., Sabotin, I., Drešar, P., Valentinčič, J.: AWJ cutting process control by means of process visualisation. Procedia Eng. **149**, 224–228 (2016)

Investigations on the Influence of Small Impact Angles for Abrasive Water Injector Jet Cutting of High-Performance Workpieces

Waldemar Reder[1(✉)] and Eckart Uhlmann[1,2]

[1] Institute for Machine Tools and Factory Management (IWF), Technische Universität Berlin, Berlin, Germany
reder@iwf.tu-berlin.de

[2] Fraunhofer-Gesellschaft, Institute for Production Systems and Design Technology IPK, Berlin, Germany

Abstract. State of the art high-performance materials, like titanium alloy or nickel-base alloy, are used in high stressed applications. Still, conventional machining is struggling to machine these materials cost-efficiently due to high costs and wear of tools. Trimming, a multilevel near-net-shape pre-contouring with multistage abrasive water injector jet cutting (AWIJC), could solve these problems. While machining three-dimensional components, process parameters such as jet impact angle α, cutting length l_d and standoff distance s may change continuously. Especially acute impact angles α were not investigated in the past and are a major question for the workpiece angle γ. Experiments were carried out with stainless steel plates of the type X5CrNi18 (EN 1.4301), while varying the jet impact angle α and the feed speed v_f within three levels. To realize comparability, the thickness of the plates was adjusted to the jet impact angle α to achieve a constant cutting depth l_d of 40 mm. The resulting workpiece angle γ was measured, and a prediction model based on regression analysis was established. An influence of the jet impact angle α and feed speed v_f on the workpiece angle γ was detected, especially for very small angles. Furthermore, the model was used for predicting the workpiece angle γ for Ti6Al4V and Inconel 718, which are commonly used high-performance alloys. The predicted values fit the measured workpiece angle γ appropriately. The investigations offer fundamental knowledge for water jet cutting with small impact angles α, which is essential for water jet trimming.

Keywords: Waterjet Cutting · Pre-contouring · Near-net-shape · Trimming

1 Introduction

1.1 High-Performance Materials

High-performance materials such as titanium- or nickelbase-alloys are widely used in the industry today which points out a wide range of applications of these materials. The majority of these applications is used in the aerospace industry. Within this industry the

L. Sitek et al. (Eds.): Advances in Water Jetting II, LNME, pp. 169–181, 2025.
https://doi.org/10.1007/978-3-031-72778-8_14

high material and manufacturing costs are economically acceptable due to fuel savings achieved by weight reduction. Inconel 718 is an example for a high-performance material in the aerospace industry. The development and initial use of Inconel 718 can be dated to the early 1960s and has since become the most widely used material for passenger aircraft engines, accounting for over 30% of the mass of a modern engine [1]. Still an economical machining operation that expands the range of applications for these high-performance materials has not been developed yet. Manufacturing technology constantly needs to adapt to the challenging material innovations in order to allow economical production.

The machining of high-performance materials, such as composites, ceramics, and hardened metals has been a major challenge in manufacturing industries. Conventional machining such as grinding, milling, and drilling show limitations in machining high-performance materials like titan- and nickel-based-alloys. These methods are limited by the hardness, toughness and low ductility of the materials, leading to high tool wear and inadequate workpiece quality. Furthermore, the heat load in the cutting area increases due to the low thermal conductivity which can cause thermal damage to the tool and the workpiece. Additionally, the material properties can be compromised by high thermal loads [2].

Abrasive waterjet cutting is a thermal cold machining process with low process forces due to the decoupling of tool and workpiece. Additionally, the tool wear is independent of the machined material. Abrasive waterjet cutting can be used to machine a wide variety of materials such as metal, fiber composites, ceramics and even high-performance materials. The process is also frequently used for cutting titanium and nickel-based alloys, making it capable for pre-contouring of difficult to machine workpiece materials [3].

1.2 Abrasive Water Injector Jet Cutting

There are several waterjet cutting applications that are characterized by different process strategies including an abrasive medium in contrast to the use of a pure waterjet. The latter is used for cutting soft materials such as polymers and food. Adding abrasive media makes it possible to also cut hard materials, such as steel, stone, ceramics and composites. Industrially, AWIJC is preferred in manufacturing and production applications due to the uninterrupted processing and the possibility of high-frequency actuation of the cutting valve. Currently, AWIJC is used to process a wide range of materials in a variety of applications. However, AWIJC is mainly limited to the machining of two-dimensional geometries due to the high-energy abrasive waterjet. After cutting through the workpiece and exiting at the bottom of the workpiece, a high residual energy remains in the jet. This remaining energy of the abrasive waterjet needs to be converted appropriately. Usually a water tank is used, where the remaining energy is transferred into heat [3]. Therefore, it is essential that the abrasive waterjet is directed towards the water tank. Due to this restriction AWIJC is mainly used for two-dimensional or two-and-a-half-dimensional machining, like cutting of sheets, plates and curved panels as well as the chamfering of the aforementioned components.

Currently, there are only few three-dimensional machining operations performed which are mostly concentrated in research and aerospace applications, justifying an extensively required manual effort [4]. Taking into account relevant literature, no detailed description of the manufacturing steps for three dimensional components were presented in publications. One example of a three-dimensional geometry manufactured by AWIJC is a surgical clamp from a round stainless-steel profile with a diameter of $d = 9$ mm [5]. Another example is the fabrication of chess pieces which were manufactured by AWIJC [6, 7]. Additionally, efforts have been described to avoid a water basin while implementing complex machining with a robot-guided cutting head. Hashish [8] adds a catcher arm to the cutting head, where a catcher cup is attached. The residual energy of the abrasive water jet is converted within the catcher cup as it exits the bottom of the machined workpiece.

Another approach to three-dimensional pre-contouring using AWIJC is cutting with a controlled cutting depth [9]. The aim of the process is to remove entire volume segments by superimposing a number of individual cuts without cutting the entire material. Figure 1 illustrates the process of three-dimensional pre-contouring utilizing distinct processing forms. Using abrasive water jet milling, volume segments can be removed from a workpiece by superimposing two kerfs from different sides of the workpiece. This is particularly relevant if it is not desired to cut through the entire workpiece. Applying cutting with a defined kerf depth to cylindrical geometries is called abrasive water jet groove turning. By superimposing of two kerfs, entire ring segments can be removed [9, 10].

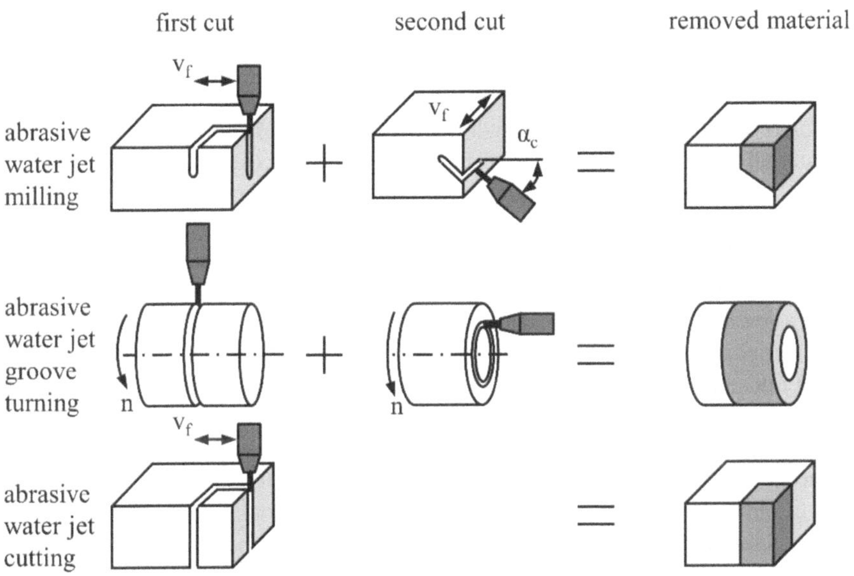

Fig. 1. Difference between abrasive water jet milling, groove turning and cutting for AWIJC and the removed material according to the different approaches.

Moreover, there is a novel approach for three-dimensional machining with AWIJC, referred to as trimming. This process variant describes a multi-stage two-dimensional cutting process. Within the process, outer volume segments are separated and the workpiece gradually converges to the target geometry. For trimming, available knowledge and process parameters for cutting with AWIJC from literature can be applied [4, 11].

1.3 Trimming

Pre-contouring three-dimensional components by trimming can reduce the challenges of conventional machining of high-performance materials and additionally utilize the benefits from of AWIJC. Furthermore, AWIJC can be used to separate a large amount of the material which has to be removed with a low number of single cuts. After pre-contouring with AWIJC, conventional milling can be performed to meet the required component specifications. With this novel process chain the volume to be machined by conventional machining is minimized and thus the tool wear of conventional milling.

Trimming is a multi-stage two-dimensional machining process with defined rotations of the workpiece in between individual AWIJC-steps which results in three-dimensional machining. Figure 2 demonstrates pre-contouring by trimming of a turbine blade, where a bar stock is used as the initial material geometry. After the first cut using AWIJC, a defined rotation is performed followed by a second cut. After a second rotation and the third cut, the pre-contouring by trimming is finished producing a near-net-shape workpiece. Subsequently, the remaining material can be removed, for example, by a milling process. By trimming, 93.3% of the volume which needs to be removed is separated by AWIJC and the remaining 6.3% needs to be milled conventionally [11].

Without trimming all material would have to be completely removed by conventional milling. In contrast, using trimming, entire volume segments are removed by single cuts. Thus, pre-contouring by trimming reduces the volume of material to be removed by conventional machining based on AWIJC separation. The costs for roughing titanium-aluminide γ-TiAl with AWIJC, turning, milling and grinding were compared and the financial potential in pre-contouring high-performance materials with AWIJC was displayed. A production order was simulated by a machining operation with a reference material removal volume of $V = 50\ \text{cm}^3$ of γ-TiAl for 600 repetitions. Compared to AWIJC a resulting increase of manufacturing costs of $C_{m,t} = 33\%$ for turning, $C_{m,g} = 27\%$ for grinding and $C_{m,m} = 8\%$ for milling was observed. By reducing the material removal rate and the material removal volume, a wide range for possible use cases for pre-contouring with AWIJC could be identified [10].

The goal of trimming is to develop a flexible and automated process for pre-contouring three-dimensional geometries by AWIJC. For this purpose, the workpiece handling and the generation of cutting paths needs to be automatized [12]. Furthermore, complex geometries lead to changes within the process parameters. Additional investigations for these process parameters need to be performed. One of these parameters are acute impact angles α which are addressed within this paper.

1.4 Three-Dimensional AWIJC

As described above, AWIJC is used for cutting two- and two-and-a-half-dimensional geometries. Since trimming is a multi-stage two-dimensional cutting process, it is possible to use known process parameters. Nevertheless, machining of three-dimensional

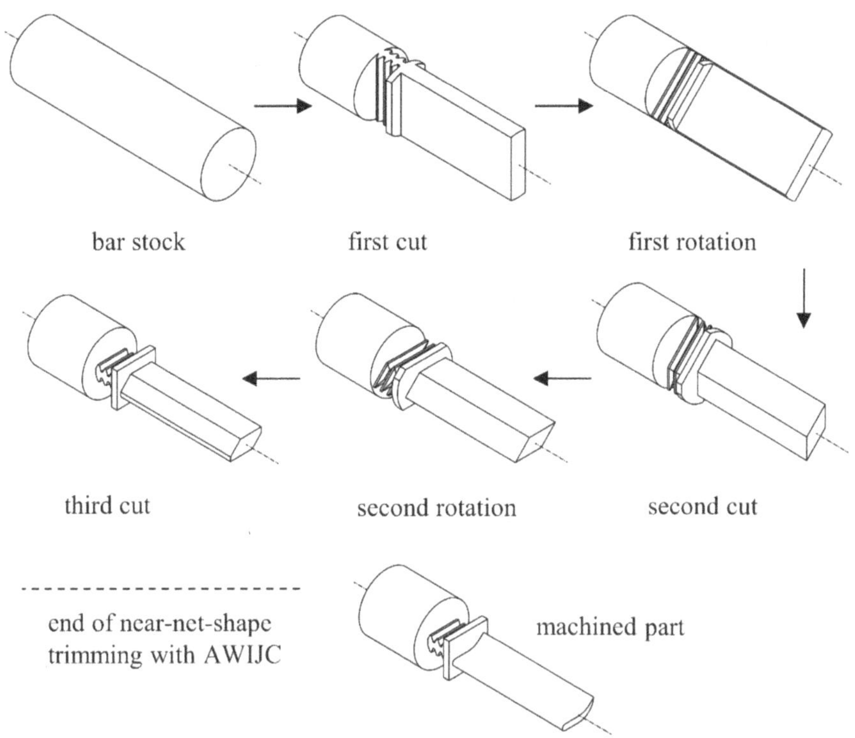

Fig. 2. Process chain of an turbine blade which is machined out of a bar stock with AWIJC-trimming within three cuts and final conventional milling operation.

geometries introduces new challenges, such as varying cutting depths l_d, varying and very small cutting angles α, and long effective distance s_{eff}. For some applications, round bars could be suitable semi-finished products for pre-contouring three-dimensional geometries using trimming. Depending on the position of the cutting head radial to the round bar, the cutting angle α changes. At the center of the rotatory axis, there is a cutting angle of $\alpha = 90°$. If the position of the cutting head changes in radial direction of the round bar, the cutting angle α also changes, see Fig. 3. Especially at the edge of the round workpiece very small cutting angles of $\alpha \leq 15°$ occur.

The geometric correlations between the jet impact angle α and the radial distance from the rotation axis a for a round bar with the radius r can be defined according to Formula (1):

$$\alpha = arccos\frac{a}{r} \tag{1}$$

In the machining of complex three-dimensional geometries, acute jet impact angles α can occur which have not been sufficiently investigated according to the current state of the art. These jet impact angles α will be addressed and investigated in more detail in the present article. The following sections will discuss the technological investigations and evaluations that were carried out in detail.

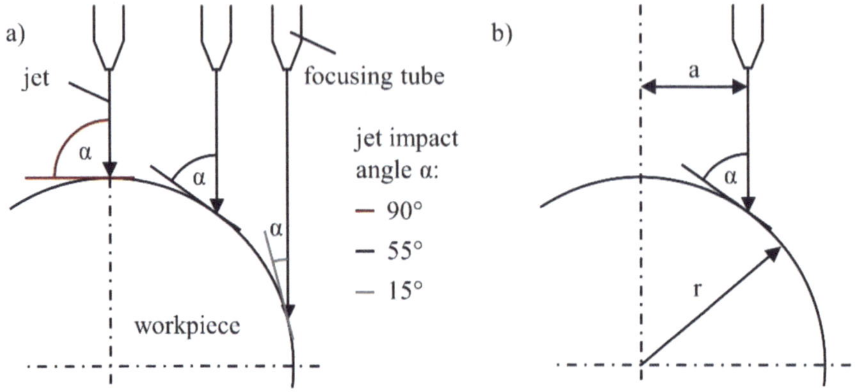

Fig. 3. The influence of the jet positioning on the jet impact angle α for complex geometries: a) different jet impact angle α for different jet positions; b) correlation between jet impact angle α, workpiece radius r and the distance between jet and center line a.

2 Investigations

2.1 Experimental Setup

The investigations were carried out on a robot-guided waterjet system WaterJet HRX160L from MaximatorJet GmbH, Schweinfurt. The machine has a positioning accuracy P of ± 0.1 mm and a repeatability R of ± 0.05 mm. To ensure a safe test setup, the samples were clamped vertically in the system to ensure that the cutting head, and thus the abrasive water jet, was always aligned in the direction of the water tank. A high-pressure pump of type UHDE HPS 6045 from Uhde High Pressure Technologies GmbH, Hagen, was used which allows the application of a maximum pressure $p_{max} =$ 600 MPa and a maximum volume flow of $\dot{m}_w = 2.5$ l/min. The technological investigations are intended to clarify which workpiece angle γ is obtained with a corresponding jet impact angle α. In particular, small jet impact angles of $\alpha \leq 20°$ may lead to an increased deflection of the jet. By that, a significant difference between workpiece angle γ and jet impact angle α is possible, Fig. 4a.

When varying the jet impact angle α, the standoff distance s must be precisely defined. The standoff distance s is defined as the perpendicular distance between the focus tube and the workpiece and ideally shows a distance of $s_{ideal} = 2$ to 3 mm [3]. By varying the jet impact angle α, the effective distance s_{eff} changes relative to the standoff distance s, compare Fig. 4b. The geometric correlation is described in Formula (2). Especially for acute jet impact angles of $\alpha \leq 15°$, the effective distance s_{eff} increases significantly. A compensation of the effective distance s_{eff} to the standoff distance s is technically not possible due to the geometrical shape of the focus tube.

$$s_{eff} = \frac{s}{sin\,\alpha} \tag{2}$$

Changing the jet impact angle α also changes the cutting depth l_d for workpieces with the same thickness. To ensure comparability of the cutting results with different jet

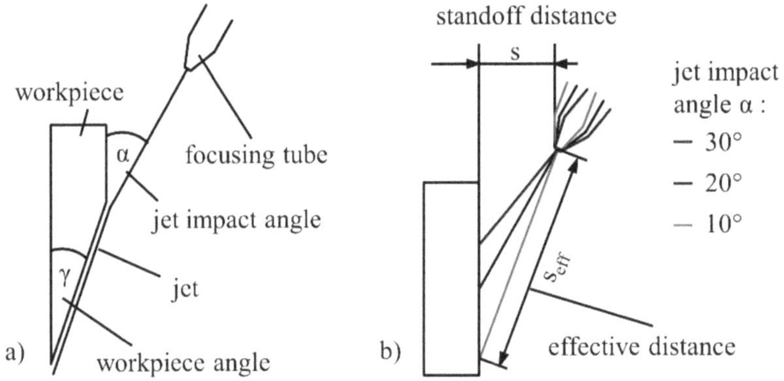

Fig. 4. Schematic presentation of the experimental setup: a) jet alignment and display of jet impact angle α and workpiece angle γ; b) display of standoff distance s and effective distance s_{eff} for various jet impact angles α.

impact angles α, the workpiece thicknesses b had to be adjusted. A cutting depth of $l_d = 40$ mm was defined for all investigations. Therefore, a jet impact angle of $\alpha = 30°$ results in a workpiece thickness of $b = 20$ mm, a jet impact angle of $\alpha = 20°$ results in a workpiece thickness of $b = 13.68$ mm, and a jet impact angle of $\alpha = 10°$ results in a workpiece thickness of $b = 6.95$ mm. The workpiece thickness was prepared by conventional milling to obtain a constant boundary conditions and to produce two plane-parallel reference surfaces.

The investigations were carried out using stainless steel X5CrNi18, since this is a suitable material in the corrosive environment of the waterjet system. The obtained results were transferred to high-performance materials in the second step of the investigations. The two most common high-performance materials Ti6Al4V and Inconel 718 were chosen for this study. With the pre-contouring of three-dimensional geometries, an economically attractive process chain is meant to be developed, therefore the process parameters were adjusted to a maximum cutting performance. A summarized overview of the process parameters is given in Table 1. A pressure of $p = 380$ MPa, which is frequently used in industry, and garnet sand with 80 Mesh were selected. The abrasive mass flow rate was defined as $\dot{m}_A = 500$ g/min and the standoff distance $s = 2$ mm. Furthermore, a nozzle with a diameter of $d_O = 0.35$ mm and a focus tube with a diameter of $d_F = 1.02$ mm were selected.

The jet impact angle α was varied between 10°, 20° and 30°. In literature the jet impact angle $\alpha \leq 45°$ is not described appropriately. To close the gap a wide range of impact angles α was chosen. Also, the feed speed v_f was varied in a wide range. The slowest feed speed of $v_f = 15$ mm/min can produce a high quality cut and the highest feed speed of $v_f = 45$ mm/min produces a separation cut with a rough surface for the selected stainless steel. A full factorial experimental design was chosen by varying both parameters in three steps. All tests were repeated twice with a total of 81 tests that were carried out.

Table 1. Parameter for the experimental investigations with stainless steel X5CrNi18.

parameter	symbol	unit	value
pressure	P	MPa	380
garnet sand		Mesh	80
abrasive mass flow	$\dot{m}_A$	g/min	500
standoff distance	s	mm	2
orifice	d_O	mm	0.35
focus tube	d_F	mm	1.02
cutting depth	l_d	mm	40
jet impact angle	α	°	10…30
feed speed	v_f	mm/min	15…45

2.2 Measuring Setup

In order to metrologically detect the workpiece angle γ, the contour of the cut workpieces was measured with an optical sensor, see Fig. 5a and Fig. 5b. The determined profile was further processed with a Matlab script in order to obtain the workpiece angle γ, Fig. 5c. The contour of the separated workpieces was measured using an optical roughness tester of the type FRT Mark III from FORMFACTOR FRT METROLOGY, formerly FRT GMBH, Bergisch Gladbach.

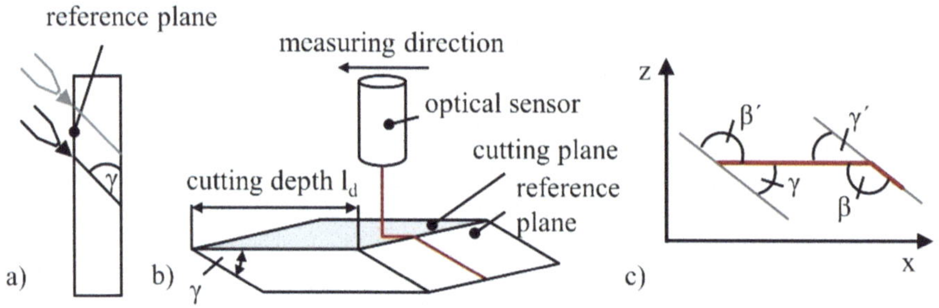

Fig. 5. Schematic description of the measuring setup; a) cutting the workpiece with the intended impact angle α; b) measuring the profile of the workpiece with an optical sensor; c) calculating the workpiece angle γ from the profile.

By analyzing the processed data, an aggregate contour profile can be extracted based on the cutting plane profile as well as the reference plane profile. Due to the geometrical arrangement of the individual profiles, the angle β can be calculated. The reference planes of the workpiece were machined by conventional milling before the cutting experiments and are parallel to each other. Therefore, the adjacent angle can be calculated according to Formula (3):

$$\gamma' = 180^\circ - \beta \tag{3}$$

Due to the plane-parallelism of the reference planes, this yields the required workpiece angle $\gamma = \gamma'$, see Fig. 5c.

2.3 Results and Discussion

In this section the results are presented in two parts. First, an evaluation of the investigations and measurement results for stainless steel X5CrNi18 is presented. The data was analysed using linear regression to identify a correlation between the jet impact angle α, feed speed v_f and workpiece angle γ. This correlation was transferred in the second part for the investigation of AWIJC of the high-performance materials Ti6Al4V and Inconel 718.

Analysis for stainless steel X5CrNi18. The obtained measurement results of the workpiece angle γ are shown in Fig. 6 for the investigated jet impact angle α in relation to the feed speed v_f. For a jet impact angle of $\alpha = 30°$, a deviation of the workpiece angle of $\Delta\alpha = 0.68°$ was determined only at a feed speed of $v_f = 45$ mm/min. For the other feed speeds v_f, the workpiece angle γ corresponds to the jet impact angle α. At a jet impact angle of $\alpha = 20°$, the workpiece angle γ matched only at a feed speed of $v_f =$ 15 mm/min. As the feed speed increases, the workpiece angle γ decreases continuously. By that, the deviation at a feed speed of $v_f = 45$ mm/min measures $\Delta\alpha =$ -1.9°. The workpiece angle γ corresponds with a jet impact angle of $\alpha = 10°$ at a feed speed of v_f = 30 mm/min. In this case, a linear correlation between the workpiece angle γ and the feed speed v_f can be determined. The largest difference of $\Delta\alpha =$ -2.7° was measured at a jet impact angle of $\alpha = 10°$ and a feed speed of $v_f = 45$ mm/min. Thus, a dependence of the workpiece angle γ on the jet impact angle α as well as on the feed speed v_f can be determined.

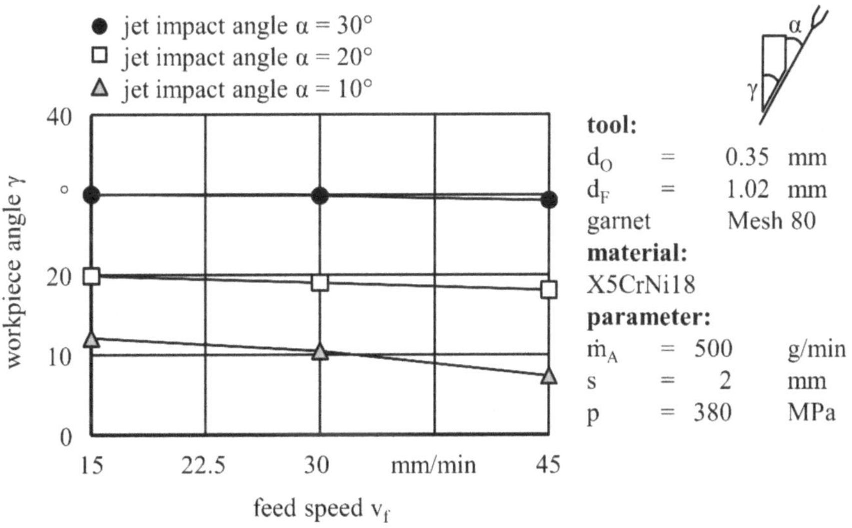

Fig. 6. Measured workpiece angle γ for different jet impact angle α and feed speeds v_f for stainless steel X5CrNi18.

In order to identify the interaction between the workpiece angle γ with the jet impact angle α and the feed speed v_f precisely, the obtained measurement data was implemented in the statistical software Minitab of the company MINITAB INC., Pennsylvania, USA. A linear regression analysis was performed and the following correlation was calculated:

$$\gamma_p = 6.21^\circ + 0.789 \cdot \alpha + 0.234 \frac{\min \cdot ^\circ}{mm} \cdot v_f + 7.32 \cdot 10^{-3} \frac{\min}{mm} \cdot \alpha \cdot v_f \tag{4}$$

The workpiece angle γ is dependent from the jet impact angle α and the feed speed v_f as can be seen in Formula (4). Both factors influence the workpiece angle γ. Furthermore, an interaction between the jet impact angle α and the feed speed v_f on the workpiece angle γ was determined.

Transfer to high-performance materials. After building a linear regression model, the predictability of the model for high-performance materials like Ti6Al4V and Inconel 718 was tested. Hence, the investigations and measurement were repeated with both materials. Only the feed speed v_f was adjusted to the material according to known cutting parameters for two-dimensional AWIJC. By that, the quality of the cut surface was similar to the quality of stainless steel. With the lower feed speed v_f a high surface quality was realized and with the higher feed speed v_f a separation cut with high surface roughness was performed. The chosen feed speed v_f for both materials are listed in Table 2. All other parameters were identical for the investigations with stainless steel.

Table 2. Feed speed for high-performance materials.

material	parameter	symbol	unit	value
Ti6Al4V	feed speed	v_f	mm/min	17.5 … 52.5
Inconel 718	feed speed	v_f	mm/min	12.5 … 37.5

For titanium Ti6Al4V, a jet impact angle of $\alpha = 30^\circ$ and a feed speed of $v_f = 17.5$ mm/min resulted in an exact match of the workpiece angle γ. When the feed speed v_f was increased, a slight deviation of the workpiece angle γ from the jet impact angle α was observed. Overall, the measurement results for the titanium specimens fit with the predicted workpiece angle γ_p by Formula (4) for a jet impact angle of $\alpha = 30^\circ$. A similar behavior is observed for a jet impact angle of $\alpha = 20^\circ$. At a feed speed of $v_f = 17.5$ mm/min, the workpiece angle γ coincides with the jet impact angle α. Furthermore, the workpiece angle γ decreases with increasing feed speed v_f. The measured workpiece angle γ fits with the predicted workpiece angle γ_p for a jet impact angle of $\alpha = 20^\circ$. The largest deviation of the measured workpiece angle γ from the predicted workpiece angle γ_p is observed for a jet impact angle of $\alpha = 10^\circ$. The largest deviations were determined at a feed speed of $v_f = 35$ mm/min with $\Delta\gamma_p = 2.1^\circ$ as well as $\Delta\gamma_p = 2.0^\circ$ at a feed speed of $v_f = 17.5$ mm/min. The results are presented in Fig. 7.

Similar results were obtained for Inconel 718 as for Ti6Al4V, see Fig. 8. At a jet impact angle of $\alpha = 30^\circ$, a similar workpiece angle γ was determined for the feed speed of $v_f = 12.5$ mm/min and $v_f = 25$ mm/min. At a feed speed of $v_f = 37.5$ mm/min,

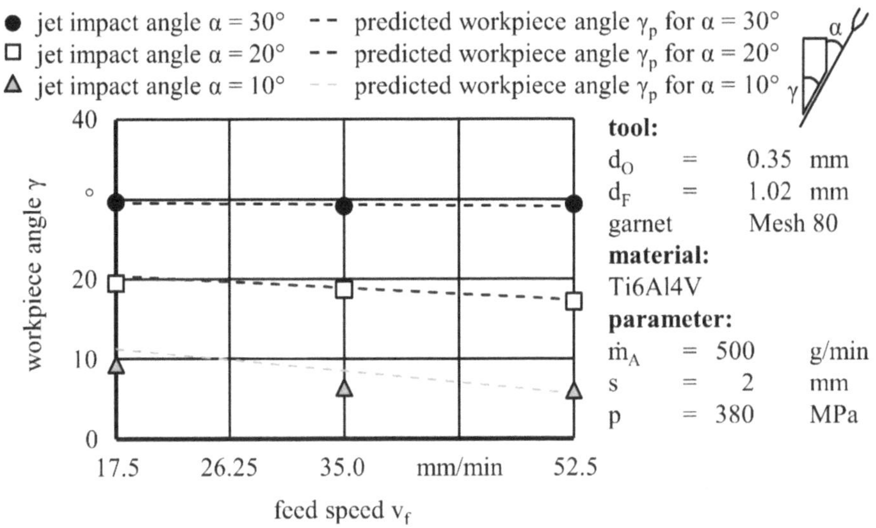

Fig. 7. Comparison between measured workpiece angle γ and predicted workpiece angle γ_p for different jet impact angle α and feed speeds v_f for Ti6Al4V.

the workpiece angle deviates by $\Delta\alpha = 0.83°$. Also, the measurement results fit with the predicted workpiece angles γ_p at a jet impact angle of $\alpha = 30°$. As in the case of Ti6Al4V and stainless steel, the deviations of the workpiece angle γ from the jet impact angle α increases as the jet impact angle α decreases. In the case of Inconel 718, the deviation of the workpiece angle γ is also higher at a jet impact angle of $\alpha = 20°$ and highest at $\alpha = 10°$. For a jet impact angle of $\alpha = 20°$, the deviation of the workpiece angle γ increases when the feed speed v_f increases and reaches $\Delta\alpha = 1.3°$ at a feed speed of $v_f = 37.5$ mm/min. However, the deviations from the predicted workpiece angle γ_p increasingly deviate at a jet impact angle of $\alpha = 20°$. The highest deviations are detectable at a jet impact angle of $\alpha = 10°$. The highest deviation with $\Delta\gamma_p = 2.9°$ was measured at a feed speed of $v_f = 25$ mm/min.

The highest deviation of the measured workpiece angle γ from the predicted workpiece angle γ_p was found for Inconel 718. Here, even at a jet angle of $\alpha = 20°$, a larger deviation was detected compared to the results regarding Ti6Al4V. This can be explained by the varying feed speed v_f for the different materials. Linear regression is limited in its ability to extrapolate results. This effect is also evident for the results in this case.

3 Summary and Outlook

In summary, pre-contouring of three-dimensional geometries with AWIJC of high-performance materials is a promising approach with significant economic potential. This approach offers significant cost savings by reducing tooling costs associated with conventional machining. However, the implementation of three-dimensional cutting with AWIJC leads to complex problems in determining the optimum cutting parameters. In

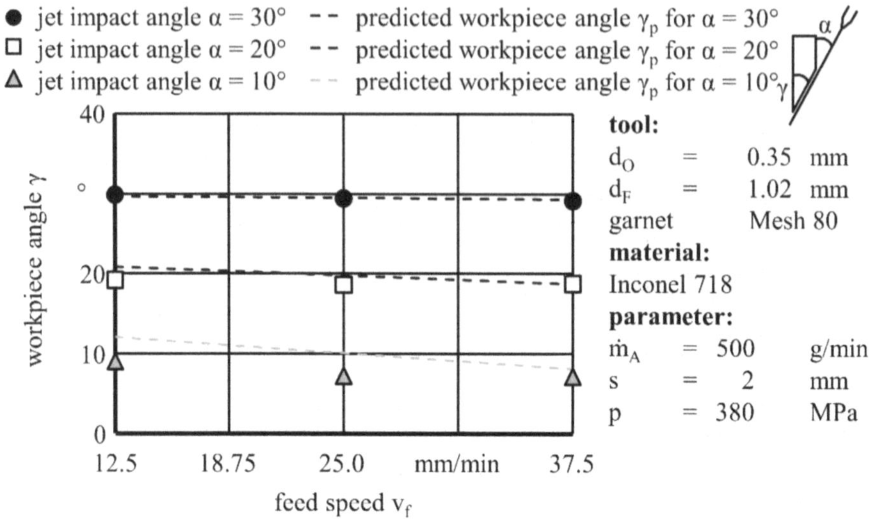

Fig. 8. Comparison between measured workpiece angle γ and predicted workpiece angle γ_p for different jet impact angle α and feed speeds v_f for Inconel 718.

particular, acute impact angles α result in different workpiece angles γ for different feed speed v_f, making it difficult to achieve consistent results.

Within the presented paper, acute impact angles α were investigated. A method to measure acute workpiece angles $\gamma \leq 20°$ was described. Furthermore, workpiece angles γ for stainless steel were measured and this data was successfully integrated into a predictive model based on linear regression. This integration allowed a prediction of workpiece angles γ_p for high-performance materials Ti6Al4V and Inconel 718. The results of this study not only provide a cost-effective way to investigate acute impact angles α, but also provide a valuable framework for three-dimensional pre-contouring with AWIJC. With this novel approach the applications for high-performance materials can be increased and lead to reduced manufacturing costs.

Further investigations will concentrate on the expansion of the prediction model for impact angles of $\alpha \leq 10°$. Additionally, the investigations will be extended by including various high-performance materials. With the achieved results and database, an accurate compensation of errors while cutting with acute jet impact angles α can be implemented into machine systems.

Acknowledgements. This paper is based on results acquired in the project JetTrim (ZF4067143LL9), which is kindly supported by the Zentrales Innovationsprogramm Mittelstand (ZIM) of the Federal Ministry for Economic Affairs and Climate Action (BMBF).

References

1. Qi, H.: Review of INCONEL 718 Alloy. J. Mater. Eng. **2**, 92–100 (2012)
2. Klocke, F., König, W.: Fertigungsverfahren 1: Drehen, Fräsen, Bohren. Springer Berlin, Heidelberg (2007)
3. Kolb, M.: Wasserstrahlschneiden, Materialbearbeitung mit einem Hochdruckwasserstrahl. Die Bibliothek der Technik, Band 295 (2006)
4. Uhlmann, E., Anders, S., Jaczkowski, R., Reder, W.: Near-net-shape trimming process by abrasive water jet cutting of high-performance workpieces for the aerospace industry. SSRN J. **35**, 93–98 (2020)
5. Liu, H.-T., Olsen, J.H.: Application of abrasive-waterjet for 3D machining. In: WJTA-IMCA Conference and Expo (2013)
6. Hashish, M.: Energy Fields in waterjet machining. In: Zhang, W. (ed.) Intelligent Energy Field Manufacturing: Interdisciplinary Process Innovations, pp. 141–171. CRC Press (2018)
7. Liu, H.-T.: "7M" Advantage of abrasive waterjet for machining advanced materials. J. Manuf. Mater. Process. **1**, 1–19 (2017)
8. Hashish, M.: Current and potential robotic abrasive waterjet systems. In: WJTA-IMCA Conference and Expo (2013)
9. Faltin, F.: Endkonturnahe Schruppbearbeitung von Titanaluminid mittels Wasserabrasivstrahlen mit kontrollierter Schnitttiefe. Hrsg., Uhlmann E., Dissertation. Technische Universität Berlin (2018)
10. Uhlmann, E., Männel, C.: 3D-Vorkonturierung mittels Wasserabrasivstrahl. Zeitschrift für wirtschaftlichen Fabrikbetrieb **113**, 479–483 (2018)
11. Reder, W., Uhlmann, E., Anders, S.: Mit dem Wasserabrasivstrahl in eine neue Dimension. VDI-Z **163**, 48–50 (2021)
12. Reder, W., Uhlmann, E.: Enabling of automatically generation of cutting paths for three-dimensional pre-contouring with waterjet trimming. J. Mach. Eng. **23**(2), 66–76 (2023)

Preliminary Investigation of the Cutting Edge Preparation of Carbide Tools Using the Abrasive Water Injection Jet Technology

Nermin Redžić[1](✉), Florian Morczinek[1], Sven Winter[2], Joachim Regel[1], and Martin Dix[1,2]

[1] Institute for Machine Tools and Production Processes IWP, Chemnitz University of Technology, Reichenhainer Str. 70, 09126 Chemnitz, Germany
nermin.redzic@mb.tu-chemnitz.de

[2] Fraunhofer Institute for Machine Tools and Forming Technology IWU, Reichenhainer Str. 88, 09126 Chemnitz, Germany

Abstract. In order to increase the tool life of carbide tools by applying a defined mechanically stable cutting edge geometry and removing micro defects and cracks, a cutting edge preparation process can be applied. As a novel technology, the high-pressure abrasive water injection jet technology can be used. The small jet diameter and high abrasive particle velocity result in an efficient material removal localized directly on the cutting edge. In this research work, the application of this technology for preparing carbide punching tools is analyzed and discussed. To gain insights into this novel preparation procedure, an experimental analysis using a 6-axis abrasive water injection jet test stand with a jet diameter of 0.3 mm applying water pressures up to 300 MPa was conducted. The aim of this analysis was to conduct a preliminary investigation of the application of this technology as a cutting edge preparation process, as well as describing the influence of process parameters on the prepared cutting edge radius.

Keywords: Abrasive Water Injection Jet · Cutting Edge Preparation · Punching · Carbide Tools

1 Introduction

Cemented carbide is a sintered material that consists of hard carbide particles (usually tungsten carbide) and a binder phase (usually cobalt). Its high hardness, compressive strength and wear resistance make it suitable for manufacturing tools for various machining processes, forming tools as well as tools for punching of high-strength sheet metals [1]. Because the shape of these tools is achieved by applying the grinding technology, the cutting edges are typically characterized by local micro cracking, chipping and burr formation. These micro defects result in high edge chipping and lead to the instability of the cutting edge. This property has a negative effect on both cutting tools, as well as tools for the punching process. In order to increase the stability of the cutting edge and thereby extend the tool life, various cutting edge preparation processes can

L. Sitek et al. (Eds.): Advances in Water Jetting II, LNME, pp. 182–193, 2025.
https://doi.org/10.1007/978-3-031-72778-8_15

be used [2]. The main objective of these processes is to precisely prepare and round the cutting edge in order to increase its mechanical stability and tool life. This ensures that the tool remains in a stable operating condition for a longer time and increases the overall process stability. Applying blasting and jetting technologies for the cutting edge preparation has been proven to have a positive effect on the cutting edge stability in machining technologies before and after coating [3–5], as well as for preparing tools for punching of sheet metal [6]. The high flexibility of these technologies can be applied to prepare complex asymmetric cutting edge shapes [7, 8]. Conventional cutting edge preparation technologies are generally used up to a pressure value of 1.5 MPa with an inner nozzle diameter of 6–8 mm [4, 7, 9]. In comparison with the technologies based on dry or wet blasting and jetting, an increase in process precision and efficiency can be achieved by applying the high-pressure abrasive waterjet technology [10, 11]. When compared to high-productive processes, such as grinding or laser treatment, the abrasive water injection jet technology exhibits the important advantage of introducing no thermal effects into the material structure [12]. The application of this technology for cutting edge preparation was described in the works of Lei et al. [13] up to a pressure of 120 MPa applying a rectangular abrasive nozzle and Sauer et al. [10] using a water pressure of 200 MPa and a focusing tube with an inner diameter of 0.3 mm. However, all of the aforementioned research focuses on the preparation of cutting tools and the treatment of linear cutting edges. Furthermore, almost all of the research focuses on low-pressure abrasive jet technologies.

In this study, a preliminary investigation into the cutting edge preparation process of carbide punching tools using the abrasive water injection jet technology was conducted. The goal of this research is to analyze the abrasive waterjet preparation process and define the general process conditions for preparing a rounded, mechanically stable cutting edge to increase tool life and process stability.

2 Experimental Analysis

In this test series, a cutting head with a focusing tube with an inner diameter of $d_f =$ 0.3 mm was used for the cutting edge preparation. The inner diameter of the applied water orifice was $d_w = 0.1$ mm. Sieved Kominex garnet sand with particle size between 63 and 90 μm was used as abrasive material. The water pressure was generated using a BFT Servotron 40.45+ high-pressure intensifier pump. The tests were performed on ground cylindrical samples made of ISO K40 carbide with an outer diameter of $D =$ 5 mm. The material properties are presented in Table 1.

The preliminary tests conducted within this research work included varying the jet exposure time per 1 mm length t_e, the standoff distance and applying different water pressures to achieve different material volume removal rates. The tested values are presented in Table 2.

The experimental analysis was conducted on a 6-axis abrasive water injection jet test stand. The test stand consists of a 5-axis cutting head, which was used to position the abrasive water jet for treating the samples, as well as an additional rotary axis in which the samples were clamped and which enables the feed motion. The test setup is presented in Fig. 1.

Table 1. Properties of the tested sample materials.

Property	Value
Carbide type	WC-Co
Carbide grain size	Fine (0.8 - 1.3 μm)
Cobalt content	11.8%
Hardness (ISO 3878)	1400 HV10
Transverse rupture strength (ISO 3327)	3200 MPa

Table 2. Process parameters varied within the experimental analysis.

Property	Value
Water pressure p	200 and 300 MPa
Standoff distance x	10 and 20 mm
Jet exposure time t_e per 1 mm length	0.1 - 1.5 s/mm
Waiting time at the start point t_w	0.5 - 1.5 s

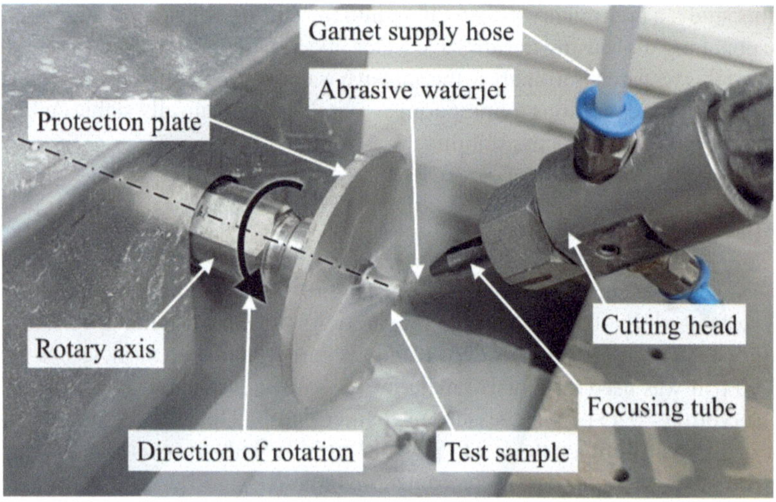

Fig. 1. Setup for the cutting edge preparation tests using the abrasive water injection jet.

In the first step, the focusing tube is positioned with an angle set up so that the symmetry line of the abrasive waterjet corresponds to the bisecting line of the tool wedge angle. During the preparation process, the cutting head remains steady in that position, while the rotation of the rotary axis enables the relative feed motion between the cutting edge and the abrasive waterjet. The cutting edge preparation procedure is presented in Fig. 2.

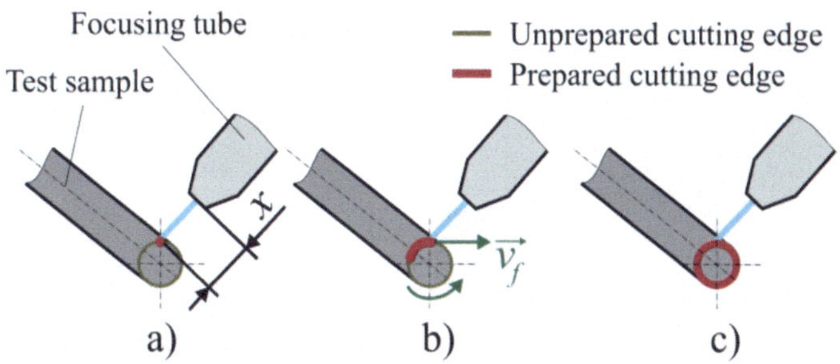

Fig. 2. Procedure of the cutting edge preparation: **a** Activation of the abrasive waterjet with a defined waiting time t_w at a standoff distance x; **b** Cutting edge preparation with a defined rotation speed of the sample which results in a relative feed motion with a feed rate v_f; **c** End of the process after one full rotation of the sample.

The waiting time t_w at the start point influences the material removal at the beginning of the process. Afterwards, the rotational speed of the rotary axis defines the feed rate v_f in mm/s at which the cutting edge is treated with the abrasive waterjet. For a defined rotational speed of the rotary axis n in min^{-1}, the feed rate at which the preparation process is conducted can be calculated as:

$$v_f = \frac{D \cdot \pi \cdot n}{60} \tag{1}$$

The preparation procedure ends with turning off the waterjet after one full rotation of the test sample, so that the whole cutting edge is treated. The cutting edge was observed before and after the preparation process using the optical 3D surface measurement GFM MikroCAD microscope. The radius measurement was conducted at three different measuring locations on the cutting edge within a length of 1.2 mm. Within this length, the software analyzes and fits circles within 400 cross section areas. The value of the radius that describes the measured length of the cutting edge is calculated as the mean value of these 400 radius values, neglecting 25 extreme values within the data set. The measuring procedure is repeated three times along the cutting edge. The representative radius value for a single cutting edge is calculated as the mean value of these three measurements, neglecting the values at the starting point of the waterjet treatment, which is discussed separately in Sect. 3.4.

3 Results and Discussion

3.1 Sample Properties Before and After the Cutting Edge Preparation

In the initial state, which is defined by the grinding process, the cutting edges on the punching tools are affected by micro cracks and defects, which can be seen in Fig. 3a. After applying the abrasive waterjet preparation process, most of the defects can be removed and the sharp edge rounded. An example of a cutting edge prepared using the abrasive waterjet technology is shown in Fig. 3b.

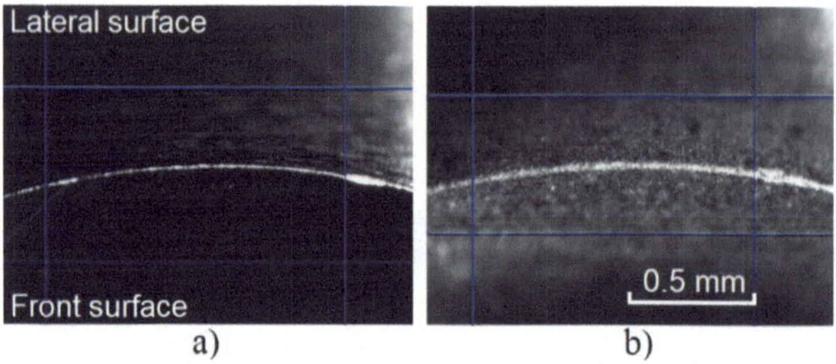

Fig. 3. Comparison an untreated cutting edge (**a**) and a cutting edge prepared using the abrasive water injection jet technology (**b**).

The cutting edge preparation process results in a rounding with a larger radius in comparison with a ground cutting edge. The radius values for the test sample in Fig. 3 are measured before and after the cutting edge preparation on the defined measuring length of l = 1.2 mm and presented in Fig. 4. Within this measuring length, a mean radius values of r_o = 11.2 µm with a standard deviation of the measured values of $\sigma = 1.79$ was measured on the unprepared ground cutting edge. The prepared cutting edge shows a more uniform radius along the cutting edge, a mean radius value of r = 25.7 µm and a standard deviation of $\sigma = 2.01$. In order to compare these values, the coefficient of variation (CV) for both sets was calculated and equals $CV = 0.16$ for the untreated and $CV = 0.08$ for the prepared cutting edge, meaning that the radius for the prepared cutting edge has a lower dispersion of the measured values than the untreated cutting edge. Based on the lower CV value, it can be concluded that the prepared cutting edge exhibits a better uniformity along the whole edge in comparison with the untreated cutting edge.

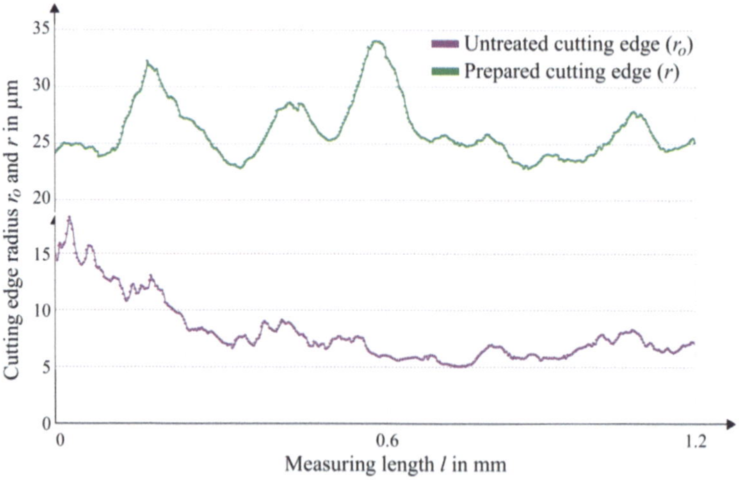

Fig. 4. Comparison of cutting edge radius values on a measurement length of l = 1.2 mm for an untreated and a prepared cutting edge using the abrasive waterjet.

3.2 Influence of the Jet Exposure Time

According to the works of Cortés Rodríguez [14], the material removal rate during the cutting edge preparation process can be described by using the term of the nominal removed area A_o, which defines the cross-section area at the tool edge described by the initial and the prepared cutting edge radius. The definition of the nominal removed area A_o is presented in Fig. 5.

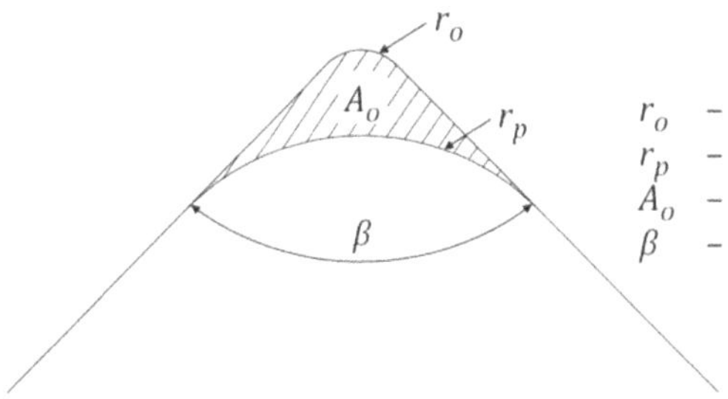

Fig. 5. Definition of the nominal removed area A_o, according to [14].

The nominal removed area can be calculated using the Eq. (2) [10]:

$$A_o = \left(r^2 - r_o^2\right) \cdot \left(\cot\left(\frac{\beta}{2}\right) - \frac{\pi}{2} + \frac{\pi \cdot \beta}{360^\circ}\right) \tag{2}$$

Since the wedge angle for the cylindrical punching tools analyzed in this research work equals $\beta = 90^\circ$, Eq. (1) can be simplified as:

$$A_o = \frac{3\pi}{4} \cdot \left(r^2 - r_o^2\right) \tag{3}$$

The material volume removal rate for achieving a defined value of A_o can be expressed using the Eq. (4) [15]:

$$\dot{V}_M = A_o \cdot v_f \tag{4}$$

From this equation, the nominal removed area of the cutting edge is inversely proportional to the applied feed rate and can be presented as:

$$A_o = \dot{V}_M \cdot \frac{1}{v_f} \tag{5}$$

Since the function inverse to the feed rate represents the jet exposure time t_e per 1 mm length, it can be stated that the area A_o is proportional to this value:

$$A_o = \dot{V}_M \cdot t_e \tag{6}$$

Substituting Eq. (6) into Eq. (3) and introducing the jet exposure time t_e as the function inverse to the feed rate, the prepared radius can be defined using Eq. (7):

$$r^2 = \frac{4}{3\pi} \cdot \dot{V}_M \cdot t_e + r_o^2 \tag{7}$$

This implies that the prepared cutting edge radius r can be described by using a square root function, which can be presented with Eq. (8):

$$r = \sqrt{\frac{4}{3\pi} \cdot \dot{V}_M \cdot t_e + r_o^2} \tag{8}$$

This correlation corresponds to the observations made in the works of Cortés Rodríguez [14] and Sauer et al. [10]. In order to analyze the influence of the jet exposure time t_e per 1 mm length on the prepared cutting edge radius r, an experimental analysis using varying the jet exposure time and water pressure was conducted. This analysis was conducted with a constant standoff distance of $x = 10$ mm. The results of this analysis are presented in Fig. 6.

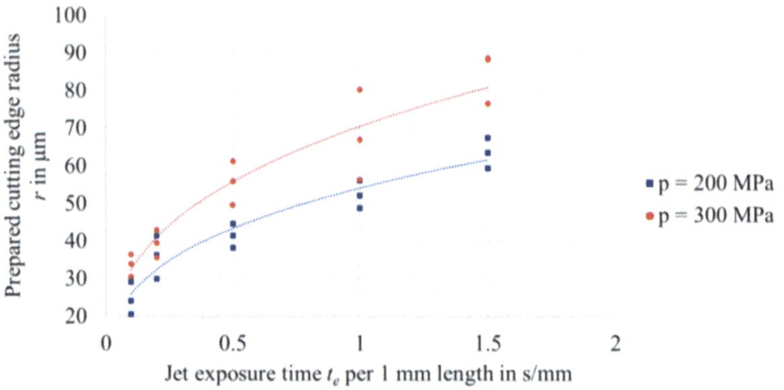

Fig. 6. Influence of the jet exposure time t_e and the water pressure p on the value of the prepared cutting edge radii for a constant standoff distance of $x = 10$ mm.

According to these results, the value of the radius of the prepared cutting edge r increases with an increase of the jet exposure time t_e. In addition to this, the increase of the material removal rate caused by higher particle velocity for the higher water pressure values p results in larger cutting edge radii. The presented results show that for a constant material removal rate value $\dot{V}_M$ a square root function describes the correlation between the applied jet exposure time t_e per 1 mm length and the resulting radius value r, which corresponds to the mathematical description shown in Eq. (8). Changing the value of $\dot{V}_M$ by applying a different water pressure value p, the slope of the describing square root function is changed, which can be observed in Fig. 6.

3.3 Influence of the Standoff Distance on the Prepared Cutting Edge Radii

Increasing the standoff distance, the jet exposure area increases [16], hence decreasing the local material removal rate which occurs on the cutting edge. This means, that the increase of the standoff distance results in smaller cutting edge radii, while increasing the area influenced by the abrasive waterjet. This was experimentally verified by applying the same process parameters presented in Fig. 6. Increasing the standoff distance to $x =$

20 mm, lower radius values r were prepared for all parameter sets. By applying a water pressure of $p = 200$ MPa, the mean radius value varied between $r = 21.5$ µm for $t_e = 0.1$ s/mm and $r = 48.0$ µm for $t_e = 1.5$ s/mm. Similarly, for the water pressure $p = 300$ MPa, the mean radius value observed for $t_e = 0.1$ s/mm was $r = 29.2$ µm and for $t_e = 1.5$ s/mm it equaled $r = 63.4$ µm. The value of the prepared cutting edge radii within the conducted analysis is presented in Fig. 7. It can be observed, that a similar correlation between the variables and the resulting radius r exists for the increased standoff distance. However, compared to the results shown in Fig. 6, a lower slope of the function can be observed, which results from the lower material volume removal rate $\dot{V}_M$ caused by the higher standoff distance.

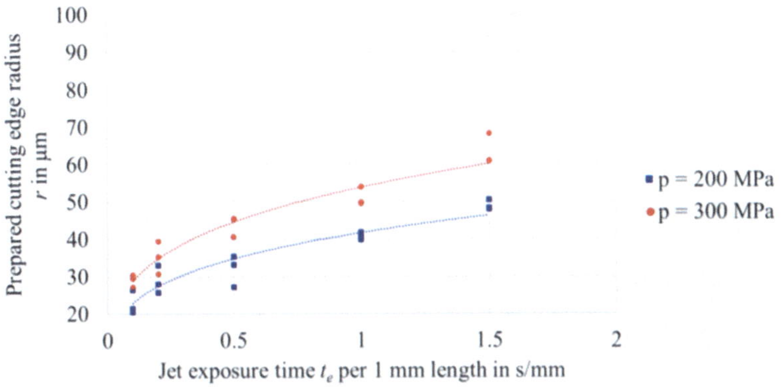

Fig. 7. Influence of the jet exposure time t_e and the water pressure p on the value of the prepared cutting edge radii for a constant standoff distance of $x = 20$ mm.

3.4 Analysis of the Start Phase on the Cutting Edge Preparation Procedure

The waiting time t_w at the starting point has a high importance on the uniformity of the prepared rounding along the cutting edge. The waiting time after starting the pure waterjet is necessary in order to ensure the suction of the abrasive flow from the supply unit through the hose into the waterjet. In addition, the start of the material removal at the cutting edge has to be ensured. However, a too long waiting time results in a larger edge rounding at the starting point in comparison with the rounding prepared during the continuous process with a defined feed rate. An example of this phenomena for a waiting time $t_w = 1.0$ s during a preparation process with a jet exposure time of $t_e = 0.5$ s/mm is presented in Fig. 8.

In order to analyze the influence of the waiting time t_w on the occurrence of this difference of the radius value, an experimental analysis with different values of the waiting time was conducted. For this experiment, the radius value at the starting point r_s and the radius of the prepared cutting edge r were measured for different values of the waiting time t_w. The locations of the measured radii are presented in Fig. 9.

The results of this analysis are summarized in the diagram presented in Fig. 10. It can be observed that a lower waiting time t_w results in a lower radius at the starting point

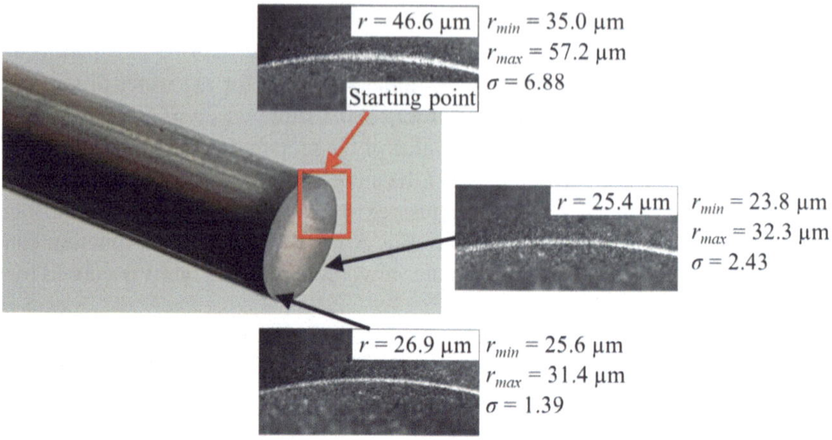

Fig. 8. Difference of the mean values of the prepared cutting edge radius r and their range, as well as the standard deviation σ at the starting point and along the cutting edge.

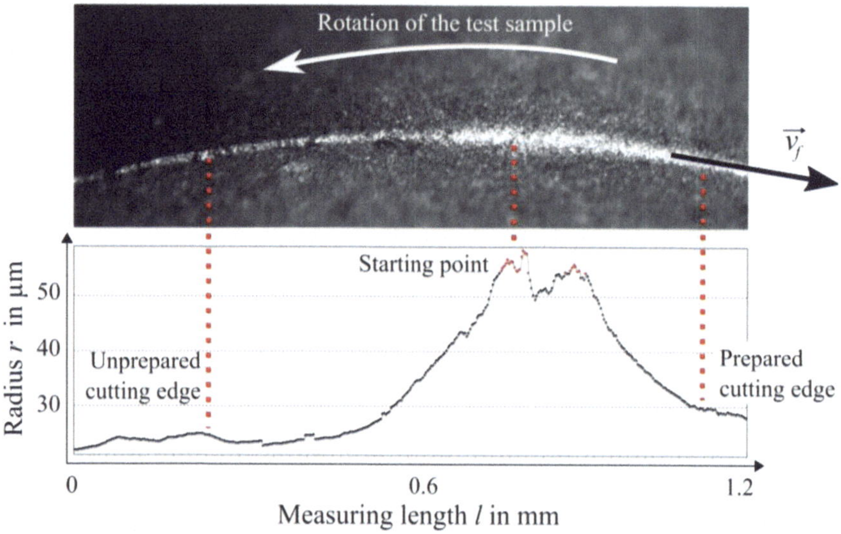

Fig. 9. Increase of the prepared cutting edge radius due to a high waiting time at the start point.

r_s. In the case of preparing the cutting edge with a jet exposure time of $t_e = 0.5$ s/mm, an equal rounding at the starting point and along the cutting edge was achieved for a waiting time of $t_w = 0.5$ s. The influence of the waiting time t_w is machine specific and depends on other process parameters (standoff distance x, jet exposure time t_e, etc.) as well as the abrasive supply hose length, which implies that the values depicted in Fig. 10 present preliminary results for one selected parameter set and the specific test stand presented in Fig. 1. On basis of these results, further experiments can be conducted to gain a more detailed insight into the influence of the beginning phase of a cutting edge preparation process on the prepared cutting edge rounding.

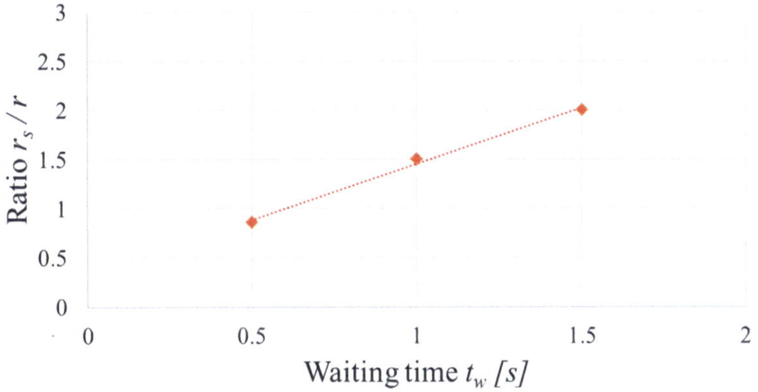

Fig. 10. Influence of the waiting time on the ratio of the mean values of the cutting edge radius at the start point and along the cutting edge for a jet exposure time of $t_e = 0.5$ s/mm, a standoff distance of $x = 20$ mm and a water pressure of $p = 200$ MPa.

4 Conclusion

In this research work, first insights into the application of the high-pressure abrasive water injection jet technology for the cutting edge preparation of carbide punching tools are presented. Within the conducted experimental analysis, it was shown that the prepared cutting edge radius r can be described as a function of the jet exposure time t_e for a constant set of other process parameters. The preliminary experiments show that this relation approaches the square root function calculated mathematically for an ideal preparation process described in Eq. (8).

The results gained within the conducted experimental analysis show that an increase in water pressure p results in an increase of the prepared cutting edge radius r. This occurs due to the fact that a higher water pressure results in a higher particle velocity and therefore a higher kinetic energy of the particles, which results in an increase of the material removal rate. In addition to that, it was shown that an increase of the standoff distance results in lower radius values of the prepared cutting edge, due to the fact that this increases the jet exposure area and hence results in a decrease of the local material removal rate. To summarize, the results presented in Figs. 6 and 7 indicate that the correlation between the prepared cutting edge radius r and the applied jet exposure time t_e for a fixed set of other process parameters is described by a square root function described in Eq. (7). The slope of the function is defined by the material volume removal rate $\dot{V}_M$, which is a function of the applied process parameters, such as the water pressure, standoff distance, properties of the used abrasive material and its mass flow rate, the water orifice diameter, etc. Within this analysis, this effect was observed by applying two different water pressure values $p = 200$ MPa and $p = 300$ MPa, as well as two different standoff distances $x = 10$ mm and $x = 20$ mm.

Analyzing the cutting edge preparation procedure, it was observed that the waiting time t_w at the beginning of the preparation process causes higher local radius values. Within a preliminary analysis values for t_w were defined which result in a uniform rounding along the whole cutting edge. Since the necessary waiting time t_w is not only a function of process parameters, but also in relation to machine properties, such as the abrasive supply feed system and the supply hose length, further investigation is required in order to completely define its influence on the preparation procedure.

Within this research work, cutting edge radii with values $r \geq 20$ µm were prepared. The preparation of smaller radius values, as well as the preparation of asymmetric cutting edge shapes, which are also of interest for the punching technology and of high importance for tools in machining technologies, are to be considered in future research work.

Acknowledgement. The authors would like to thank the Research Association for Steel Application (FOSTA) for their support in the project "ScherWas" (IGF Project no. 22508 BG), which is supported by the German Federal Ministry for Economic Affairs and Climate Action (BMWK) within the framework of the Industrial Collective Research (IGF) on the basis of a decision by the German Bundestag.

References

1. Winter, S., Richter, K., Galiev, E., Nestler, M., Psyk, V., Kräusel, V.: Punching of ultra-high-strength spring strips: evolution of cutting edge radius up to 1,000,000 strokes for three punch materials. J. Manuf. Mater. Process. **6**(2), 38 (2022). https://doi.org/10.3390/jmmp6020038
2. Denkena, B., Biermann, D.: Cutting edge geometries. CIRP Ann. **63**(2), 631–653 (2014). https://doi.org/10.1016/j.cirp.2014.05.009
3. Biermann, D., Aßmuth, R., Schumann, S., Rieger, M., Kuhlenkötter, B.: Wet abrasive jet machining to prepare and design the cutting edge micro shape. Procedia CIRP **45**, 195–198 (2016). https://doi.org/10.1016/j.procir.2016.02.071
4. Bouzakis, K.D., et al.: Effect of cutting edge preparation of coated tools on their performance in milling various materials. CIRP J. Manuf. Sci. Technol. **7**(3), 264–273 (2014). https://doi.org/10.1016/j.cirpj.2014.05.003
5. Wang, W., Biermann, D., Aßmuth, R., Arif, A.F.M., Veldhuis, S.C.: Effects on tool performance of cutting edge prepared by pressurized air wet abrasive jet machining (PAWAJM). J. Mater. Process. Technol. **277** (2020). https://doi.org/10.1016/j.jmatprotec.2019.116456
6. Ott, A., Biermann, D.: Experimental investigation of the cutting edge microshape to improve the wear resistance of punch and die tools for sheet metal punching. In: Schmitt, R., Schuh, G. (eds.) Advances in Production Research. Springer International Publishing, Cham, pp. 391–401 (2019). https://doi.org/10.1007/978-3-030-03451-1_39
7. Aßmuth, R.: Schneidkantenpräparation durch Druckluft-Nassstrahlspanen mit Industrierobotern (Dissertation). In: Schriftenreihe des ISF, no. 99. Vulkan-Verlag, Essen (2019). ISBN: 978-3-8027-8915-1
8. Segebade, E., Zanger, F., Schulze, V.: Asymmetrische Schneidkantenmikrogeometrien - Bedeutung, Charakterisierung, potenzielle Anwendungen und Heraus-forderungen. In: Spanende Fertigung: Prozesse, Innovationen. In: Biermann, D. (ed.) Werkstoffe, 7. Edition., pp. 79–85. Vulkan Verlag, Essen (2017). ISBN: 978-3-8027-2989-8

9. Kötter, D.: Herstellung von Schneidkantenverrundungen und deren Einfluss auf das Einsatzverhalten von Zerspanwerkzeugen (Dissertation). In: Schriftenreihe des ISF, no. 36. Vulkan Verlag, Essen (2006). ISBN 3-8027-8736-6, 978-3-8027-8736-2
10. Sauer, K., Witt, M., Putz, M.: Influence of cutting edge radius on process forces in orthogonal machining of carbon fibre reinforced plastics (CFRP). Procedia CIRP **85**, 218–223 (2019). https://doi.org/10.1016/j.procir.2019.09.042
11. Morczinek, F., Putz, M., Dix, M.: Comparison of abrasive water jet technologies in terms of performance and kerf geometry accuracy for cutting ceramics. Int. J. Sustain. Manuf. **4**(2–4), 201–215 (2020). https://doi.org/10.1504/IJSM.2020.107134
12. Redžić, N., Kieweg, P., Regel, J., Dix, M.: Effects of abrasive waterjet cutting on surface properties of hardened steel. Procedia CIRP **118**, 620–625 (2020). https://doi.org/10.1016/j.procir.2023.06.106
13. Lei, Y., Wu, Y., Yang, W., Zhang, Y.: Analysis of the effect of water jet parameters on the performance during preparation of cutting tool edge by fine abrasive water jet. In: WJTA Conference and Expo. New Orleans, Louisiana, USA (2019)
14. Cortés Rodríguez, C. J.: Cutting Edge Preparation of Precision Cutting Tools by Applying Micro-Abrasive Jet Machining and Brushing (Dissertation). Kassel University Press, Kassel (2009). ISBN: 978-3-89958-712-8
15. Sauer, K.: Qualitätsorientierte Prozessgestaltung der Bohrungsfertigung in kohlenstofffaserverstärkte Kunststoffe (Dissertation). In: Berichte aus dem IWU, no. 129. Verlag Wissenschaftliche Skripten, Auerbach/Vogtland (2022). ISBN: 978-3-95735-162-3
16. Momber, A.W., Kovacevic, R.: Principles of Abrasive Water Jet Machining. Springer, London (1998). https://doi.org/10.1007/978-1-4471-1572-4

Influence of Phase Composition in an Injection-Type Abrasive Slurry Jet (iASJ)

Manuel Schüler[1](✉), Ralph Lindken[2], and Brendan Görres[1]

[1] HydroMill GmbH, Arndtstr.30, 52064 Aachen, Germany
manuel.schueler@hydromill.de

[2] Hochschule Bochum, Am Hochschulcampus 1, 44801 Bochum, Germany

Abstract. Conventional manufacturing methods face challenges with hard, brittle, or thermally sensitive materials. Abrasive Water Jet (AWJ) methods offer advantages when cutting those high-performance materials. However, milling with AWJ requires refinement to gain comprehensive knowledge of the process design. This study explores an experimental comparison of commonly used AWJ and an injection-type Abrasive Slurry Jet (iASJ). The gaseous phase in the jet tool can be minimised using iASJ. Because it uses a regular AWJ system, the iASJ allows a direct comparison of the results. Preliminary investigations focused on optimising slurry jet stability, while the central experiments studied the influence of the jet's phase composition on the erosion process. Here, a dense-sintered technical ceramic made of zirconia was used. Moreover, the study suggests that the absence of air suppresses high-frequency droplet erosion, minimising material outbreaks of brittle-hard materials.

Keywords: Waterjet Machining · Slurry Injection Abrasive Waterjet · Ceramics · Surface Finishing

1 Introduction

Materials like zirconia are commonly used in prosthetics in the biomedical sector due to their high strength and biocompatibility, making them suitable for various manufacturing approaches based on specific needs [1]. Similarly, in the energy industry, Thermal Barrier Coatings (TBC) are utilised for applications requiring high heat resistance, such as steam turbine blades made of Yttria-stabilised Zirconia (YSZ) [2].

Conventional processes often prove unsuitable for processing high-performance materials required for high-technology products [3]. Non-conventional processes usually rely on mechanical contact between the tool and the workpiece, introducing heat and stress into the material [4]. The need for flexible, non-conventional techniques in production arises since the demand for unique materials across all industrial sectors is increasing [5]. Biomedical and energy generation industries utilise high-performance and hard-to-machine materials like high-strength technical ceramics combined with non-conventional approaches to match specific needs [6].

L. Sitek et al. (Eds.): Advances in Water Jetting II, LNME, pp. 194–208, 2025.
https://doi.org/10.1007/978-3-031-72778-8_16

Abrasive water jet (AWJ) machining offers high cutting performance and comparable advantages to other processes, including the ability to work with brittle and ductile materials with shallow temperature rise. Its versatility in creating complex shapes [7] could make it a viable candidate for machining oxide ceramics [8]. Material removal in ductile materials is achieved by introducing the water jet's kinetic energy to the workpiece, which then deforms plastically [9]. Introducing kinetic energy to brittle materials such as zirconia and glass promotes microcrack formation, which limits its application. Typically, the waterjet applied is adjusted on the machining task. Besides industrially standardised AWJ, as an add-on of a regular plain waterjet, there were other derivates of waterjets (Fig. 1). This preliminary work led to the different waterjet configurations, categorised by their phase composition, though some were limited to the laboratory state [10].

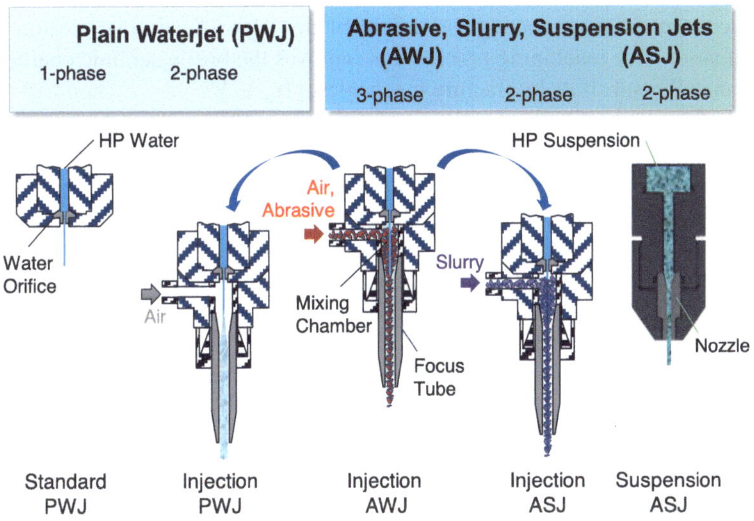

Fig. 1. An alternative categorisation of waterjet systems by the jet's phase composition.

AWJ cutting allows high performance, negligible heat generation, and machining flexibility. AWJ can also be applied to machining to a defined depth, like the conventional milling technique [11]. However, a considerable number of variables are related to material removal mechanisms. Compared to classic through-cutting, the loads acting on a defined cutting depth differ in water jet milling. Due to the predominantly orthogonal direction of kinetic engagement of the jet tool, especially at shallow cutting depths, the cutting front is generally flat [12]. The corresponding conditions suppress material removal mechanisms under a low angle [13].

Abrasive Water Jet Milling (AWJM), also known as Controlled Depth Machining (CDM), is a modified approach of AWJ cutting, focusing on controlled material removal rather than cutting through the material [11]. Surface quality directly impacts the machining result in a time-critical manner so that a better surface can generally be achieved with a higher feed rate. Note that this phenomenon is contrary to typical striation marks that occur in AWJ cutting, which are mitigated by reducing the traversing speed. As the process control is significantly more dynamic, special requirements are placed on a stable and continuous waterjet tool - Fluctuations in the fluid mixing lead to problematic instabilities that directly affect the processing result. A different abrasive jet phase composition altered the material removal rate. Studies have shown that changing the air content of the jet tool impacts machining characteristics and erosion effects [14]. Moreover, the literature examined the influence of abrasive properties on erosion in AWJ machining, with findings indicating that particle shape and size impact surface roughness and material removal rate (MRR) [15].

Pure or Plain Water Jet (PWJ) utilises a pressurised water jet without abrasive injection into the mixing chamber. Its homogenous fluid composition without chaotic mixing is ideal for processing soft, ductile materials. Under certain circumstances, AWJ may be substituted with PWJ when the target material is sensitive to breaking or composed of pores. It can remove material selectively from multi-composites or coating systems such as ceramic thermal barrier coatings on turbine blades. Here, the droplet erosion leads to a contamination-free machining process that removes the brittle ceramic coatings from a ductile metallic substrate by fracture mechanics. [16, 17].

The abrasive suspension jet (ASJ) head is like that of PWJ and employs a pressurised mixture of abrasive and water to be squeezed through a nozzle in the absence of a gaseous carrier phase. However, the systemic effort is higher due to the pressurised vessel containing the abrasive to form the suspension, so this system is not well established for production and is currently used for extraordinary applications. Due to the direct acceleration through a nozzle without a post-mixing process, a higher cutting performance can be achieved [18]. Studies on composites confirm that ASJ improves machining quality due to the lack of air present in the jet [19].

Slurry Injection Abrasive Water Jet (iASJ) is a modification of traditional AWJ involving a slurry composed of abrasives, water, and additives injected into the mixing chamber. Contrary to the common ASJ, this derivate uses the slurry under atmospheric conditions. At the cost of reduced cutting performance due to the higher mass of sucked liquid to be accelerated, it enables the processing of fine abrasives without clogging, a common issue using fine and dry abrasives [20]. Previous studies have explored iASJ finishing for surface improvement, utilising multiple passes and different mixtures of the slurry for multi-stage processes [6, 21]. Further research investigates different abrasives and materials like ceramics and glass, identifying areas for improvement in iASJ setups [22, 23].

Literature shows various settings affect the waterjet machining process [24, 25] Klicken oder tippen Sie hier, um Text einzugeben.. Multiple response optimisation is a common approach due to the interaction of parameters. Variables include machine parameters, material-specific factors like abrasive type and target material properties that affect output variables like surface texture or MRR. Despite these fundamental complexities, there is a lack of process understanding of the jet characteristics. Due to the altered engagement for milling operations, there are critical effects in machining brittle-hard materials [27]. Different wear mechanisms were associated with the flow regime in the process zone [14].

Existing knowledge indicates a gap in understanding the water jet milling process. Despite the known process understanding transferred from AWJ cutting, there are differences in material removal related to the waterjet tools used. This paper focuses on the fragmentation effects of brittle-hard materials due to the dynamic interaction of the jet phases induced droplet impact and the gas phase of the jet tool. This phenomenon appears to be particularly important in perpendicular conditions of tool engagement, as is the case with water jet milling brittle-hard materials. A direct comparison between a two-phase and a three-phase AWJ according to the injector principle has not yet been considered to investigate the influence of the jet phases on the machining result of increased complexity in the development of special AWJ applications.

2 Experiments

This study employed a two-phase injection-type abrasive slurry jet (iASJ) to compare the systemic impact when machining brittle-hard materials. A comparison was made under identical conditions comparing iASJ with a three-phase injection AWJ. The two phases in iASJ are abrasives as a dispersed solid phase in a water carrier phase (the density of the water is increased by adding a thickener), forming a slurry, and water as a continuous phase. The slurry and the high-velocity water from the pump are mixed in the mixing chamber. Then, the water in the slurry and the pure water flowing through the high-pressure pump become one phase. The three phases in the AWJ are abrasives as dispersed solid phases in an air carrier phase. At this stage, the air is in a continuous phase. A high-velocity water phase sucks in the air/solid mixture. During this process, due to the significant difference in density, i.e. a difference in inertia, the solid moves from the air phase to the water phase, creating a three-phase flow with a continuous water phase and dispersed abrasive solid phase in water and dispersed air as a gaseous phase in water phase. In the experiments of this study, two principal materials were machined: soda-lime glass (SiO_2) and dense-sintered zirconium dioxide (ZrO_2), also known as zirconia ceramic. Both materials exhibit brittle characteristics besides other characteristics. Due to the low cost and ability to observe variability in the process results by visual inspection, glass served as an analogy material. Slurry mixtures with varying viscosities were applied and analysed according to the process's stable conditions. Slurry viscosity control was achieved by adding a soluble dietary fibre as a thickener.

For the experiments, a particular waterjet machine, H.G. Ridder type WARICUT E-11045, in combination with a UHDE Type HPS 4022 intensifier pump, was used, capable of producing water pressure up to 400 MPa. The cutting head was configured with a focus tube diameter of 0.54mm and an orifice diameter of 0.12 mm for all experiments, conducted with a standoff distance to the workpiece surface of 2 mm. Garnet abrasive was sieved to match a grit size between 75 μm and 125 μm. An advanced microscopy 3D measurement system, Alicona InfiniteFocus G5, supplemented with a Laser scanning microscope Keyence type VK-X100, was used for surface analysis according to ISO 25178. Preparation of the materials included vibrating sieving to match abrasive grain sizes and ultrasonic cleaning to remove residual particles from the workpiece surface before inspection.

The study can be divided into two experimental sections. The fluid properties were investigated in the preliminary experiments to define the main experiment parameters.

Parameters include machine parameters, abrasive concentrations, viscosity, and flow rate to explore the jet stability and erosion characteristics for machining. In the central part, experiments with varying feed rates, water pressure, and abrasive flow were conducted to compare the process results between AWJ and iASJ.

The preliminary experiments focused on the fluid properties of the slurry to generate the iASJ. Here, a corresponding abrasive flow rate entering the mixing chamber is based on the amount of abrasive determined by the slurry flow rate and its abrasive concentration. This calibration produces a particle load comparable to that of the AWJ. In the experimental setup, the flow rate could be controlled by using machine parameters and specific ball valve opening positions. Furthermore, the jet exiting the focusing tube was observed regarding changes induced by different slurry mixtures in the mixing chamber. Images were taken during the slurry flow rate experiment of iASJ and evaluated with computer vision tools to assess jet characteristics inspired by related work [28]. The static stability of the slurry was judged according to its time-dependent settlement of abrasive still in suspension, hindering the passage of light in time intervals of different thickener concentrations (Table 1).

Table 1. Parameters for preliminary experiments of slurry injection.

Parameters: Slurry flow rate	Levels	Low Level	High Level
System pressure	4	75 MPa	200 MPa
Abrasive flow equivalent	3	7 g/min	28 g/min
feed rate	5	250 mm/min	1250 mm/min
Thickener concentration	4	0 g/l	2.25 g/l
Time interval (grit settlement)	4	1 min	10 h

Experiments for the slurry characteristics fed into the iASJ, or dynamic stability, were initially conducted using only water, revealing a narrow stability window. Subsequent investigations involved slurry additives and experiments to determine the stability characteristics based on the flow restrictions by the valve position. The thresholds for a stable flow of the slurry were defined by adjusting the valve position from fully closed to fully open, which was recorded as a percentage value indicating valve openness. Supplementary experiments were conducted to judge the jet stability of processing glass specimens.

3 Results

The initial phase of the study, as described in Table 1, revealed a dependency of the flow rate on the slurry viscosity (Fig. 2). In a pure water jet, the flow rate exhibited an almost linear increase relative to pressure, followed by a gradual stabilisation. Similar behaviour is observed when the slurry contains low concentrations of thickener, but the stabilisation occurs at higher pressures. Conversely, higher slurry viscosity resulted in a relatively constant flow rate throughout the experiment, indicating reduced flowability and a less dependent characteristic on the applied pressure.

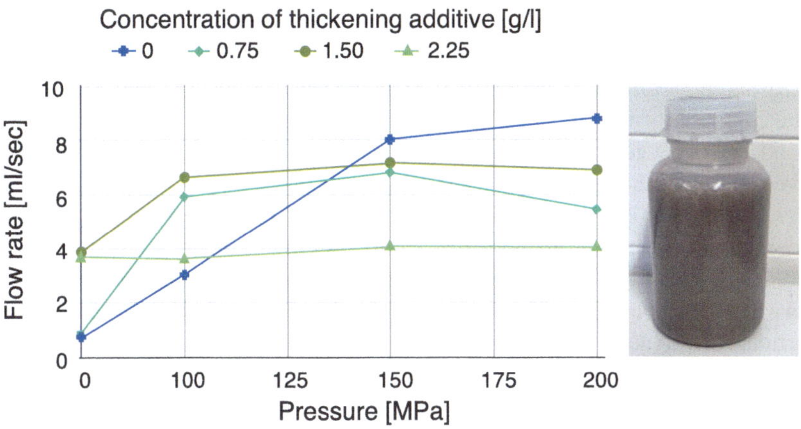

Fig. 2. Slurry flow rate in dependence on system pressure based on different concentrations of the thickener.

Another study concluded that using additives minimally impacts machining results, suggesting little effect on the MRR or profile geometry [29]. The data collected from the jet-shape analysis of preliminary experiments concluded a linear increase in the jet divergence without significant changes in jet diameter for batches of different concentrations of the thickening additive. However, a minor rise in jet diameter was observed when comparing no and maximum thickener concentrations. During erosion analysis, other variables, such as the homogeneity of the jet mixture, could cause more significant changes than those observed from the free-air jet.

In connecting fluid characteristics of the slurry and perceived erosion consistency, the abrasive settlement was crucial in understanding slurry behaviour when introducing thickener into the mixture. The primary objective was to track the time of abrasive particles' settling. Six batches were prepared, and photos were taken periodically to document the settling process. Adding thickener to the mixture prolonged the abrasive settlement time due to a change in viscosity. The latter must be matched with the gravitational effect acting on different particle sizes. For example, a significant portion of the bottle's abrasive solution settled in the first eight minutes, which contained larger particles and a low thickener concentration. Notably, the results suggested an upper threshold of viscosity beyond which the settlement time dramatically exceeds the experimentation time. Furthermore, this experiment shed light on how grain size influences settlement time, indicating that finer abrasives remain in suspension longer than larger-grained counterparts. This insight is valuable for understanding slurry behaviour and optimising the viscosity of the slurry composition for industrial applications.

The investigation conducted in this work highlights the crucial importance of jet stability in preparation for the main experiments of a systematic comparison. Stability was attained through a series of experiments involving varying valve opening positions. An area of stability could be observed around 20…50% restriction of the flow intake by the regulation valve along the investigated viscosities, with a narrow spot of exception for injection of plain water (Fig. 3). The graph delineates the maximum valve opening (opening threshold) before instability occurred and the minimum valve opening necessary for a stable jet (restricted threshold). Deviations from these thresholds result in inhomogeneous slurry flow or even pulsating of the jet. The results indicated that viscosity and flow limitation hydraulic parameters can stabilise the slurry flow.

The computational analysis of the images taken regarding the jet shape showed an independence of jet geometry from the fluid's viscosity. In contrast, jet pulsation can be observed on the machined glass pieces in Fig. 4, where a stable jet produced continuous grooves. It is imperative to distinguish between the two types of jet instability observed. In one scenario, excessive valve opening led to very low-frequency vibrations of great amplitude, visible on the machine as vibration of the slurry hose. Conversely, instability occurs when the valve is closed beyond the threshold, resulting in insufficient slurry flow proportional to the vacuum inside the mixing chamber or the build-up of gaseous bubbles. The experiments could not differentiate between the pressure-dependent expansion of microscopically tiny air bubbles or a drop below the vapour pressure of the slurry. However, the gaseous volume separated the slurry flow, leading to a pulsating jet that inconsistently dispenses abrasive material.

An appropriate configuration for the iASJ system was identified through both static and dynamic stability of the slurry feed and its effect on the machining result. As an intermediate result, a concentration of 2.5 g/l of the thickening additive was chosen for the following experiments. This concentration yielded favourable abrasive settlement times and stability in the slurry feed or jet shape.

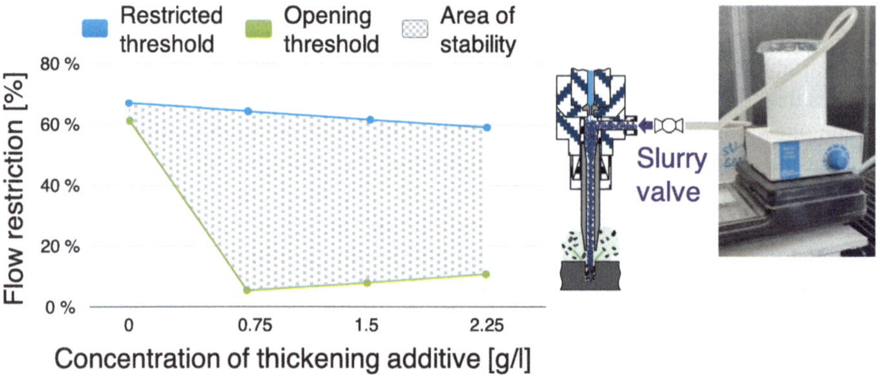

Fig. 3. Determined jet stability based on the valve position and viscosity of the slurry by adding thickening additives.

Milling with water jets requires good controllability of the removal depth. Here, the workpiece's absolute depth values were analysed for an initial assessment. The average depth values in each line were analysed based on the input variables, showing an increase proportional to the abrasive concentration and system pressure. As expected, a reduction in erosion depth was observed with increased feed rates. Preliminary experiments focused on the formation of a stable jet. The average values of erosion depth and and the standard deviation of the depth values were taken into account to judge over jet stability. Here, the erosion depth reliably indicated the material removal capacity in each slurry mixture. As slurry concentration and pressure increased, groove depth also

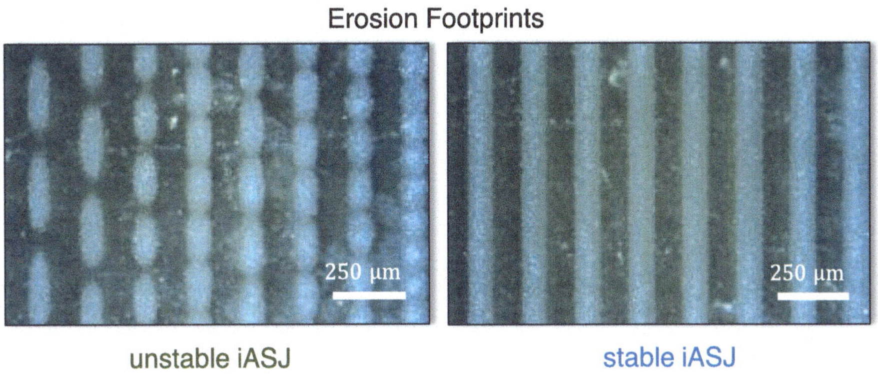

Fig. 4. Unstable (left) and stable (right) jet material-removing characteristics demonstrated on glass test pieces.

rose. An unstable removal process could be assumed for pressure levels of 100 MPa and 200 MPa, as the depth values fluctuate comparatively more. In contrast, values at 150 MPa show less variation overall. This was accompanied by the corresponding standard deviation values at this pressure level, particularly with an abrasive concentration of 5. However, this pressure may not be optimal for stable results with an abrasive concentration of 7.5%, but it yielded satisfactory results of 10%. Different abrasive concentrations were investigated to explore the optimal range within the investigation. The most promising results were obtained at 150 MPa pressure and abrasive concentrations of around 5%.

Finally, the depth consistency was enhanced with the analysis of surface texture. Here, the maximum height Sz of the filtered profile was evaluated within an area of 33% of the profile's bottom centre line. Following the analysis of the depth values, an abrasive concentration of 5% exhibited better surface quality values than others, indicating an optimum abrasive concentration and higher feed rate could help improve surface quality within a certain level. However, excessively high abrasive concentrations can deteriorate these values, with comparably good results for abrasive concentrations between 7.5% and 10%.

Further evaluation was made to obtain the erosion performance by the material removal rate (MRR) condensed precisely by the integration of the depth measurements from the 3D footprint at the zirconia specimens. The literature [21] postulated a change of wear mechanism that occurs at a critical speed. This change from sliding wear to impact wear at higher velocities can be stated as a slope in Fig. 5 due to a change in the angle of the AWJ cutting front. Although a sudden change in slope could be observed with iASJ in Fig. 6, the MRR characteristics show distinctive differences.

Surface Roughness measurements of the kerf are visualised over the feed rate in Figs. 7 and 8. As can be seen, the differences between AWJ and iASJ are minor, and surface roughness behaviour is comparable. In both cases, there is a decrease in surface roughness with increasing feed rate until a minimum at feed rates of 750 mm/min is reached.

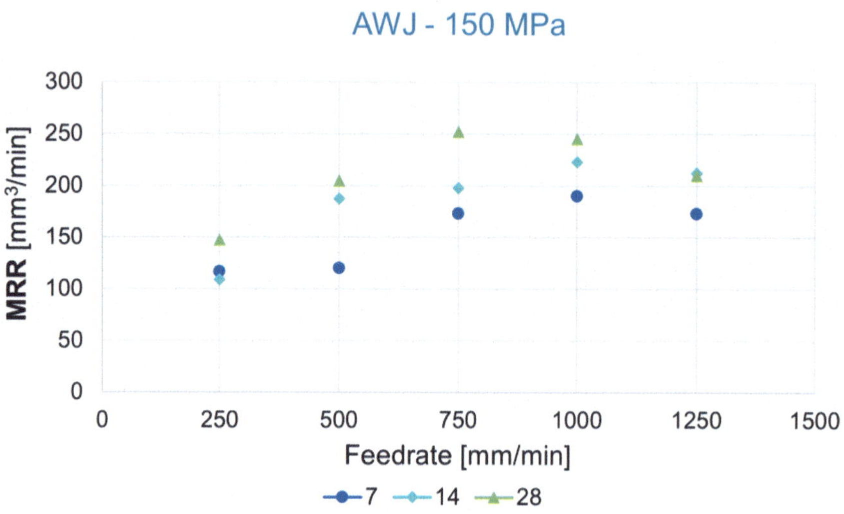

Fig. 5. AWJ erosion performance – Material Removal Rate (MRR) in dependence on the feed rate.

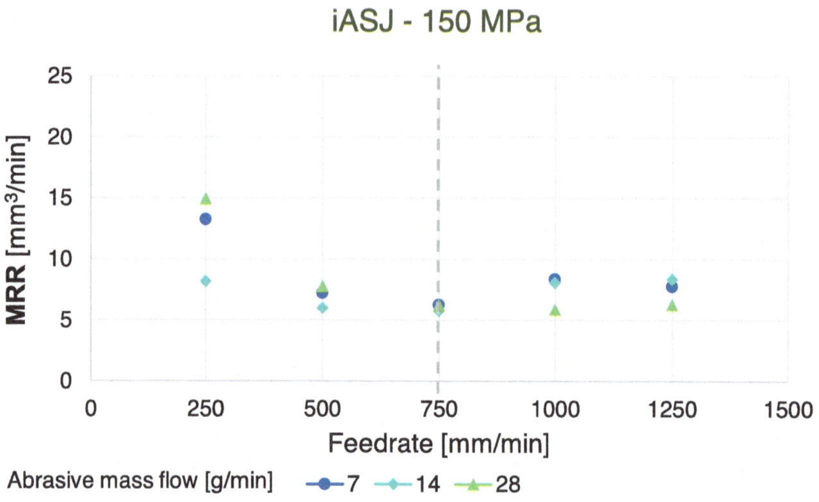

Fig. 6. iASJ erosion performance – Material Removal Rate (MRR) in dependence on the feed rate.

In addition, the cut edge profiles were also analysed in this study (Figs. 9, 10, 11, and 12), which indicates that significantly more material fractures occurred when processing the brittle-hard ceramic with an AWJ compared to the iASJ configuration. The analysis of the results obtained on soda-lime glass clearly showed that the effect of the gaseous phase of the AWJ led to more significant dynamic mechanical stress on the material compared to a two-phase iASJ. The edge zone of the profiles processed using AWJ showed larger dimensions of the individual chipping, interspersed with recognisable cracks or brittle

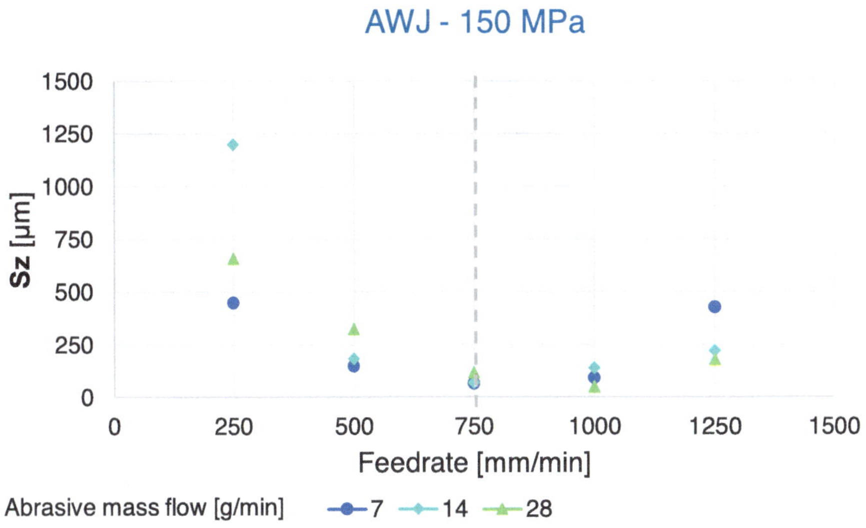

Fig. 7. AWJ surface texture – maximum surface height (Sz) in dependence on the feed rate.

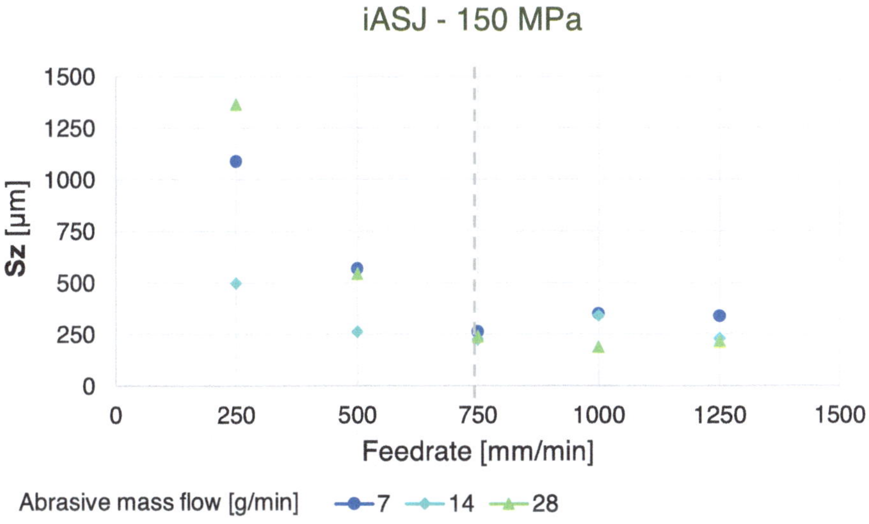

Fig. 8. iASJ surface texture – maximum surface height (Sz) in dependence on the feed rate.

breakouts. In comparison, at a system pressure of 150 MPa, the effective loads on the component surface on the iASJ side appeared to be significantly lower but more evenly distributed in frequency. This impression was equalised at an increased system pressure of 200 MPa, which could indicate further fluidic phenomena of the turbulent mixing of an injection jet head, e.g. cavitation formation. These results were obtained across multiple pressure levels, abrasive concentrations, and travel speed ranges.

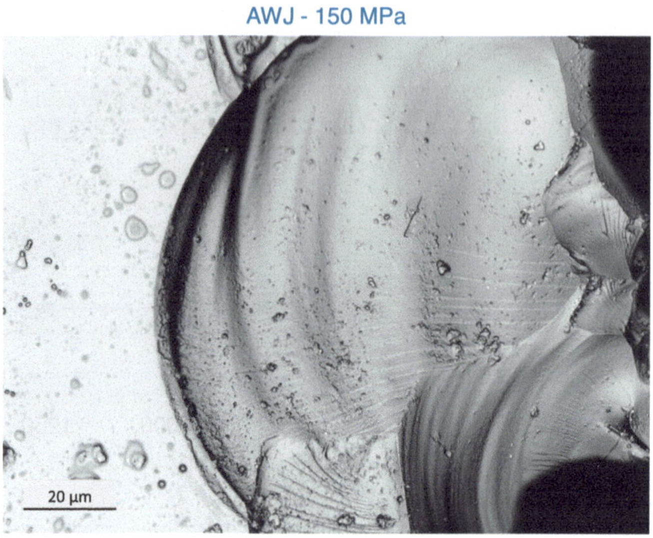

Fig. 9. Top view on the profile edge: breakout characterisation of AWJ erosion on glass material machined with 150 MPa.

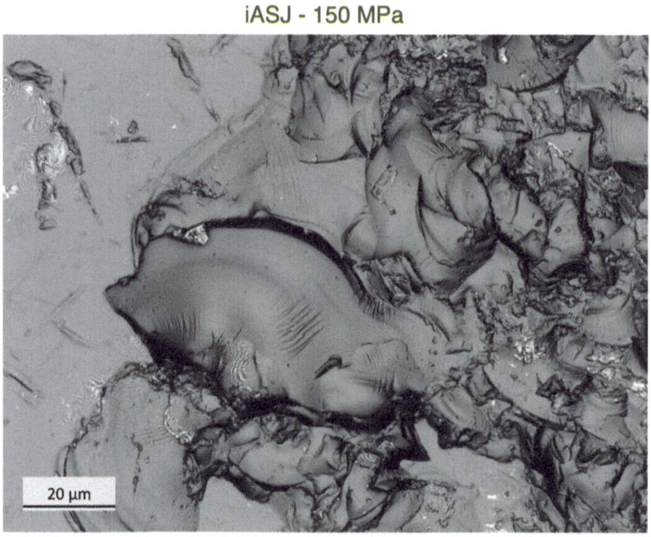

Fig. 10. Top view on the profile edge: breakout characterisation of iASJ erosion on glass material machined with 150 MPa.

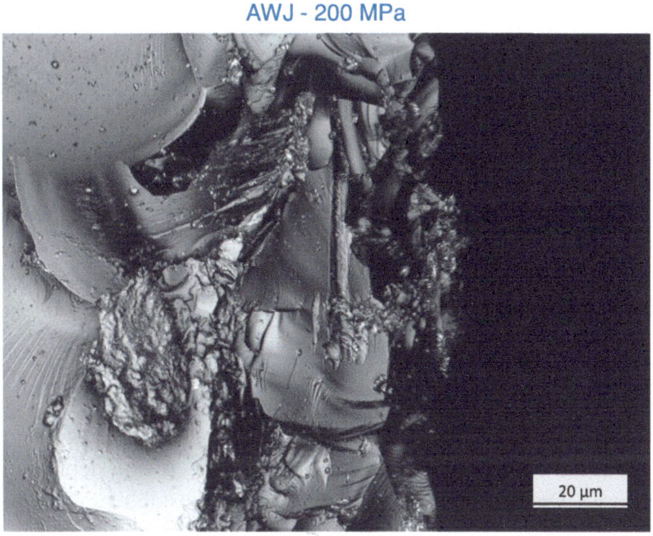

Fig. 11. Top view on the profile edge: breakout characterisation of AWJ erosion on glass material machined with 200 MPa.

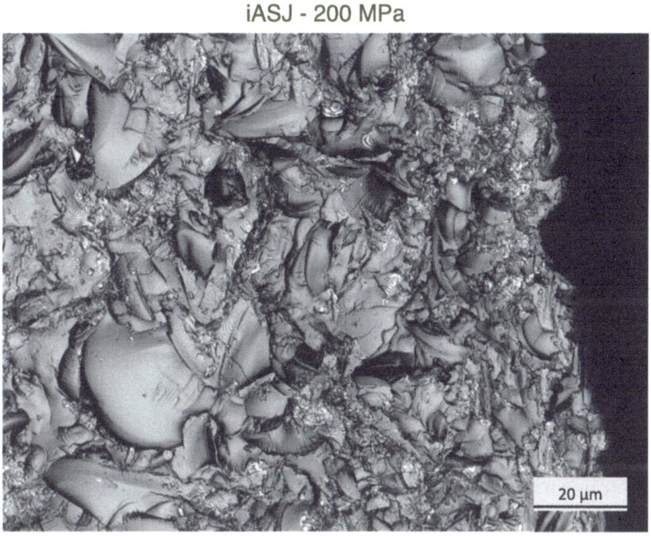

Fig. 12. Top view on the profile edge: breakout characterisation of iASJ erosion on glass material machined with 200 MPa.

4 Conclusion

This work focussed on the influence of the jet composition on the processing result regarding the physical phase components. The industrially established abrasive water jet (AWJ) as a three-phase jet tool was compared with the two-phase injection abrasive slurry jet (iASJ) variant. The latter used the identical jet head, but instead of an abrasive-laden air stream, a suspension without an air phase was sucked in. In the first part of the experimental investigations, the fluidic variables influencing the jet formation of the iASJ were analysed in terms of stability. Subsequently, the process comparison focussed on the influence of the air phase on the fragmentation effects that occur during the processing of brittle-hard materials; in addition to a densely sintered technical zirconium oxide ceramic (ZrO_2), a soda-lime glass was used as an analogue material for simplified analysis.

In the experimental part of this study on the optimisation of iASJ, the dependence of the flow rate and its influence on the stability of the jet tool on the viscosity of the slurry was shown. Adding thickeners prolonged the sedimentation time of the abrasive particles and allowed first estimations towards an optimal viscosity for industrial applications. The stability of the jet tool could be controlled by the fluid's viscosity and the slurry flow's hydraulic limitation.

The eroded profiles showed increased removal depth proportional to the abrasive concentration and the system pressure. As expected, the erosion depth decreased with increasing feed rate. Due to fluctuating depth values, an unstable removal process could be assumed for both materials in the waterjet system at a system pressure between 100 MPa and 200 MPa. In contrast, values at 150 MPa showed less variation overall. Here, the standard deviation of depth values at 150 MPa and an abrasive concentration of 5% was low and stable. However, this pressure may not be optimal for stable results at an abrasive concentration of 7.5%, but it gives satisfactory results at 10% abrasive content of the slurry.

A final comparison of the erosion performance based on the material removal rate (MRR) showed apparent differences between the jet systems. On the one hand, a change in the removal mechanisms at a critical feed rate of around 750 mm/min is assumed in both variants but with opposing curves. A supplementary visual analysis of the profile edge zone revealed significantly more material fractures with an AWJ than with the iASJ configuration. Likely, the influence of the waterjet's composition on the gas phase could play an essential role in the fragmentation and the breakout of brittle-hard material agglomerates. The surface texture analysis showed similar behaviours for both AWJ and iASJ results based on the values of surface height Sz. The surface parameters were also dependent on the feed rate. For AWJ and iASJ, the surface roughness decreased with increasing feed rate until a local minimum was reached around 750 mm/min feed rate.

Despite the benefits of iASJ for machining sensitive materials, its industrial use still lacks experience, underscoring the necessity for additional research in this area. Future studies should focus on understanding the erosion mechanism regarding the AWJ's phase composition so that the material outbursts can be used deliberately in a potential multi-stage process for efficient material removal for special applications. Furthermore, a focus is on developing automated systems to monitor jet stability and control process variables. Additionally, exploring the fluid dynamics could help to characterise and optimise AWJ milling with appropriate waterjet systems.

References

1. Emelogu, A., Marufuzzaman, M., Thompson, S.M., Shamsaei, N., Bian, L.: Additive manufacturing of biomedical implants: a feasibility assessment via supply-chain cost analysis. Addit. Manuf. **11**, 97–113 (2016). https://doi.org/10.1016/j.addma.2016.04.006
2. Borrmann, J.P., Dadgar, M., Döring, J.-E., Herrig, T., Bergs, T.: Empirical modeling of abrasive waterjet process for controlled depth machining of dense segmented ceramic thermal barrier coatings. Procedia CIRP **102**, 355–360 (2021). https://doi.org/10.1016/j.procir.2021.09.061
3. Axinte, D.A., Karpuschewski, B., Kong, M.C., Beaucamp, A.T., Anwar, S.: High Energy Fluid Jet Machining (HEFJet-Mach): from scientific and technological advances to niche industrial applications. CIRP Ann. Manuf. Technol. **2**, 751–771 (2024). https://doi.org/10.1016/j.cirp.2014.05.001
4. Grote, K.-H., Antonsson, E.K.: Springer Handbook of Mechanical Engineering. Springer, New York (2009)
5. Alsoufi, M.S.: State-of-the-art in abrasive water jet cutting technology and the promise for micro- and nano-machining. IJMEA **5**(1), 1 (2017). https://doi.org/10.11648/j.ijmea.20170501.11
6. Schüler, M., Herrig, T., Bergs, T.: A study on abrasive waterjet multi-stage machining of ceramics. Procedia CIRP **108**, 770–775 (2022). https://doi.org/10.1016/j.procir.2022.03.119
7. Technological Analysis of Different Abrasive Water Jet Machining Strategies for Blisk Roughing
8. Saurabh, S., Tiwari, T., Nag, A., Dixit, A. R., Mandal, N., Das, A. K., Mandal, A., Kumar Srivastava, A.: Processing of alumina ceramics by abrasive waterjet-an experimental study. Materials Today: Proceedings (2018)
9. Hassan, A. I., Kosmol, J.: Dynamic elastic-plastic analysis of 3D deformation in abrasive waterjet machining. Journal of Materials Processing Technology (2001)
10. Bergs, T., Schüler, M., Dadgar, M., Herrig, T., Klink, A. (Hrsg): Investigation of Waterjet Phases on Material Removal Characteristics. Elsevier BV (2020)
11. Fowler, G.: Abrasive Waterjet - Controlled Depth Milling of Titanium Alloys, University of Nottingham (2003)
12. Zhang, S., Wu, Y., Wang, S.: An exploration of an abrasive water jet cutting front profile. Int. J. Adv. Manuf. Technol. (2015)
13. Wang, J.: The effect of jet impact angle on the cutting performance in AWJ machining of alumina ceramics. Key Eng. Mater. **238–239**, 117–122 (2003)
14. Nguyen, T., Wang, J.: A review on the erosion mechanisms in abrasive waterjet micromachining of brittle materials. Int. J. Extrem. Manuf. **1**(1), 12006 (2019). https://doi.org/10.1088/2631-7990/ab1028
15. Schüler, M., Dadgar, M., Herrig, T., Klink, A., Bergs, T.: Experimental investigation of abrasive properties in waterjet machining. Procedia CIRP **101**, 210–213 (2021). https://doi.org/10.1016/j.procir.2021.03.128
16. Bergs, T., Borrmann, J. P., Schüler, M., Herrig, T., Döring, J.-E. (Hrsg): Pure waterjet controlled depth machining for stripping ceramic thermal barrier coatings on turbine blades. Elsevier BV (2019)
17. Zhao, K., Gao, C., Liu, Z., Guo, C.: Investigation of removing thermal barrier coatings from Nickel based super-alloy using abrasive water jet. IOP Conf. Ser.: Mater. Sci. Eng. **452**, 22112 (2018). https://doi.org/10.1088/1757-899X/452/2/022112
18. Putz, M., Dix, M., Morczinek, F., Dittrich, M.: Suspension Technology for Abrasive Waterjet (AWJ) cutting of ceramics. Procedia CIRP **77**, 367–370 (2018). https://doi.org/10.1016/j.procir.2018.09.037

19. Ramesha, K., Santhosh, N., Kiran, K., Manjunath, N., Naresh, H.: Effect of the process parameters on machining of GFRP composites for different conditions of abrasive water suspension jet machining. Arab. J. Sci. Eng. **44**(9), 7933–7943 (2019). https://doi.org/10.1007/s13369-019-03973-w
20. Tazibt, A., Parsy, F., Abriak, N.: Theoretical analysis of the particle acceleration process in abrasive water jet cutting. Comput. Mater. Sci. **5**(1–3), 243–254 (1996). https://doi.org/10.1016/0927-0256(95)00075-5
21. Nouraei, H., Kowsari, K., Papini, M., Spelt, J.K.: Operating parameters to minimise feature size in abrasive slurry jet micro-machining. Precis. Eng. **44**, 109–123 (2016). https://doi.org/10.1016/j.precisioneng.2015.10.008
22. Kowsari, K., Schwartzentruber, J., Spelt, J.K., Papini, M.: Erosive smoothing of abrasive slurry-jet micro-machined channels in glass, PMMA, and sintered ceramics: Experiments and roughness model. Precis. Eng. **49**, 332–343 (2017). https://doi.org/10.1016/j.precisioneng.2017.03.003
23. Kowsari, K., Sookhaklari, M.R., Nouraei, H., Papini, M., Spelt, J.K.: Hybrid erosive jet micro-milling of sintered ceramic wafers with and without copper-filled through-holes. J. Mater. Process. Technol. **230**, 198–210 (2016). https://doi.org/10.1016/j.jmatprotec.2015.11.027
24. Hashish, M.: Optimisation Factors in Abrasive - Waterjet Machining. J. Eng. Ind. **113** (1991)
25. Perec, A., Pude, F., Kaufeld, M., Wegener, K.: Obtaining the selected surface roughness by means of mathematical model based parameter optimization in abrasive waterjet cutting. SV-JME **63**(10), 606–613 (2017). https://doi.org/10.5545/sv-jme.2017.4463
26. Perec, A., Pude, F., Kaufeld, M., Wegener, K.: Obtaining the selected surface roughness by means of mathematical model based parameter optimization in abrasive waterjet cutting. J. Mech. Eng. (2017)
27. El-Domiaty, A.A., Abdel-Rahman, A.A.: Fracture mechanics-based model of abrasive waterjet cutting for brittle materials. Int. J. Adv. Manuf. Technol. **13**(3), 172–181 (1997). https://doi.org/10.1007/BF01305869
28. Zeleňák, M., Říha, Z., Jandačka, P.: Visualisation and velocity analysis of a high-speed modulated water jet generated by a hydrodynamic nozzle. Measurement **159**, 1–9 (2020). https://doi.org/10.1016/j.measurement.2020.107753
29. Kowsari, K., James, D.F., Papini, M., Spelt, J.K.: The effects of dilute polymer solution elasticity and viscosity on abrasive slurry jet micro-machining of glass. Wear **309**(1–2), 112–119 (2014). https://doi.org/10.1016/j.wear.2013.11.011

Experimental Characterisation of the Secondary Jet in Abrasive Water Jet Machining on the Example of an Oxide Ceramic

Manuel Schüler(✉) and Manuel Walotka

HydroMill GmbH, Arndtstr. 30, 52064 Aachen, Germany
info@hydromill.de

Abstract. In this work, the secondary abrasive water jet was investigated experimentally. The secondary jet is the deflection of the abrasive water jet in the workpiece engagement. The focus was on analysing the direction and damage potential of the secondary jet when machining an oxide ceramic. To better understand the deflection of the abrasive water jet, the formation of the cutting front was first examined utilising analogy tests on transparent glass material. In the subsequent analysis of the secondary jet direction, a comparison was made between the analogy material and an oxide ceramic. For this purpose, the average exit and opening angles of the secondary jet were evaluated. The investigation of the damage potential of the secondary jet followed this.

Keywords: Secondary Water Jet · Abrasive Water Jet (AWJ) · AWJ Milling

1 Introduction

The increasing use of high-performance materials characterises modern production. At the same time, the entire manufacturing environment is changing. High-performance materials are often difficult to process conventionally. Alternative processes usually do not have the necessary level of development, and the performance of the process is typically in competition with an upcoming orientation of sustainability. Abrasive water jet (AWJ) machining is a high-performance method of great flexibility, ranging from cutting to shape or surface machining of high-performance materials. Therefore, AWJ is a promising alternative for roughing and micro-machining difficult-to-cut materials. Ceramic materials have great potential in various applications and can be efficiently machined using AWJ, contrary to other methods. However, this technology has not yet been fully developed compared to conventional processes.

A significant challenge in the practical application of this technology is the time- and location-critical effect of the AWJ tool engaging the workpiece. Due to many influencing variables, the implementation of the AWJ in computer-aided manufacturing (CAM) has not been realised. A fundamental challenge for shape machining, or milling with AWJ, is the non-solid characteristics of a continuously changing nature. Reflections of the jet lead to unwanted contour damage because the jet's energy is not fully transferred onto primary

L. Sitek et al. (Eds.): Advances in Water Jetting II, LNME, pp. 209–227, 2025.
https://doi.org/10.1007/978-3-031-72778-8_17

contact and releases its energy at secondary contact. A reliable basis for integrating the AWJ characteristics in computer-aided manufacturing standards that cover the secondary jet's direction and effect is needed to deal with this challenge.

2 State of the Art

Due to the flexibility of the process and the low thermal or mechanical loads on the workpiece, high-pressure water jets are used in many industrial sectors [13]. Efforts were made to transfer the process benefits towards water jet milling [6, 10, 14, 17]. High-strength materials, like water jet cutting, can be efficiently machined in various applications. Water jet machining for defined shape removal is divided into two approaches. On the one hand, there is an approach of cutting through the entire workpiece. Conversely, there is a milling approach of cutting to a defined depth. Typically, the 3D composition of the tool path is similar to that of conventional milling [17].

2.1 Microscopic Mechanism of the Abrasive Water Jet

Abrasive particles are mandatory when machining high-strength materials such as alloys or ceramics. For this purpose, the impulse of the pure water jet is transferred to the abrasive particles, initiating erosion in terms of micro-abrasions. The sum of the impacts leads to a macroscopically observable material removal [3].

Finnie and Bitter carried out a fundamental analytical description of the removal mechanisms by individual particle impacts. According to Finnie, the particle mass, the yield point of the target material, the force ratio in the horizontal and vertical direction, the removal depth, the impact angle, and the particle velocity influence the material removal. When evaluating Finnie's calculated results, there was good agreement between the results for shallow impact angles below 45° on a ductile target material. The removal process changes on materials with brittle material properties and steeper impact angles, and the calculated values differ from the experimental values due to crack formation and accumulation. In addition, disintegration processes occur that cannot be described by the path curve of a particle [7–9].

Bitter's studies on erosion mechanisms are founded on the energetic consideration of the erosion process of freely moving particles. Categorisations were made on two wear mechanisms - sliding wear along the surface and impact wear. Both wear mechanisms coincide during contact between the particle and the target material, depending on the impact angle and material properties. At steep angles of particle impact, sliding wear is the predominant mechanism, while at shallow angles, impact wear accounts for most of the erosive material removal. Similar to the findings of Finnie, particle velocity significantly influences material removal [1, 2].

The interaction between particles and target material is also differentiated according to the types of contact at the cutting front (Fig. 1). Henning defines three kinds of particle interaction: Impact, sliding and rolling along the cutting front. In the impact process, the solid particle transfers part of its kinetic energy to the cut front, thereby generating material removal. Sliding and rolling cause material removal due to the normal force of the abrasive particles in the direction of the contact surface. After primary contact, the particles' direction of movement and rotation change. Different removal mechanisms are used when contact is made again [12].

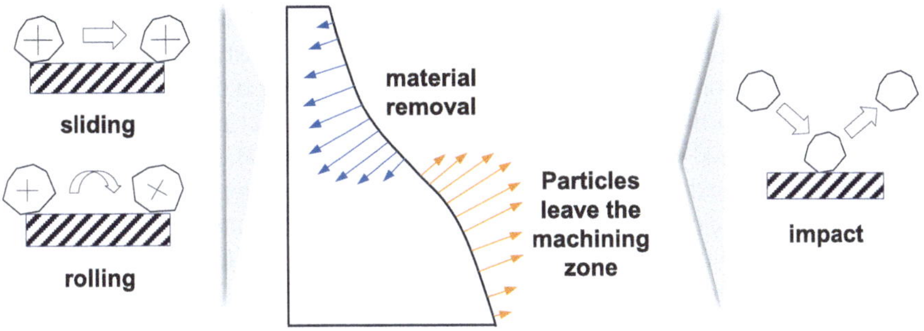

Fig. 1. Particle contact possibilities [12].

Secondary particle contact in the form of sliding and rolling generally only occurs on concave cutting front areas, as the normal force of the particle depends on the centrifugal force of the abrasive water jet (Fig. 1). On convex surfaces within the machining zone, the abrasive water jet is not in contact with the material surface. Examples of convex surfaces are steps on the cutting front [12].

2.2 Macroscopic Mechanism of the Abrasive Water Jet

Hashish carried out a macroscopic approach to describe the shape of the AWJ tool. The temporal development of kerf formation was analysed using transparent polymethyl methacrylate (PMMA). Hashish divided the removal mechanism into three horizontal process phases and three vertical zones. Each zone characterises the existing mechanism within a cycle. Bitter defined wear mechanisms to describe the removal mechanism in the respective area. In the first zone, the particles impact the cutting front at a sharp angle. The present removal mechanism is sliding wear. In the second zone, the AWJ loses intensity, and the curvature at the cutting front increases with the growing depth of the kerf. An increase in curvature causes a step where the abrasive particles hit the surface with increased impact wear. As the process continues, the impact angle of the AWJ increases until a maximum kerf depth is reached. In the third zone, the cutting process develops into a drilling process. Impact wear dominates during the drilling process. The individual phases describe the phenomena of the removal process in the feed direction [11].

Zeng and Kim investigated the deflection of particles at the steps of the cut front of aluminium oxide ceramics and stainless steel. For this purpose, kerf cuts with sudden jet interruption were carried out. The cut front was then examined using scanning electron microscopy and divided into a primary and a secondary zone of influence. In the primary area of influence, the particles hit the workpiece directly and cause material removal. In the secondary location of influence, the secondary jet mainly determines the material removal. The kerf formation process takes place in five sections (Fig. 2).

The abrasive particles hit the surface directly in the first section of the cutting front until the abrasive water jet is entirely on the workpiece surface (Sect. 1). The resulting jet deflection causes the material removal to change abruptly. As a result, the curvature at the transition to the secondary influence zone adjusts. This results in the first micro-step (Sect. 2). A deflection of the jet at the foremost jet front protects the newly initiated step. The protective reflection causes the step to grow continuously. The step only grows further into the material due to the increasing material removal at the step edge (Sect. 3). Simultaneously with the formation of the step, the primary jet penetrates further into the material in the feed direction and is deflected again (Sect. 4). Another micro-step is formed. However, the deflected jet initially suppresses the development of the new step. Due to the prevalence of an optimal erosion state, there is no change in the superior curvature. The first formed step and subsequent micro-steps continue their growth into the material to a specific depth. When the optimum erosion condition is no longer present at the cutting front, the final macroscopic step with a particular curvature and depth results (Sect. 5). This process is continuous and independent of the material [19].

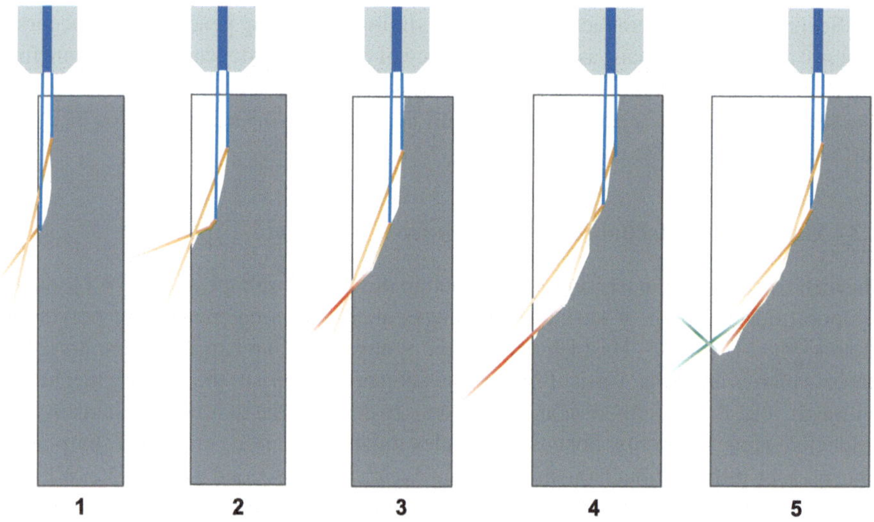

Fig. 2. Step formation process according to [19].

Experimental investigations by Männel and Uhlmann on the opening angle at the exit of the secondary jet on concave and convex workpiece samples illustrated an extensive range of influence of the damage potential caused by the secondary jet. The opening angle α (Fig. 3) correlates with the remaining jet intensity. The larger the opening angle of the secondary jet, the lower the secondary jet intensity. Due to the focusing of the secondary jet at small exit angles, the jet intensity remains high. Large opening angles mainly occur with large impact angles of the primary jet, as with pure impact wear. The opening angle is smaller for steeper impact angles of the primary jet [18].

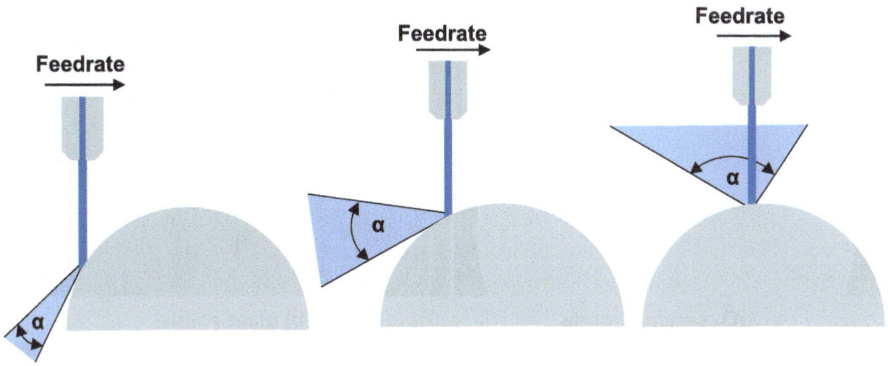

Fig. 3. Opening angle of the secondary jet according to [18].

3 Experimental Setup

3.1 Basic Investigations on the Kerf Geometry

Fundamental knowledge of the damage potential as a function of the secondary jet direction is essential for the process design of material removal by the abrasive water jet. Dense-sintered yttrium-stabilised zirconia (3Y-TZP) was used. In addition to a plate, a ronde of 100 mm diameter was used to visually examine the direction of the secondary jet at different jet exit angles (Fig. 4). A soda-lime glass was used to substitute the opaque ceramics that may not be optically inspected while processing. The main component in this glass group is silicon dioxide (SiO_2) [4, 5]. Notably, most physical properties, like the density of this glass material (2.5 g/cm^3), were lower than that of zirconia (6 g/cm^3).

Similarly, the Mohs hardness or the compressive strength is way lower for glass. Although the erosion results are different, both materials show predominantly brittle properties. Finally, the remaining erosion characteristics of the secondary jet were analysed by the erosion footprint on a synthetic polymer (Fig. 4d). A transparent Polymethylmethacrylate (PMMA) material was selected to reveal a more sensitive behaviour than the original zirconium dioxide sample. The PMMA surface was tinted black to detect scattering areas of the secondary abrasive water jet.

A preliminary parameter study was conducted to ensure the analogy in material behaviour. For this purpose, the cutting fronts on the oxide ceramic and the glass material were analysed to compare the macroscopic kerf formation. At the kerf bottom, the related angles of the cutting front (CF) were measured for both target materials (Fig. 5). CF is defined relative to the workpiece surface. The kerf depth is the perpendicular distance of the reference plane to the kerf base. A particular freeze of continuous erosion was achieved with a sudden shut-off of the jet while it passed over the sample. This procedure ensures that the cutting front retains its actual gradient. Table 1 overviews the experimental design and the chosen process parameters. A successive approach was used with pressure and feed rate adjustment to compromise erosion performance and process stability.

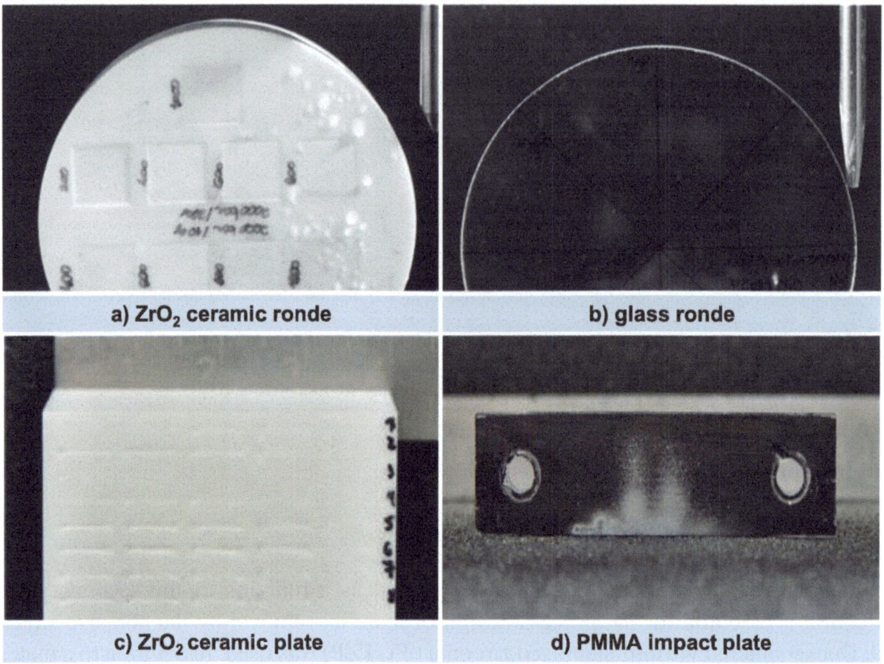

Fig. 4. Sample material.

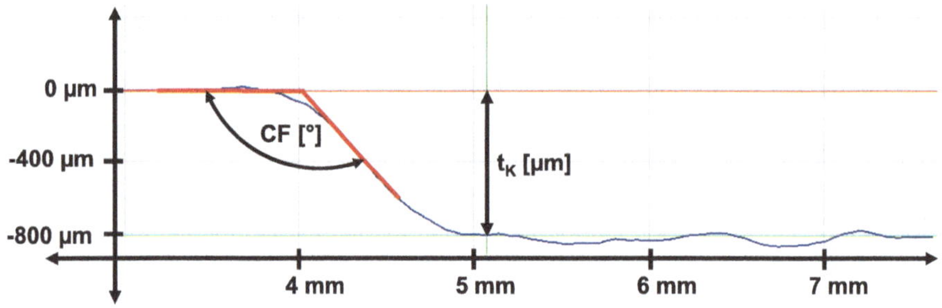

Fig. 5. Measurement of the cutting front angle CF and the kerf depth t_K.

The abrasive used for all tests is white fused alumina, consisting of over 99% pure aluminium dioxide (Al_2O_3). Compared with standard garnet abrasive, the physical properties of this sharp-edged abrasive are improved. It shows a higher hardness of about 9 Mohs and a raised melting point of the primary material around 2050 °C. This synthetic abrasive was sieved between 200 µm and 212 µm to maintain a narrow variation of grit size [15, 16].

Table 1. Preliminary test parameters.

Constant Parameter	Jet Angle [°]		90	
	Orifice diameter [mm]		0,12	
	Focus tube diameter [mm]		0,54	
	Massflow [g/min]	20	40	60
Parameter Glas Trial 1	Pressure [MPa]	50	100	150
	Feedrate [mm/min]	1000	1500	2000
Parameter Glas Trial 2	Pressure [MPa]	100	125	150
	Feedrate [mm/min]	2000	2500	3000
Parameter ZrO_2 Trial 1	Pressure [MPa]	100	150	200
	Feedrate [mm/min]	500	1000	1500
Parameter ZrO_2 Trial 2	Pressure [MPa]	100	150	200
	Feedrate [mm/min]	200	600	1000

3.2 Investigation of the Secondary Abrasive Water Jet Direction

A circular disk was used to investigate the secondary jet direction based on the entry angle (Fig. 6). Machining the circular disk along the outer contour ensured a smooth transition of all disk angles (γ) from 15° to 165°. The angle range from 15° to 90° represents backward machining by the abrasive water jet, while the content from 90° to 165° equals forward machining. The tests were conducted on glass and oxide ceramic disks to verify the secondary jet's behaviour (Fig. 4). This approach allowed for visual examination through optical images and yielded material-specific results on the oxide ceramic. The glass disk underwent visual evaluation by comparing exit angles at different kerf depths, which were increased by elevating the jet intensity. Tests on the zirconia ceramic followed a two-stage test plan, varying parameters such as pressure, feed rate, and mass flow. This approach identifies the influencing variables on the mean secondary jet exit angle (β) within a limited number of test runs.

3.3 Investigation of the Effect of the Secondary Abrasive Water Jet

A test setup was developed that characterises the secondary jet on an impact plate positioned at a defined distance of 5 mm. The impact plates are aligned parallel to the kerf and pressed perpendicularly onto the oxide ceramic surface to prevent the secondary jet from flowing underneath the sample. The test utilised PMMA plates (Fig. 4d) for their plastic properties, providing the necessary sensitivity to measure damage potential across various abrasive water jet intensities. The sample surface was tinted black to visualise the overall effect of the secondary jet contact on the impact plates, enhancing the visibility of even slight contacts outside the central damage zone. The secondary jet is generated solely through jet deflection in the processing zone.Consequently, the primary abrasive

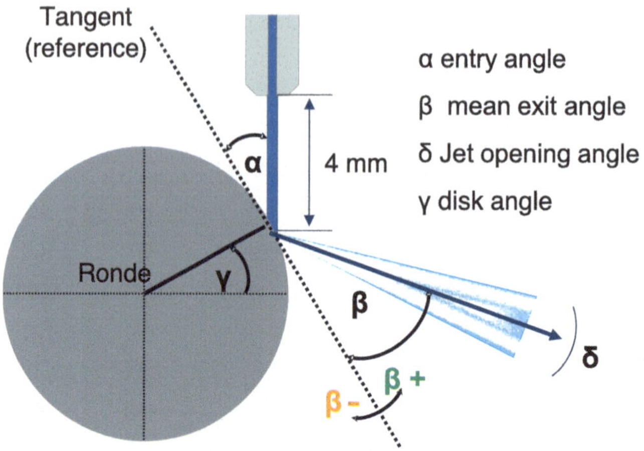

Fig. 6. Definition of the jet angles.

water jet was applied to a planar oxide ceramic sample (Fig. 4c) at different angles (Fig. 7). A design of experimental approaches on the oxide ceramic involves varying pressure, feed rate, mass flow, and jet angle to determine the influencing variables on the damage potential of the secondary jet. The design used two levels for the pressure, feed rate and mass flow parameters. The jet angle was divided into five levels per run - 15°, 30°, 45°, 60° and 75°. This division of jet angle levels results from the preliminary tests and the directional investigation of the secondary jet. The selected parameter levels lead to eight distinct parameter combinations with five angle settings. A total of 40 kerfs were created to investigate the damage potential.

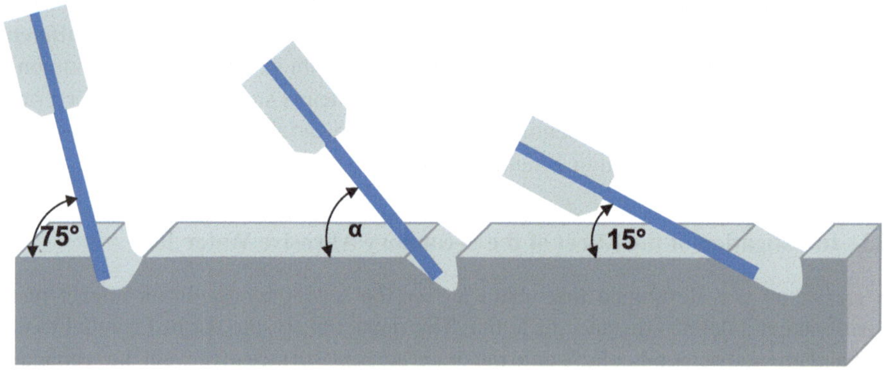

Fig. 7. Impact angle variation.

4 Results and Discussion

4.1 Basic Investigations on the Kerf Geometry

The preliminary tests show that the kerf angles CF are comparable for zirconium dioxide and glass with the same kerf depth (Fig. 8). The measurement points of the kerf angle CF are plotted on the ordinate. The kerf depth values are entered on the abscissa. Furthermore, the abscissa is scaled logarithmically to display the shallower kerf depth ranges. A trend in the kerf angles with increasing kerf depth can be seen. In the field from 20 µm to 80 µm, there are only measured values for kerfs on the zirconium dioxide ceramic. The kerf angles of all measuring points are between 172° and 175°. The lowest kerf depth on glass is 100 µm, with a kerf angle of around 170°.

Further measurements on the oxide ceramic, in the kerf depth range between 85 µm and 100 µm, form a cluster around CF of the lowest kerf depths on glass. The angles on the oxide ceramic vary between 164° and 169°, decreasing with increasing kerf depths. Kerf angles between 145° and 152° were determined in the range from around 314 µm to 470 µm. On both materials, kerf angles between 145° and 149° are observed at these specific kerf depths. One individual kerf on the zirconia ceramic has a depth of about 1 mm and a kerf angle of approximately 126°. Surrounding results on the glass samples have similar CF ranging from 125° to 130°. The kerf angles do not decrease linearly with increasing kerf depth. In a range up to 2.1 mm depth, the kerf angles converge to 110°. An increased reduction in the kerf angles is primarily recognisable in this representation in a range up to a kerf depth of 300 µm. After that, the gradient of this curve decreases.

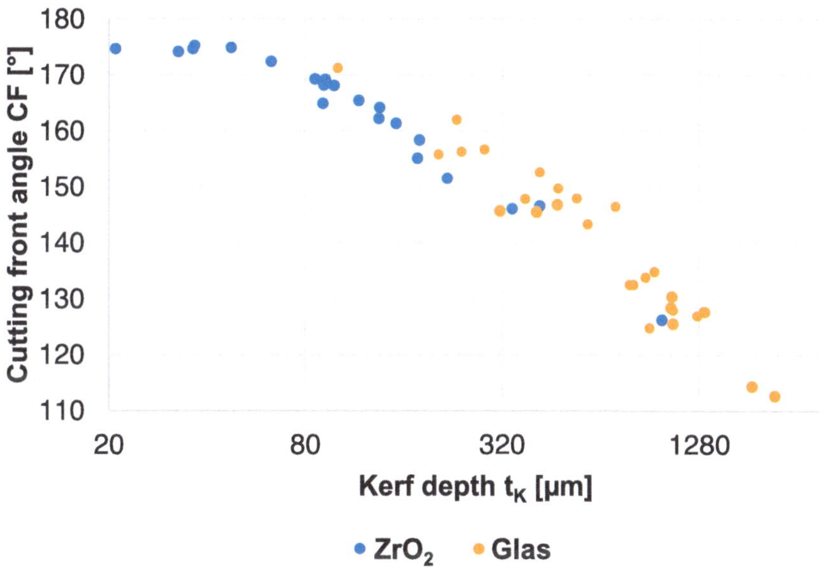

Fig. 8. Cutting front angle as a function of the kerf depth on ZrO_2 and glass.

Perpendicular machining with the abrasive water jet and knowledge of the CF can provide the actual entry angles of the primary jet along the kerf angle. There is no increased cutting angle variation for kerf depths up to 80 μm. As a result, entry angles of the primary jet around 85° can be determined (Fig. 8) - it indicates an almost perpendicular jet engagement. As the kerf depth increases, the entry angles decrease as a function of the kerf depth. In the kerf depth range between 80 μm and 300 μm, the kerf angles are between 145° and 170°. That corresponds to average entry angles from 55° to 80°. In the subsequent kerf depth section from 300 μm to 1 mm, the entry angles drop to around 35°. For the smallest measured cut front angle, the entry angle of the primary jet is around 23°. The associated kerf of the sharpest cutting front angle is the deepest measurement of the test series with a value of over 2 mm. The entire graph is approximated by a second-degree polynomial (Formula 1). The equation calculates the average kerf angle as a function of the kerf depth achieved for a specific set of parameters.

$$CF = 0,00001577 \cdot t_k^2 - \frac{187}{3125} \cdot t_k + 172,316[°] \quad (1)$$

Preliminary tests on glass were conducted as analogy tests for directional testing on a transparent material with brittle properties. The prevailing understanding indicates that the abrasive water jet primarily deflects at the CF (Sect. 2). Consequently, the cutting front geometry plays a significant role in determining the direction of the secondary jet. The assumption was made that if the cutting front angle is the same for identical kerf depths, the jet behaviour on glass would resemble that on an oxide ceramic. Plotting the CF curves at various kerf depths revealed that different parameters yield similar kerf depths with identical CF. This observation suggests that the cutting front angle depends on the kerf depth.

In general, the kerf depth can be adjusted through various process parameters. The jet intensity includes abrasive mass flow, particle velocity (pressure), feed rate, jet diameter, and exposure time (Formula 2 & 3). Demonstrating similar CF with comparable kerf depths using different parameters on two brittle materials indicates that the kerf depth can also be depicted as a function of the jet intensity.

$$CF \sim t_K \quad (2)$$

$$t_K \sim I_{AWJ} \quad (3)$$

Adjusting the cutting front angle within geometric limits is achievable by parameterising a specific kerf depth. The calculation of the cutting front angle as a function of the kerf depth was determined mathematically using a compensation curve derived from experimentally measured values. The second-degree polynomial characterised the cutting front angle on brittle materials (Formula 1), drawing from 50 individual measured values. The consistency in results between CF on the oxide ceramic and glass indicates a correlation for brittle materials.

4.2 Investigation of the Secondary AWJ Direction

Taking the secondary AWJ into account adds further geometric relationships to the CF cut front. In the second investigation, the analysis focused on the average exit angle β and the opening angle δ of the secondary jet (Fig. 6). In Fig. 9, the statistical distribution of the mean jet angle and the opening angle for different intervals of the ronde angle γ are visualised in a boxplot diagram. The results were measured for each parameter, which was a combination of pressure, mass flow, and feed rate. The disk angle γ is plotted on the abscissa. A scale from minus ten to sixty degrees is plotted on the ordinate to categorise the average exit angle β and the opening angle δ with the variation of the jet entry angle α. The mean value is indicated by a cross within the interquartile range. Similarly, the median is represented by a crossbar.

In an interval of $15° < \gamma < 30°$, the average secondary jet exit angle was independent of the parameters pressure, mass flow and feed rate. The average exit angle β is in an interval of $0° < \beta < -3.5°$ for 50% of the measured values. At a disk angle of $50° < \gamma < 60°$, the average exit angle β increases steadily to a maximum of around $\beta \approx 3°$. It is also evident that the median and the mean value of the exit angle β are below the reference tangent up to a disk angle of $40° < \gamma < 50°$ and can be read as 0° on the reference line. In the further course, the mean value and the median increase to a round angle $\gamma = 90°$ and remain in the positive range. The most significant deviation of the mean value of the measured exit angles β is at a perpendicular impact angle as $\beta \approx 5°$. Up to a 90° disk angle, the variance of the average exit angle β of the secondary jet is defined in a narrow range. The mean value of the exit angle deviates from the reference tangent by a maximum of around 5° degrees. In the case of forward machining, the mean values of the exit angle are also close to the reference tangent. Remarkable is the higher variance of the exit angle β of the secondary jet in the case of a forward-cut approach with disk angles in the range of $120° < \gamma < 150°$.

The opening angle of the secondary jet behaves similarly. The measured exit angles are focused on a disk angle of $15° < \gamma < 80°$ and only deviate slightly from each other. The interquartile range increases steadily at a slight rate as the disk angle γ increases. The mean values also increase somewhat. From a disk angle $\gamma = 90°$, the variance of the opening angle suddenly increases. With a perpendicular jet entry, the span of the interquartile range almost triples. The maximum opening angle δ is approximately $\delta \approx 28°$. In the case of forward machining with a disk angle in the field of $120° < \gamma < 130°$, the opening angle reaches a maximum of $\delta \approx 41.5°$. At the same time, the variance of the opening angle δ also reaches a maximum. The observation of the opening angles at disk angles in the range of $130° < \gamma < 150°$ show a reduction in the variance and a reduction by half of the median compared to the disk angle range of $120° < \gamma < 130°$.

Since jet intensity relies mainly on pressure, mass flow, and feed rate parameters, the investigation focused on understanding their influence on the direction of the secondary jet. Principal effect diagrams of the average exit angle of the secondary jet were utilised for the backward and forward jet entry angle range ($40° < \alpha < 50°$) on an oxide ceramic. Beginning with the examination of the central effect diagram for entry angles of 40° to 50° in the backward processing zone of the AWJ (Fig. 10), it was observed that the parameters - pressure, feed rate, and mass flow rate - generally have insignificant effects on the exit angle for backward machining within this entry angle range. This observation aligns with evaluating the average exit angles (Fig. 9). The secondary jet demonstrated a limited dependence on the parameter selection. With a backward setting of the AWJ at $40° < \alpha < 50°$, all effect differences were within an angular range of $\pm$ 2° relative to the reference tangent. A pressure elevation from 100 MPa to 300 MPa resulted in a

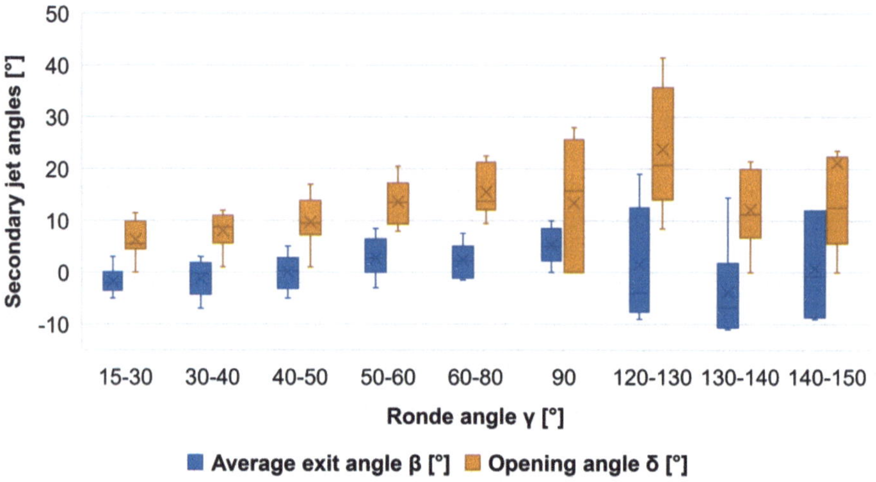

Fig. 9. Average exit angle and opening angle of the secondary jet on ZrO_2.

rise of the mean secondary jet deflection angle. Simultaneously, elevating the feed rate from 200 mm/min to 2000 mm/min slightly increased the jet angle. Conversely, raising the mass flow rate from 20 g/min to 60 g/min produced the opposite effect, causing the average exit angle of the secondary jet to decrease below the reference tangent.

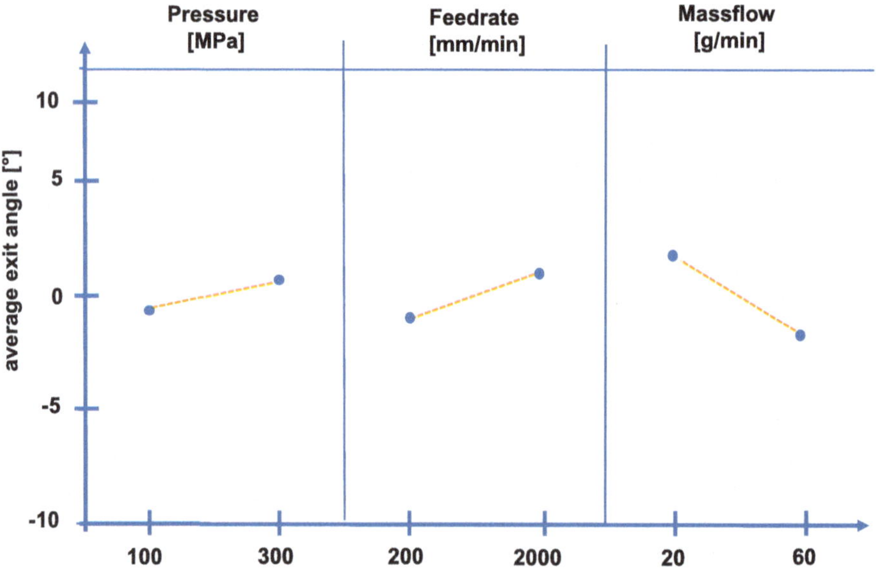

Fig. 10. Average exit angle for a backward cut angle α between 40°- 50°.

In contrast to backward machining, forward machining is significantly influenced by the main parameters (Fig. 11). A variation in pressure from 100 MPa to 300 MPa results in an average increase in the mean deflection angle from $\beta = -7.5°$ to a positive angle of $\beta = 3.5°$ relative to the reference tangent. Conversely, the effect of the feed rate is opposite to that of the pressure. An increase in the feed rate leads to a reduction in the mean angle of deflection. At a 200 mm/min feed speed, the mean jet angle is approximately $\beta = -0.8°$. Increasing the feed speed to 2000 mm/min decreases the exit angle by $\Delta\beta = 6°$. The mass flow rate reveals a more pronounced effect than the one during backward machining. At a 20 g/min mass flow rate, the average exit angle is $\beta \approx 2°$. Varying the mass flow from 20 g/min to 60 g/min causes the average exit angle of the secondary jet to decrease to a value of $\beta \approx -10°$ relative to the reference tangent.

During backward machining, a concave cut front is formed under any angle variation of the jet entry. The high velocity of the jet, coupled with the curvature at the cutting front, induces a dominant centrifugal force on the secondary jet (Fig. 1). This force bundles the secondary jet within the kerf geometry, offering insights into the mechanisms of secondary jet decay. Primarily observed at low jet intensities, jet decay occurs more prominently in shallower kerfs, resulting in larger CF. As per Henning's observations, the generation of the centrifugal force along the concave cutting front may be prevented, leading to particle reflection in random directions upon initial impact (Fig. 1).

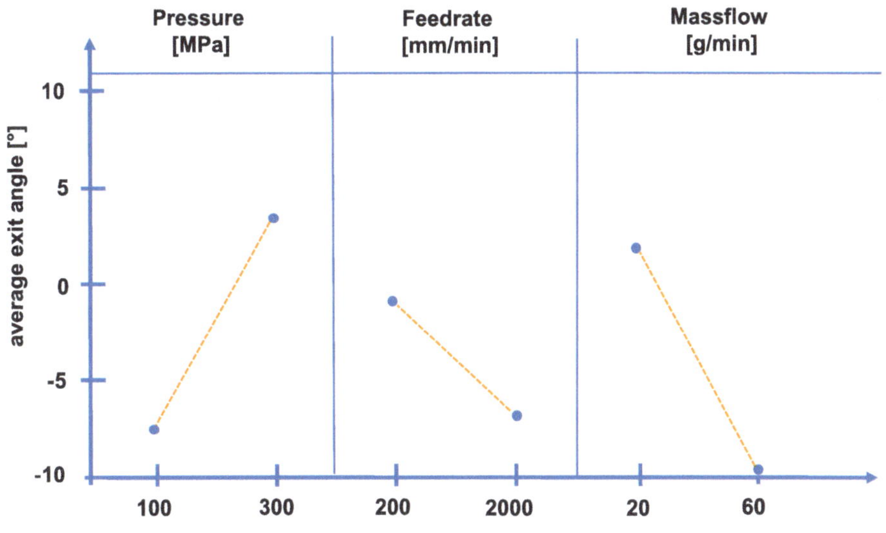

Fig. 11. Average exit angle for a forward cut angle α between 40°- 50°.

Deeper kerf machining can lead to macroscopic steps (Fig. 2), causing the secondary jet to deflect at the bottom of the step and exit at the step's edge. This deviation shifts the average exit angle from the reference tangent in the kerf bottom, causing the secondary jet to deflect from the kerf geometry. The cyclical process of the abrasive water jet leads to alternating exit angles, impacting process stability during material removal. Uncontrolled movements of the secondary jet may induce contour damage in distant component areas, particularly as a still-focused secondary jet remains bundled over longer distances.

Moreover, the main effects of the parameters on the opening angle are not negligible (Fig. 12). The opening angle in degrees is plotted on the ordinate. The varied factors of the parameters - pressure, feed rate, and mass flow - are plotted along the abscissa. Among these factors, the pressure of the AWJ demonstrated the most significant effect on the opening angle. They vary the pressure from 100 MPa to 300 MPa resulting in an average increase in the opening angle from $\delta \approx 10°$ to $\delta \approx 17.4°$. Changing the feed rate from 200 mm/min to 2000 mm/min results in a more minor increase in the opening angle compared to pressure variations. At 200 mm/min, the average opening angle is $\delta \approx 12.5°$, and increases by an average of $\Delta\delta \approx 2.5°$ to $\delta \approx 15°$. The mass flow rate generally has a more negligible effect on the opening angle, with a rise from 20 g/min to 60 g/min resulting in a decrease by $\Delta\delta \approx 1°$ from an average of $\delta \approx 14.2°$ to around $\delta \approx 13.3°$.

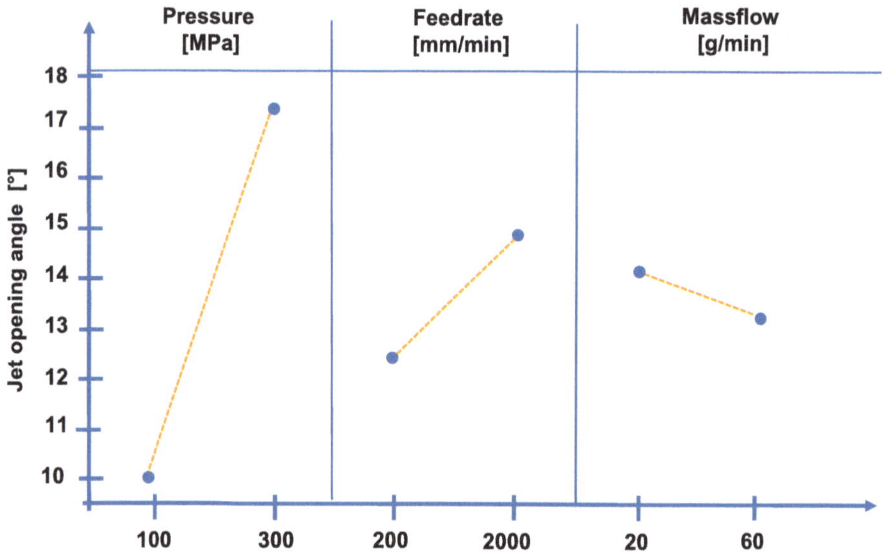

Fig. 12. Main effects on the opening angle of the secondary jet.

The results regarding the opening angle of the secondary jet indicate its limitation to a smaller interval during backward machining compared to perpendicular or forward machining (Fig. 9). Zeng and Kim's investigation of step formation justifies the increased opening angle (Fig. 2). During step formation, the leading jet stream deflects first, crossing the rear jet stream area (Fig. 13, left). This results in the weakening of the material removal capacity at the rear. The abrupt change in material removal causes the rear jet area to deflect in a different direction, resulting in overlapping exit angles of the front and rear jet areas. This phenomenon was also observed in tests, where the exit angle increased as the rear jet area crossed the previously deflected jet front and formed the lower jet exit angle (Fig. 13, right). Without direct contact with the workpiece, the rear stream flows above the front jet stream, only reflected by the upper edge of a step. This explains the often clearly bounded lower part of the secondary jet. In the trailing area at higher jet intensities, the rear jet flowed almost tangentially out of the machining zone along the kerf front. As sliding wear dominates the removal mechanism, a smooth kerf surface is produced without disturbing irregularities. The lack of obstacles at the bottom of the kerf prevented contact between the rear primary jet stream or upper secondary jet stream and the target material. The kerf geometry emerges as one of the main factors influencing the opening angle. Pressure also influences the opening angle (Fig. 12). An increase in pressure affects the kerf geometry, causing disturbing irregularities due to the cyclical material removal characteristic of the water jet (Sect. 2). These irregularities can deflect the secondary jet.

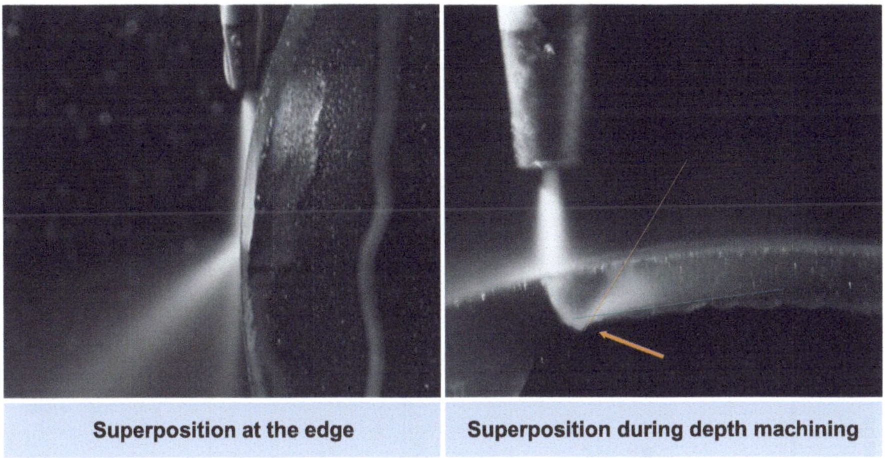

Fig. 13. Superposition of the front and rear jet.

4.3 Experimental Investigation of the Secondary Abrasive Water Jet Effect

To analyse the remaining effect of the secondary jet, damage depth (tD) and damage height (hD) measurements were conducted on an impact plate, considering pressure, feed rate, mass flow, and impact angle. First, the damage depth on the PMMA impact plate was shown in a corresponding main effect plot (Fig. 14). A variation of pressure and feed rate exhibits a moderate dependency. Here, a rise in pressure from 100 MPa to 200 MPa results in an average penetration depth change of around $\Delta tD \approx 225$ μm. This magnification represents an increase of approximately 167% compared to the previous value of $tD \approx 135$ μm at 100 MPa. A change in the feed rate is more modest than a variation in pressure. Here, it reduces the average damage depth from $tD \approx 332$ μm to around half the previous value of $tD \approx 173$ μm. The mass flow rate has a minor effect on damage depth, with a variation from 20 g/min to 60 g/min, causing a growth in mean damage depth from $tD \approx 242$ μm to $tD \approx 280$ μm. The jet impact angle has the most significant effect among the parameters on the damage depth of the secondary jet. A sharp jet angle of $\alpha = 15°$ results in an average indentation of $tD \approx 900$ μm. Raising the impact angle to $\alpha = 30°$ decreases the damage depth by factor three to an average of approximately tD ≈ 320 μm.

Further elevation to jet angles of $\alpha = 45°$ results in an average damage depth of t_D ≈ 100 μm. Neglectable damage of less than 15 μm is evident at jet angles α between 60° and 75° due to the disintegration of the secondary jet (Sect. 4.2). Catching up on the studies by Uhlmann and Männel (Fig. 3), a focused secondary jet exhibits high-energy characteristics. According to Zeng and Kim, the preservation of high kinetic jet energy was explained by the absence of macroscopic step formation due to the predominance of the optimum erosion state along the cutting front (Sect. 2). Sliding wear dominates along the cutting front in the forward orientation, while the backward direction minimises additional contact with the target material. It can be said that only a fraction of the kinetic energy is transferred to the target material. Compared to the jet angle, the parameters pressure, feed rate, and mass flow have a more negligible effect on the damage depth.

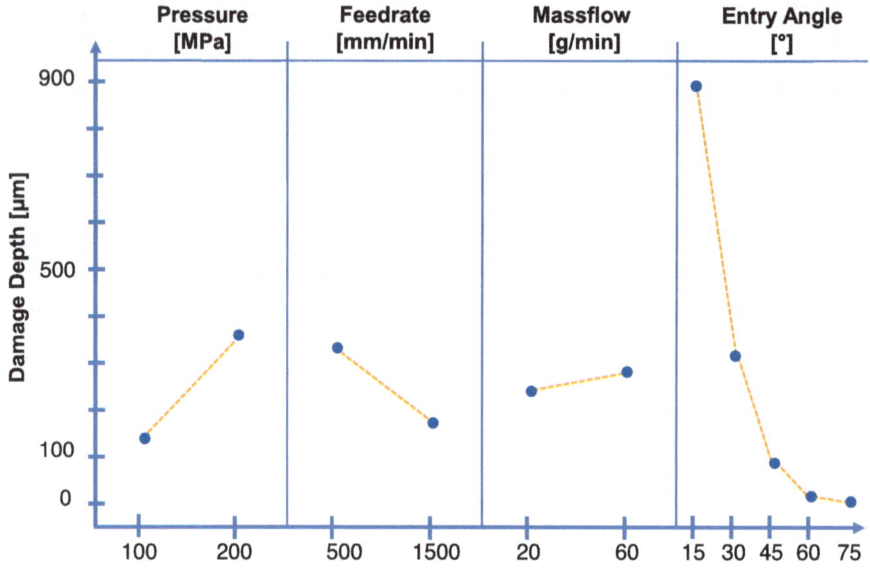

Fig. 14. Main effects on the damage depth t_D of the secondary jet.

The variables influencing the damage height were also examined using a central effect diagram (Fig. 15). By doubling the pressure from 100 MPa to 200 MPa, the average damage height decreases from h_D ≈1320 μm to h_D ≈ 1280 μm by an amount of Δh_D ≈40 μm. The pressure parameter has no significant influence on the damage height. The mass flow rate has a comparably low impact on the damage height of the secondary jet. At a mass flow rate of 20 g/min, the damage height of the secondary jet is approximately h_D ≈1220 μm. On average, the damage height increases by 130 μm when the mass flow rate is 60 g/min. The feed rate has an increased influence on the damage height. By increasing the feed rate from 500 mm/min to 1500 mm/min, the average damage height of the secondary jet reduces by half from h_D ≈1650 μm to h_D ≈814 μm.

Similar to the damage depth, the jet entry angle also significantly influences the effect of the secondary jet regarding the damage height. At a jet entry angle of 15°, the damage height reaches a value of $h_D \approx 1300$ µm. Raising the jet angle to 30° magnifies the damage height to around $h_D \approx 1800$ µm. The maximum damage height is at a jet angle of around $\alpha = 45°$ with an average of $h_D \approx 1900$ µm. If the jet angle is increased further to $\alpha = 60°$, the average damage height decreases significantly to a value of $h_D \approx 1114$ µm. The peak of the damage height at a jet entry angle of $\alpha = 45°$. Due to the previously defined jet angles, a deviating optimum between $30° < \alpha < 60°$ is possible. At a jet angle of $\alpha = 75°$, no damage height is detectable within the scope of this investigation method. Therefore, the damage height is set to zero. The jet angle and feed rate significantly influence the damage height. It shows that pressure and mass flow only affect the damage height to a small extent (Fig. 15). An increase in the feed rate causes a reduction in the damage height. This effect can be explained through lower kerf waviness and more shallow kerfs at higher feed rates.

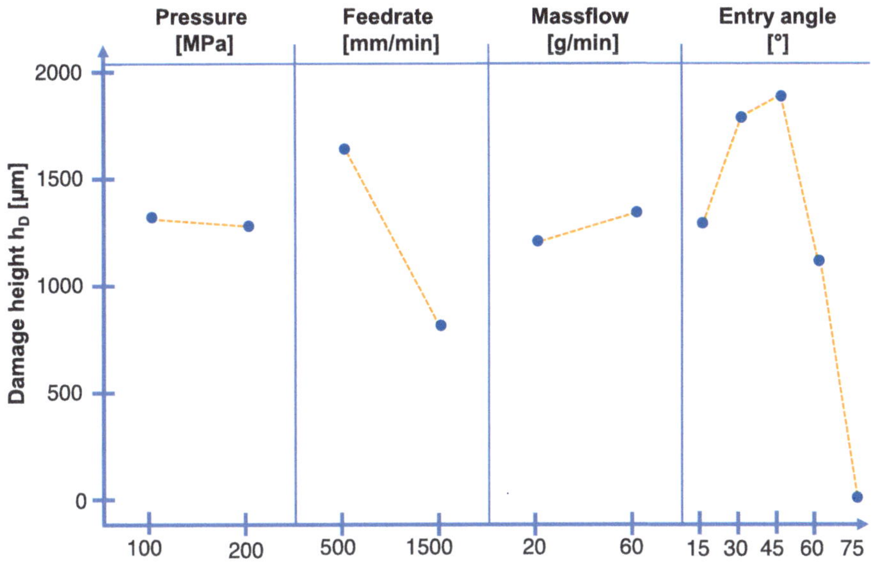

Fig. 15. Main effects on the damage height h_D of the secondary jet.

5 Conclusion

This study investigated the variables influencing the direction and the effect of the secondary jet in abrasive water jet (AWJ) machining. Being a non-solid tool, the engaged AWJ may be reflected. This so-called secondary jet could violate the original contour due to unwanted material removal. Characterising the secondary jet is the prerequisite for compensating in 3D water jet milling. In the experimental studies, a distinction was made between the direction and the erosive effect of the secondary jet. Here, a zirconia ceramic (ZrO_2) specimen was machined with AWJ. Besides zirconia ceramic as the original sample material, transparent soda lime glass was also used as workpiece material to visualise the phenomena of the secondary abrasive water jet.

Preliminary results indicated a correlation between the cutting front angle and kerf depth. A mathematical model was developed to determine the dependence of the cutting front angle on the kerf depth. The secondary jet direction analysis revealed the correlation between jet orientation (forward/backward), kerf geometry, and jet angle. In addition, the relationship between the opening angle and the achieved topography on the kerf bottom was highlighted. A particular relation between the orientation of the primary jet damage potential of its secondary counterpart was noticed.

Future work could include simulation techniques to provide analytical insights into the causes of secondary jet deflection and characterize them on an energy basis. They could also be used to quantify the damaging effects on different materials by including material properties. In addition, existing process models could be extended for a more comprehensive description and relate them to the angles of the cutting front. Integrating the secondary jet effects into waterjet-specific computer-aided manufacturing (CAM) modules has significant practical relevance, enabling the design of milling operations with improved process stability.

References

1. Bitter, J.G.A.: A Study of Erosion Phenomena. Part II. Wear 169–190 (1962). https://doi.org/10.1016/0043-1648(63)90073-5
2. Bitter, J.G.A.: A Study of Erosion Phenomena. Part I. Wear **1963**, 5–21 (1962). https://doi.org/10.1016/0043-1648(63)90003-6
3. Blickwedel, H.: Erzeugung und Wirkung von Hochdruck-Abrasivstrahlen. Fortschr.-Ber. VDI 2 (206), Düsseldorf: VDI-Verlag (1990). Dissertation, Leibnitz Universität Hannover
4. DIN Deutsches Institut für Normung e. V. (1990) Flachglas im Bauwesen. Chemische und physikalische Eigenschaften (1249 - 10)
5. DIN Deutsches Institut für Normung e. V. (2001) Glas. Teil1: Begriffe für Glasarten und Glasgruppen (1259-1)
6. Faltin, F.: Endkonturnahe Schruppbearbeitung von Titanaluminid mittels Wasserstrahlabrasivstrahlen mit kontrollierter Schnitttiefe. Dissertation, TU Berlin (2017)
7. Finnie, I.: Erosion of surfaces by solid particles. Wear **3**, 87–103 (1960). https://doi.org/10.1016/0043-1648(60)90055-7
8. Finnie, I., Levy, A.D.M.: Fundamental mechanisms of the erosive wear of ductile metals by solid particles (1979)
9. Finnie, I., McFadden, D.H.: On the velocity dependence of the erosion of ductile metals by solid particles at low angles of incidence, vol. 48 (1978)
10. Fowler, G.: Abrasive Water-jet - Controlled Depth Milling of Titanium Alloys. University of Nottingham (2003)
11. Hashish, M.: Visualisation of the abrasive-waterjet cutting process. Exp. Mech. **28**(2), 159–169 (1988)
12. Henning, A.: Modellierung der Schnittgeometrie beim Schneiden mit dem Wasserabrasivstrahl. Zugl.: Stuttgart Univ. Dissertation. IPA-IAO-Forschung und Praxis 470, Jost-Jetter, Heimsheim (2007)

13. Klocke, F., König, W.: Fertigungsverfahren 3. Abtragen, Generieren Lasermaterialbearbeitung, 4., neu bearbeitete Auflage. Fertigungsverfahren 3, Springer-Verlag Berlin, Heidelberg (2007)
14. Klocke, F., Schreiner, T., Schüler, M., et al.: Material removal simulation for abrasive water jet milling. CIRP Conference of Electro Physical and Chemical Machining (2018). https://doi.org/10.1016/j.procir.2017.12.110
15. Kuhmichel Abrasiv GmbH: Edelkorund weiß (2022). https://www.kuhmichel.com/de/produkte/edelkorund/. Accessed 05 Feb 2024
16. Kuhmichel Abrasiv GmbH: Granatsand (2022). https://www.kuhmichel.com/de/produkte/granatsand/. Accessed 05 Feb 2024
17. Laurinat, A.H.: Abtragen mit Wasserabrasivinjektorstrahlen. Dissertation, Düsseldorf, Deutschland (1994)
18. Uhlmann, E., Männel, C., Braun, T.: Efficient abrasive water jet milling for near-net-shape fabrication of difficult-to-cut materials. Int. J. Adv. Manuf. Technol. **111**, 685–693 (2020). https://doi.org/10.1007/s00170-020-06074-3
19. Zeng, J., Kim, T.J.: Development of an abrasive waterjet kerf cutting model for brittle materials. In: Lichtarowicz, A. (ed.) Jet cutting technology 13, pp. 483–501. Kluwer, Dordrecht, Boston (1992)

Use of Accelerated-Hydro-Abrasive-Cavitation-Erosion (AHACE) Method to Investigate the Erosion Resistance of Rocks to Fast Flowing Liquids

Libor Sitek[1(✉)], Petr Hlaváček[1], Petr Martinec[1], Lenka Bodnárová[2], Dagmar Klichová[1], and Josef Foldyna[1]

[1] Institute of Geonics of the Czech Academy of Sciences, Studentská 1768, Ostrava, Czechia
libor.sitek@ugn.cas.cz
[2] Faculty of Civil Engineering, Brno University of Technology, Veveří 512, Brno, Czechia

Abstract. The paper focuses on the resistance of rocks to fast-flowing liquids using a well-established method of accelerated hydro-abrasive-cavitation simulation based on special application of abrasive water jets. Erosion resistance was tested on large samples of nine rock types from various quarries in Czechia (granite, gneiss, basalt, granodiorites, metabasites and marbles). New surfaces created after the application of simulative erosion flows were analysed and their texture was described. Relationships between the erosion resistance of rocks and their selected properties (i.e. compression and flexural tensile strengths, abrasiveness and abrasivity) were studied. It has been found that erosion resistance of rocks tested on the hydro-abrasive erosion differs significantly and is closely related to the mineral composition and internal rock structure. Basalt was proved to be the most resistant, followed by metabasites, fine-grained Litice Granodiorite, coarse-grained granite, gneiss, both marbles and fine-grained Silesian Granodiorite. No relationship was found between the observed properties of rocks and their resistance to the accelerated hydro-abrasive-cavitation erosion. The method developed by the authors is unique, enabling quick analysis of erosion resistance of various materials to fast-flowing liquids. The paper provides with new knowledge on behaviour and resistance of rock interacting with fast-flowing liquids related to their structure and properties, which has not been studied so far.

Keywords: Fast Flowing Liquid · High-Speed Water Jet · Accelerated Hydro-Abrasive-Cavitation Erosion (AHACE) · Rock · Erodents

1 Introduction

Erosion wear of rock and cementitious composite structures loaded by the action of fast-flowing polluted liquids is an interesting phenomenon which cannot be easily analytically described due to the synergy of several disintegration mechanisms. Apart from the abrasion caused by the presence of small particles in the water flow (silt, sand, gravel,

L. Sitek et al. (Eds.): Advances in Water Jetting II, LNME, pp. 228–257, 2025.
https://doi.org/10.1007/978-3-031-72778-8_18

ice, or debris), disintegration can be caused by cavitation or chemical attack and bacterial action. The surface becomes porous, thus reducing its resistance to erosion. All these processes happen simultaneously and cannot be separated from each other. This combined disintegration is often observed in hydraulic rock and concrete structures and structures exposed to effects of the water flow impacting on surface, whether on a permanent or one-time basis during particular events (typically floods, etc.). The processes occurring during disintegration may be so complex that it is impossible to simulate them, and to predict possible consequences of the action of flowing water on structures. The only reliable way to describe these processes is the experimental investigation of real-life structures or test samples.

Long-lasting and often slow disintegration of real structures is an obstacle to the finding of durable materials or to the determination of suitable surface treatments of materials exposed to fast-flowing liquids. Measurable failure occurs within a few months, often after several years or decades. In addition, the damage happens on a non-regular and often one-time basis depending on the actual quantity of the liquid and its velocity (torrential rains, floods, etc.). Although it is actually good that the process of disintegration of real structures is slow, laboratory tests investigating the resistance of various materials to degradation effects of real flowing liquids in real life could be unbearably long.

Methods based on the use of high-velocity water jets represent a promising research direction of the erosion wear of rock materials, because the physical similarity enables fast and simple experimental simulation of complex real-life situations. For this purpose, the authors used their own method of the accelerated simulation of hydro-abrasive-cavitation erosion (AHACE) based on the action of an abrasive water jet on the material surface, which they had previously tested successfully in determining the erosion resistance of concrete to fast-flowing liquids.

2 State of the Art

Most of the literature dealing with erosion wear caused by flows of polluted liquids focuses exclusively on the wear of cementitious composites, as well as on recommendations from various expert associations for testing the resistance of concretes and practices to prevent concrete degradation [1]. A detailed study of hydro-abrasive erosion of fibre reinforced concrete was presented, for example, in [2]. However, due to character of the entire disintegration process and efforts to keep conditions similar to the real situation, testing methods are often very time-consuming [3]. An extreme case of long-lasting laboratory testing of hydro-abrasive erosion is reported by Binici et al. [4]. The testing lasted three years and the loss in concrete was investigated once a year. The obvious lack of practicability as well as the limited use of these methods encouraged the application of conventional and rapid laboratory methods and studies. However, they have examined individual degradation processes separately, which generally do not reflect the reality (e.g. research into abrasion [5], cavitation [6, 7], chemical attack [8, 9] or biological degradation [10]). The recommendations concerning the testing procedure for determination of abrasive resistance of concrete, for example, is the subject of some reports made by the American Society for Testing and Materials - ASTM C1138 [11], ASTM C944 [12] or ASTM C779/C779M-19 [13].

Logically, studies investigating the use of high-velocity water flows technology as a suitable and fast method of concrete disintegration appeared in the professional area. These methods are based on the similarity with the flow process of often polluted liquids with greater or smaller content of free hard particles of various sizes which may be part of the flowing liquid (i.e. abrasives), or the particles gradually eroded from the original concrete surface and driven by the liquid. Since high-velocity water flows exiting from the nozzle can reach velocities of up to hundreds of $m \cdot s^{-1}$, material erosion is fast. With the appropriate combination of water flow parameters and test setting, this method enables faithful simulation of real erosion processes. Wear particles that occurred during concrete erosion after the application of high-velocity water flow were studied, for example, by Momber and Kovacevic [14]. Hu et al. [15] carried out hydro-abrasive erosion tests on plain concrete and on steel-fibre reinforced concrete. The same authors also investigated the erosion resistance of several concrete types [16]. Momber studied the cavitation wear of concretes at laboratory conditions using a cavitation chamber [17] and he quantified the effects of abrasive flow kinetic energy and exposure time on the erosion of concretes exposed to abrasive water flow [18]. He also tried to use the acoustic emission sensing technique for an on-line monitoring of hydro-abrasive erosion process in pre-cracked concretes [19]. Behaviour of concretes eroded by water-driven particles and modelling of this process were described in other Momber's works [20] and [21].

On the other hand, erosion resistance of rocks loaded by fast-flowing liquids has not been systematically studied yet. Sporadic literature data on the interaction of high-velocity water flows with rock surfaces are focused on other research areas. For example, Hood et al. [22] investigated the erosion mechanisms during rock cutting and they presented a new physical model of erosion during water jet cutting. Surface and subsurface changes in rocks caused by pulsating water jets without erodents were studied experimentally by Dehkhoda and Hood [23]. Ozcelik et al. [24], for example, investigated the effects of water jets on surfaces of ornamental stones in order to improve their aesthetic appearance. Consequently, they compared the obtained results with outputs from other methods of stone treatment. Another study focused on treatment of ornamental stones to improve their antislip properties [25]. The properties of rock surfaces after cutting with an abrasive water jet were then analysed by Młynarczuk et al. [26].

Determination of suitable configuration and setting of parameters of a high-velocity flow to faithfully simulate the real-life erosion process is quite challenging and it usually requires an extensive experimental program with much experience in the waterjet technology. The authors of the presented study used their own long-term experience in the degradation of cementitious composites using high-velocity water flows (see, e.g., [27–30]) which they applied for an accelerated hydro-abrasive simulation of the degradation process of concrete surfaces affected by fast-flowing liquids. In [31], the authors successfully tested an original method for the accelerated hydro-abrasive-cavitation erosion of concretes using a high-velocity water flow. In [32], the method was used for determination of resistance of concretes treated with special impregnation agents which increased resistance of concrete surface layers. Tests revealed that various loading regimes of concretes can be suitably simulated at changing flow parameters (flow rate, amount of abrasive particles in the flow, particle size, direction of impact, etc. [33]). Furthermore, it was found that individual structural elements of cementitious composite were eroded at different intensities as in real-life erosion cases. Similar conclusions were reached in a study focused on the analysis of subsurface structures using micro X-Ray CT methods [34]. Moreover, in contrast to thermal loading of concretes, for example, where cracks (either separation cracks along aggregate grains or cracks in cement mortar) are generated due to different expansion of structural elements in the concrete, which results in a significant loss of concrete integrity and strength [35], the material is gradually removed from the surface exposed to simulation by high-velocity water flows (both without and with abrasives) and damage of the internal structure, as well as occurrence of new cracks or fractures are eliminated.

This paper focuses on the determination of the erosion resistance of selected types of rocks. The authors' hypothesis assumed that the resistance of tested rocks to the hydro-abrasive erosion by high-velocity abrasive water flows clearly depends on rock properties, such as strength or abrasion resistance determined by other methods.

3 Material Selection and Petrographic Description of Individual Rocks

Nine types of rocks from various quarries in Czechia were selected for testing of erosion resistance to flowing liquids. The list of rocks and their properties are presented in Table 1. All these values were measured at the Institute of Geonics of the CAS (IGN) according to appropriate standards [36–40, 42]. The test samples included igneous (basalt, granite, granodiorite) as well as metamorphic (marble, gneiss, metabasite) rocks.

3.1 Litice Granodiorite (Biotitic Granodiorite)

Granodiorite from Litice nad Orlicí represents an intrusive igneous granitoid with a granular, omnidirectional structure and massive texture (Fig. 1). In the structure, there are grains of orthoclase in the form of well-contoured outgrowths up to 1.2 mm x 0.8 mm with good cleavage according to {010}, hardness 6 (according to Mohse) and density of 2.56–2.63 g·cm^{-3}. White crystals of acid plagioclase of albite-oligoclase composition, 0.65 mm x 0.48 mm in size with good cleavage according to {010} and rims on orthoclase with hardness 6.0–6.5 and density of 2.62–2.67 g.cm^{-3} are other components evidenced in the structure. The also present biotite is a dark brownish-black flake-shaped biotite, 0.6–0.8 mm in size with excellent basal cleavage {0001}, with hardness 2.5–3 and density of 2.8–3.1 g·cm^{-3}. The last component is irregular, grey to white-grey quartz grains without crystallographic constraint, with average size of 0.52 mm x 0.38 mm, non-fissile, brittle, with hardness 7 and density of 2.65 g.cm^{-3}.

3.2 Liberec Granite (Porphyritic Granite)

The so-called Liberec Granite from Ruprechtice u Liberce represents an intrusive igneous rock – a granitoid (Fig. 2). Its structure is massive, porphyritic and hypidiomorphic with irregular grains. The structure includes grains of orthoclase (pink crystals) with well-contoured outgrowths up to 20 mm x 15 mm with good cleavage according to {010}, hardness 6 and density of 2.56–2.63 g·cm^{-3}. Acid plagioclases of albite-oligoclase

Table 1. Selected rocks and their basic properties [36–40, 42].

Rock type and origin	Bulk density [36] [kg·m^{-3}]	Density [36] [kg·m^{-3}]	Water absorption [39] [%]	Total porosity [36] [%]	Compressive strength [37] [MPa]	Flexural strength [38] [MPa]	Abrasiveness Böhme [40] [mm]	Abrasivity IGN [42] [mg·m^{-1}]	Abrasiveness IGN [42] [mg·m^{-1}]
Litice Granodiorite	2622	2668	0.32	2.26	164	13.2	20.0	4.20	0.66
Liberec Granite	2630	2670	0.14	1.5	168	10.4	17.0	2.19	0.66
Silesian Granodiorite	2610	2670	0.39	2.25	174	17.2	21.5	3.51	0.66
Potštejn Gneiss	2607	2722	0.28	0.71	145	12.0	18.0	1.34	0.66
Bílčice Basalt	2790	2810	0.18	0.71	251	23.9	18.0	1.31	0.66
Mistrovice Metabasite	2940	2989	0.3	1.64	155	14.3	13.0	1.13	2.65
Masty Metabasite	2942	3001	0.25	1.97	136	12.3	11.0	0.90	5.97
Lipová Marble	2710	2710	0.08	0.00	117	9.1	27.5	0.01	13.26
Supíkovice Marble	2697	2700	0.09	0.11	77	9.1	27.5	0.03	34.48

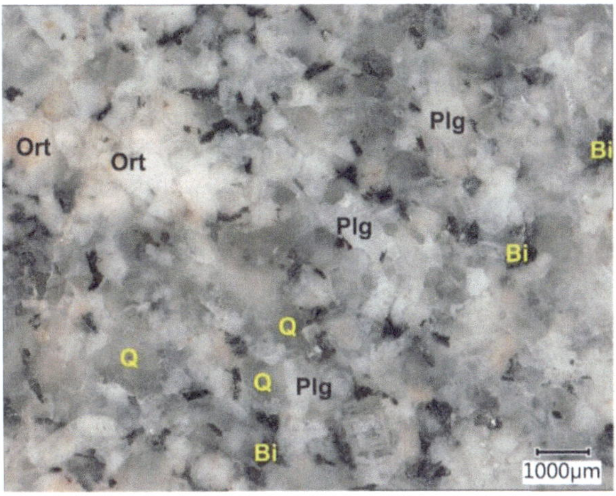

Fig. 1. A polished section of Litice Granodiorite (Q – quartz; Ort – orthoclase; Plg – plagioclase; Bi – biotite).

composition are present as smaller white crystals (about 1.5 mm) with good cleavage according to {010}, also in the form of rims on orthoclase with hardness 6.0–6.5 and density of 2.62–2.67 g·cm^{-3}. The other component is a dark brownish-black flake-shaped biotite of 1 mm in size, with excellent basal cleavage {0001}, with hardness 2.5–3 and density of 2.9–3.1 g·cm^{-3}. The structure also contains irregular, grey to white-grey quartz grains without crystallographic constraint up to 2 mm x 1 mm, non-fissile, brittle, with hardness 7 and density of 2.65 g·cm^{-3}.

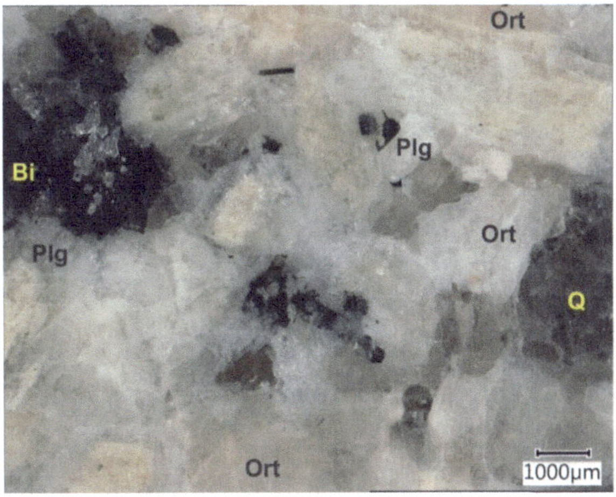

Fig. 2. A polished section of Liberec Granite (Q – quartz; Ort – orthoclase; Plg – plagioclase; Bi – biotite).

3.3 Silesian Granodiorite (Biotitic Granodiorite)

Another representative of biotitic granodiorites is the so-called Silesian Granodiorite from Žulová - Nový lom. Again, it is an intrusive igneous granitoide with a hypidiomorphic-granular, omnidirectional structure and massive texture (Fig. 3). The grains of orthoclase are hypidiomorphic constraint, with medium size of 0.75 mm x 0.55 mm, good cleavage according to {010}, hardness 6 and density of 2.56–2.63 g·cm^{-3}. Acid plagioclases of albite and oligoclase composition are present as smaller white crystals with good cleavage according to {010} and as rims on orthoclase. Their hardness is 6.0–6.5 and density 2.62–2.67 g·cm^{-3}. Dark brownish-black flake-shaped biotite of about 0.2–0.3 mm in size represents another component with an excellent basal cleavage {0001}, hardness 2.5–3 and density of 2.7–3.1 g·cm^{-3}. The last distinctive component is irregular, grey to white-grey quartz grains without crystallographic constraint, nonfissile, brittle, of medium size of 0.68 mm x 0.45 mm, with hardness 7 and density of 2.65 g·cm^{-3}.

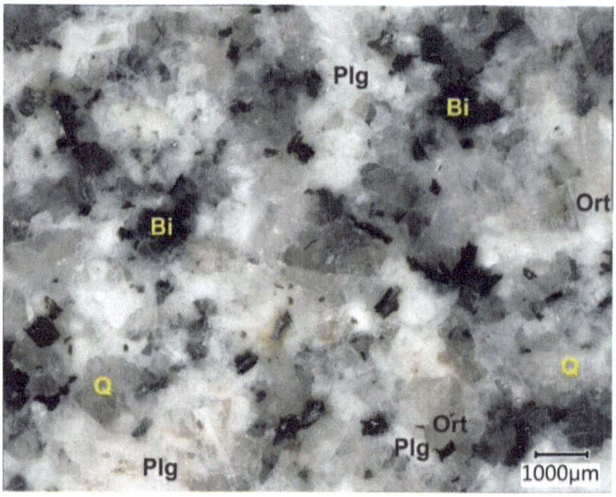

Fig. 3. A polished section of Silesian Granodiorite (Q – quartz; Ort – orthoclase; Plg – plagioclase; Bi – biotite).

3.4 Potštejn Gneiss (Granitised Gneiss)

The Potštejn Gneiss from Černá skála is a metamorphic, strongly anisotropic rock (Fig. 4). Bands of grano-lepidoblastic structure (metaster or leucosome) alternate with bands of lepido-granoblastic structure (metatekt or melanosome). The thickness of bands is variable, from 1 to 3 mm. The texture is banded (foliated) and granular. The structure contains allotriomorphic grains of orthoclase with a medium size of approx. 0.55 mm with good cleavage according to {010}, hardness 6 and density of 2.56–2.63 g·cm^{-3}. Acid plagioclases of albite and oligoclase composition in form of allotriomorphic white crystals with good cleavage according to {010} are also present. Their size is medium of 0.38 mm x 0.2 mm; hardness ranges from 6.0 to 6.5 and density is 2.62–2.67 g·cm^{-3}. Dark brownish-black flakes of biotite up to 1 mm are allotriomorphic with an excellent basal cleavage {0001}. The flakes are preferentially foliated, with hardness 2.5–3 and density of 2.7–3.1 g·cm^{-3}. The structure also contains irregular, grey to white-grey quartz grains without crystallographic constraint, non-fissile, brittle, of medium size of 0.3–0.6 mm, with hardness 7 and density of 2.65 g·cm^{-3}.

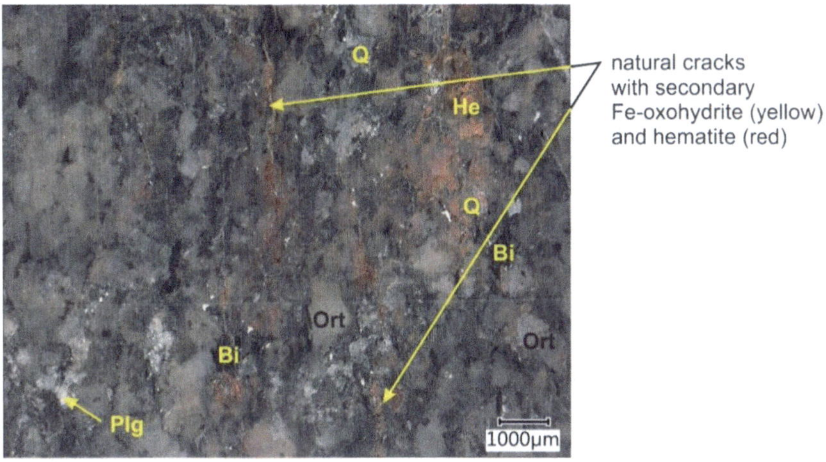

Fig. 4. A polished section of Potštejn Gneiss (Q – quartz; Ort – orthoclase; Plg – plagioclase; Bi – biotite; He – hematite).

3.5 Bílčice Basalt (Solid Type Basalt)

The olivine-pyroxenitic basalt from Bílčice is an igneous, surface and basaltoide rock (Fig. 5). Its structure is porphyritic, with a pilotaxitic texture of the basic mass. The massive texture contains gas pores and cavities up to 2 mm in size. The basal plagioclase consists of narrow lath-shaped crystals (smaller than 0.1 mm) with good basal cleavage according to {010}, density of 2.68–2.78 g·cm^{-3} and hardness 6.5. It also consists of olivine crystals of a tenth of mm with good cleavage according to {010}, density of 3.22–3.29 g·cm^{-3} and hardness 7. Glassy matrix (glass with magnetite particles) can also be determined in the structure.

3.6 Mistrovice Metabasite

The banded epidote-amphibolic metabasite (Fig. 6) is a metamorphic basal rock from Mistrovice. Its structure is microfiber granoblastic (more epidote positions alternate with more amphibole-plagioclase positions). The texture is banded (foliated) with band thickness of about 0.5–3 mm. The structure contains grains of basal plagioclase (bytownite and labradorite) in form of small lath-shaped crystals with good cleavage according to {010}, hardness 6.0–6.5 and density of 2.68–2.72 g·cm^{-3}. Next component are long crystals of

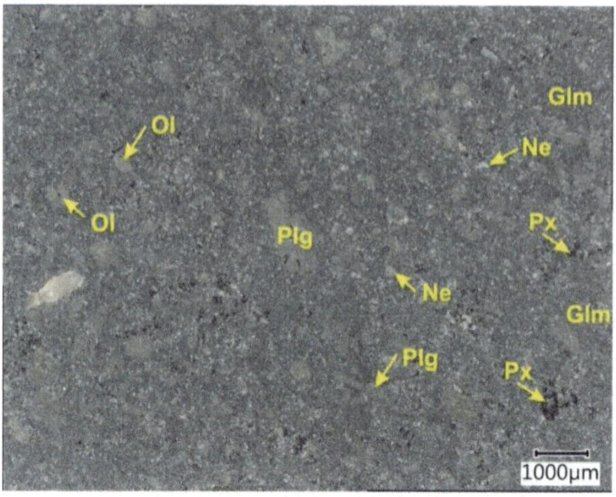

Fig. 5. A polished section of Bílčice Basalt (Plg – plagioclase; Ne – nepheline; Glm - glassy matrix with thin prismatic crystals of basal plagioclases; Px - pyroxene, Ol - olivine).

green epidote of 0.5–0.7 mm in size, with hardness 6.5 and density of 3.02–3.59 g·cm^{-3}, with perfect cleavage according to {001}. The last component is an amphibole - stretched dark grey-green crystals of hornblende of the maximum length of 1.3 mm, with hardness 5–6 and density of 3–3.5 g·cm^{-3} with perfect cleavage according to {110} and $\{10\bar{1}1\}$.

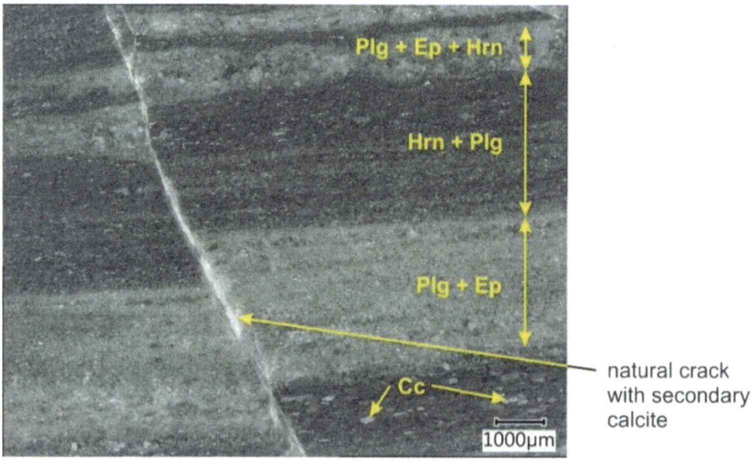

Fig. 6. A polished section of Mistrovice Metabasite (Plg – plagioclase; Cc – calcite; Ep - epidote; Hrn – hornblende).

3.7 Masty Metabasite

The second sample of metabasite with a slightly different structure comes from Masty. It is also a banded epidote-amphibole metabasite with the same composition as the previous sample. However, it contains more plagioclase laminae and the bands are more clearly defined (Fig. 7). The thickness of bands is smaller (0.5–1.5 mm) compared to the previous sample. The structure also consists of grains of basal plagioclase in form of small lath-shaped crystals with good cleavage according to {010}, hardness 6.0–6.5 and density of 2.68–2.72 $g \cdot cm^{-3}$. Next component are long crystals of green epidote of 0.5–0.7 mm in size, with hardness 6.5 and density of 3.02–3.59 $g \cdot cm^{-3}$, with perfect cleavage according to {001}. The last component is an amphibole – i.e. stretched dark grey-green crystals of hornblende of the maximum length of 1.3 mm, with hardness 5–6 and density of 3–3.5 $g \cdot cm^{-3}$ with perfect cleavage according to {110} and $\{10\bar{1}1\}$.

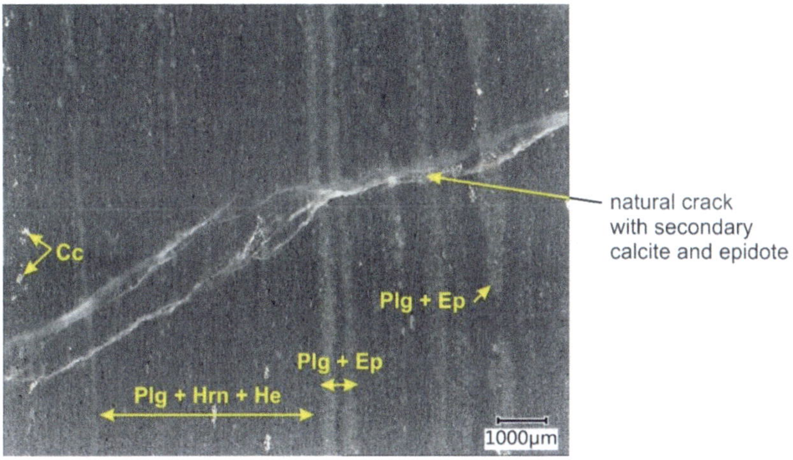

Fig. 7. A polished section of Masty Metabasite (Plg – plagioclase; Cc – calcite; Ep - epidote; Hrn – hornblende, He – hematite).

3.8 Lipová Marble (Calcite Marble)

This metamorphic rock is from Lipová. It is coarse-grained rock with calcite crystals up to 4 mm in size (Fig. 8). It is monomineralic (the rock contains exclusively calcite grains), with hardness 3 and perfect cleavage according to rhombohedrons $\{10\bar{1}1\}$. The structure is jagged, granoblastic, coarse-grained, massive and orientated.

3.9 Supíkovice Marble (Calcite Marble)

This marble is very similar to the previous sample. It is from the locality of Supíkovice. Again, it is coarse-grained rock with calcite crystals up to 4 mm in size (Fig. 9). It is monomineralic (100% of calcite). The texture is jagged and granoblastic, coarse-grained, massive and orientated. The rock mainly consists of calcite grains ($CaCO_3$) with hardness 3 and perfect cleavage according to rhombohedrons $\{10\bar{1}1\}$.

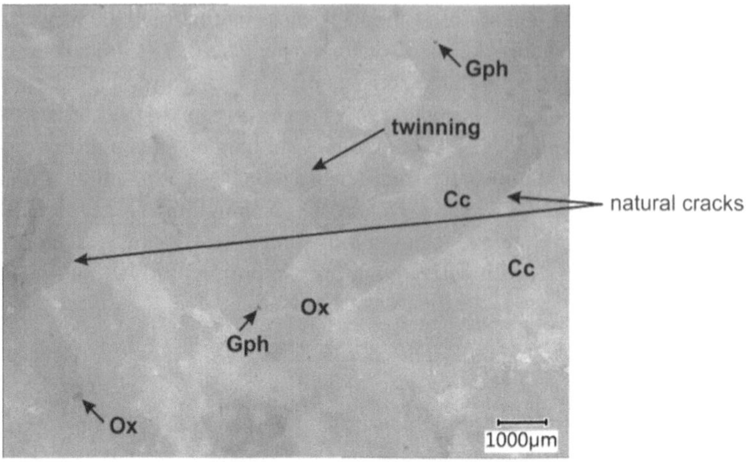

Fig. 8. A polished section of Lipová Marble (Cc – calcite; Ox - Fe-oxohydrite; Gph – graphite).

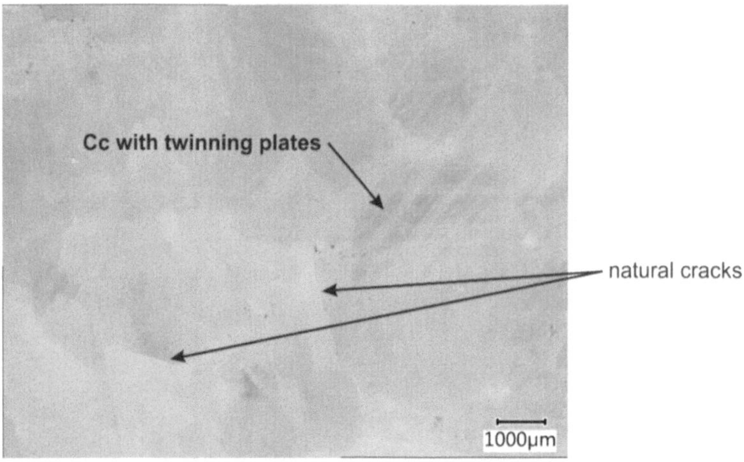

Fig. 9. A polished section of Supíkovice Marble (Cc – calcite).

4 Experimental Setup and Procedure

Erosion of rocks by flowing liquids was simulated based on the well-established *Accelerated Hydro-Abrasive-Cavitation Erosion* (AHACE) method, i.e. loading by means of a high-velocity water flow with erodent (abrasive) particles acting perpendicularly to the surface under study and moving along a predetermined trajectory (see [33]). Due to this, a larger area was affected, as the diameter of the flow was only 1 mm. One erosion cycle consisted of two mutually perpendicular zig-zag trajectories (Fig. 10). The perpendicular impact of the flow simulated real-life situation that occurs, for example, when water hits the bottom of a stilling basin of dam. In order to simulate the process as close as possible to real situation, relatively low impact velocities of the flow were applied and a small amount of abrasives was used (see, e.g. [33]). In this case, the material was not cut through and the water flow with erodents spreads at high velocity over the surface of the material after the impact, loading the surface layers with additional degradation processes (for example, shear forces, cavitation, abrasion, etc.).

The high-velocity flow was generated by a standard Slice I cutting head with a diamond water nozzle. This head is commonly used in abrasive waterjet cutting. Flow parameters in the experiments were set as specified in Table 2. The abrasive (erodent) material of the Australian GMA garnet of grain size of 80 mesh was used in experiments. Abrasive grain distribution of the Australian GMA garnet is presented in Fig. 11.

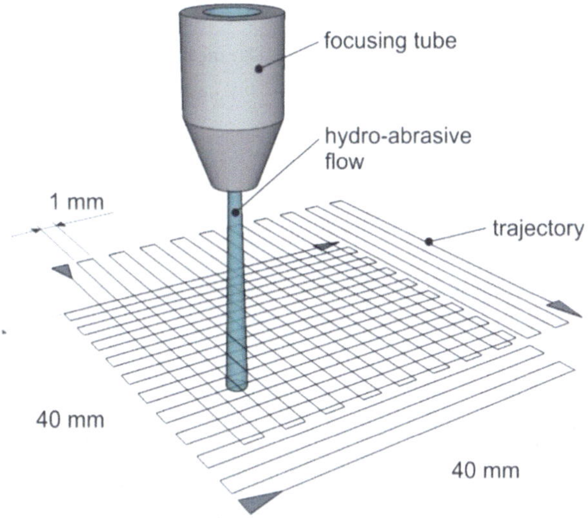

Fig. 10. Complex trajectory for one erosion cycle during surface treatment by the high-velocity water flow with erodents.

Rock samples were gradually eroded by an abrasive water flow. The flow generating nozzle was moved over the sample according to the trajectories shown in Fig. 10. One erosion cycle or two erosion cycles at the same place were carried out in each sample. The flow created some kind of eroded depressions (pockets) in test samples. The amount of material removed from these pockets V_e was a basic parameter used for the evaluation of erosion processes. This parameter was determined by a handheld 3D scanner CREAFORM GO!SCAN 3D. Once the eroded rock volume of individual pockets V_e was measured, the volumetric erosion rate V_r was specified as follows:

$$V_r = \frac{V_e}{t_e} \tag{1}$$

where t_e is the exposure time

$$t_e = N \cdot t_c \tag{2}$$

Table 2. Parameters of the applied high-velocity abrasive water flow.

Parameter	Value
upstream water pressure	40 MPa
water nozzle diameter	0.33 mm
focusing tube diameter	1.02 mm
focusing tube length	76.2 mm
stand-off distance	50 mm
impact angle of the flow	90°
traverse velocity of the flow	500 mm·min^{-1}
grain size of erodents	80 mesh
mass flow rate of erodents	20 g·min^{-1}

N is the number of erosion cycles and t_c is the time for one erosion cycle. The volumetric erosion rate V_r indicates the rate of erosion of surface layers of the test samples which better determinates the erosion abilities of the flowing water with erodents. Higher amount of eroded material, and thus higher volumetric erosion rate, indicated lower resistance to the flow applied, and vice versa.

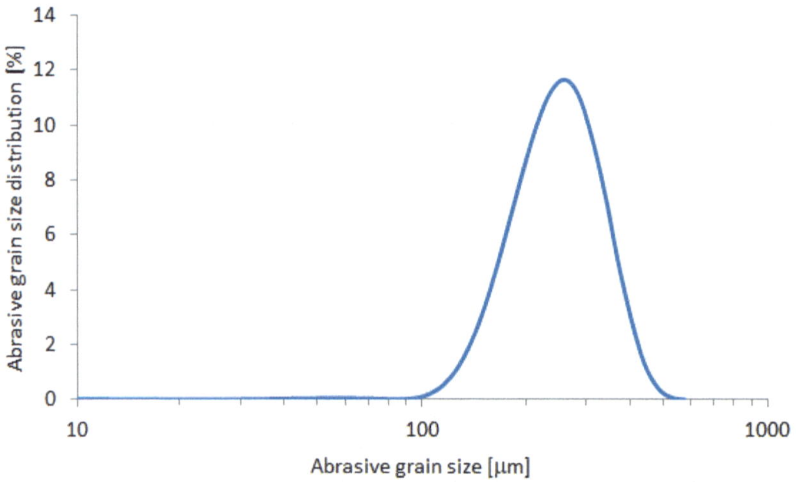

Fig. 11. Abrasive grain distribution of the Australian GMA garnet.

Apart from the material loss, new surfaces formed after the application of the AHACE method were analysed from a macro- and microscopic point of view. For microscopic analyses, a mobile digital Keyence VHX 5000 microscope was used. Newly-created surface profiles were then analysed by a FRT MicroProf optical profilometer in two, mutually perpendicular directions x and y.

In addition, abrasiveness and abrasivity according to the internal IGN methodology and the standard abrasiveness (abrasion resistance) according to Böhme [40] were measured for all selected rocks. The definitions of rock abrasiveness and rock abrasivity are similar and, therefore, very often need clarification. Even within the academic community, researchers often use the names interchangeably, which needs to be corrected. *Abrasiveness* is defined in different ways. In the case of natural and artificial materials, their susceptibility to abrasive wear is defined by abrasiveness. The abrasiveness is responsible for the extent to which the rock will be worn down as the abrasive material acts on it. In contrast with abrasiveness, *abrasivity* describes the ability of rocks to wear frictionally the surface of solid material element (primarily metal, but not only). Abrasivity is responsible for how much the element in contact with rock wears out [41].

The abrasivity of rocks according to the IGN methodology (based on the Czechoslovak ON 44 1121 – 1982 standard [42]) is described as the ability of rock to erode the surface of a solid tool in the disintegration process [43]. The test was based on determination of the weight loss of a standardized metal rod moving on a treated surface of a testing body at constant thrust in relation to the total length of the rod movement path [mg·m^{-1}]. For this testing, an abrasion meter was used, i.e. a rotating machine which had a constant rotational velocity of 100 rpm and generated a constant thrust of 100 N throughout the testing and which is enabled to register number of rotations with a settable switch-off. The abrasiveness according to the IGN methodology is then defined as the volume loss of a rock sample clamped in a grinding machine which performs a prescribed grinding path [43]. This testing was based on determination of the weight loss of a rock sample, which interacted with a test metal rod moving on a prepared surface of the test sample at constant thrust [mg·m^{-1}]. Here again, the test equipment consisted of an abrasion meter with a constant rotational velocity of 100 rpm and a constant thrust of 100 N.

5 Results and Discussion

Results of erosion tests (i.e. eroded volume V_e and volumetric erosion rate V_r) are presented in Figs. 12 and 13. Digital reconstructions of eroded pockets in each material, their profiles and detailed structure of obtained surfaces are shown in Figs. 14, 15, 16, 17, 18, 19, 20, 21, and 22.

5.1 Resistance to Erosion by a Hydro-Abrasive Flow

From the set of test samples, basalt was affected by erosion to the least extent (V_r = 8.4 mm^3·s^{-1}). Thus, it can be considered one of the most resistant aggregates to fast-flowing liquids. Metabasite from both the above-mentioned localities also shows a very good erosion resistance (V_r = 8.7 mm^3·s^{-1} and 11.8 mm^3·s^{-1}, respectively); while the resistance of metabasite with finer foliation is slightly worse. Resistance of approximately 60% compared to basalt was observed in fine-grained Litice Granodiorite (V_r = 13.4 mm^3·s^{-1}) followed by coarse-grained Liberec Granite (V_r = 14.9 mm^3·s^{-1}), Potštejn Gneiss (V_r = 15.4 mm^3·s^{-1}) and both marble types (V_r = 15.5 mm^3·s^{-1} or 15.7 mm^3·s^{-1}, respectively). The least resistant rock from the set of test samples is fine-grained Silesian Granodiorite with the resistance to the hydro-abrasive erosion at perpendicular impact angle of about 50% compared to basalt ($V_r = 16.6\ \text{mm}^3 \cdot \text{s}^{-1}$).

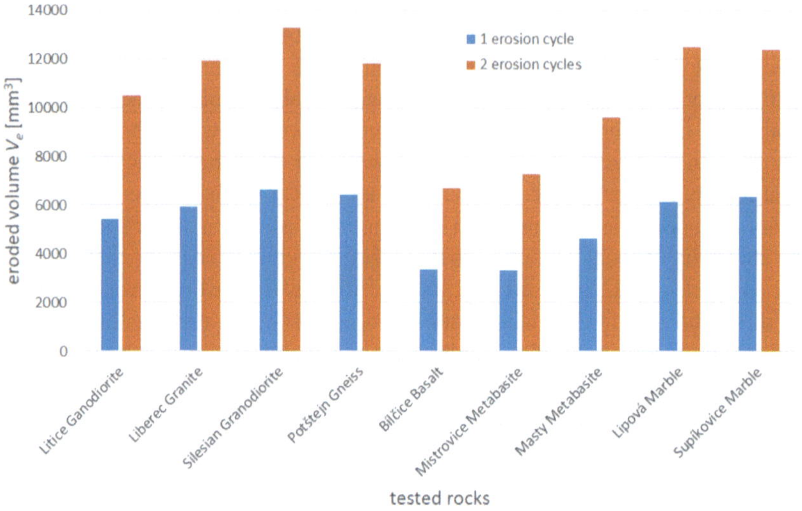

Fig. 12. Eroded volume V_e of tested rocks for one and two erosion cycles.

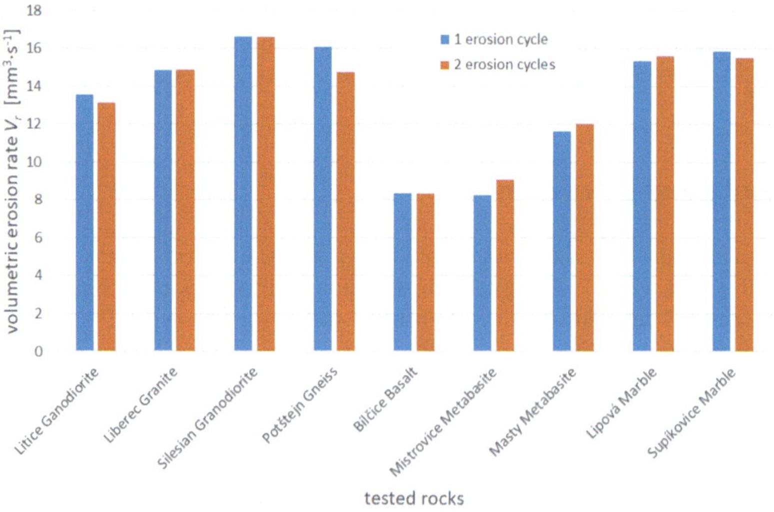

Fig. 13. Volumetric erosion rate V_r of tested rocks for one and two erosion cycles.

After the second erosion cycle applied at the same rock area loaded in the first cycle, all tested rocks showed approximately double the amount of material removal compared to the first cycle (Fig. 12). This implies that the rock samples we tested can be considered homogeneous in the macro-volume with stable strength and erosion resistance (also see Fig. 13). Indeed, the authors have previously proved that inhomogeneity in surface

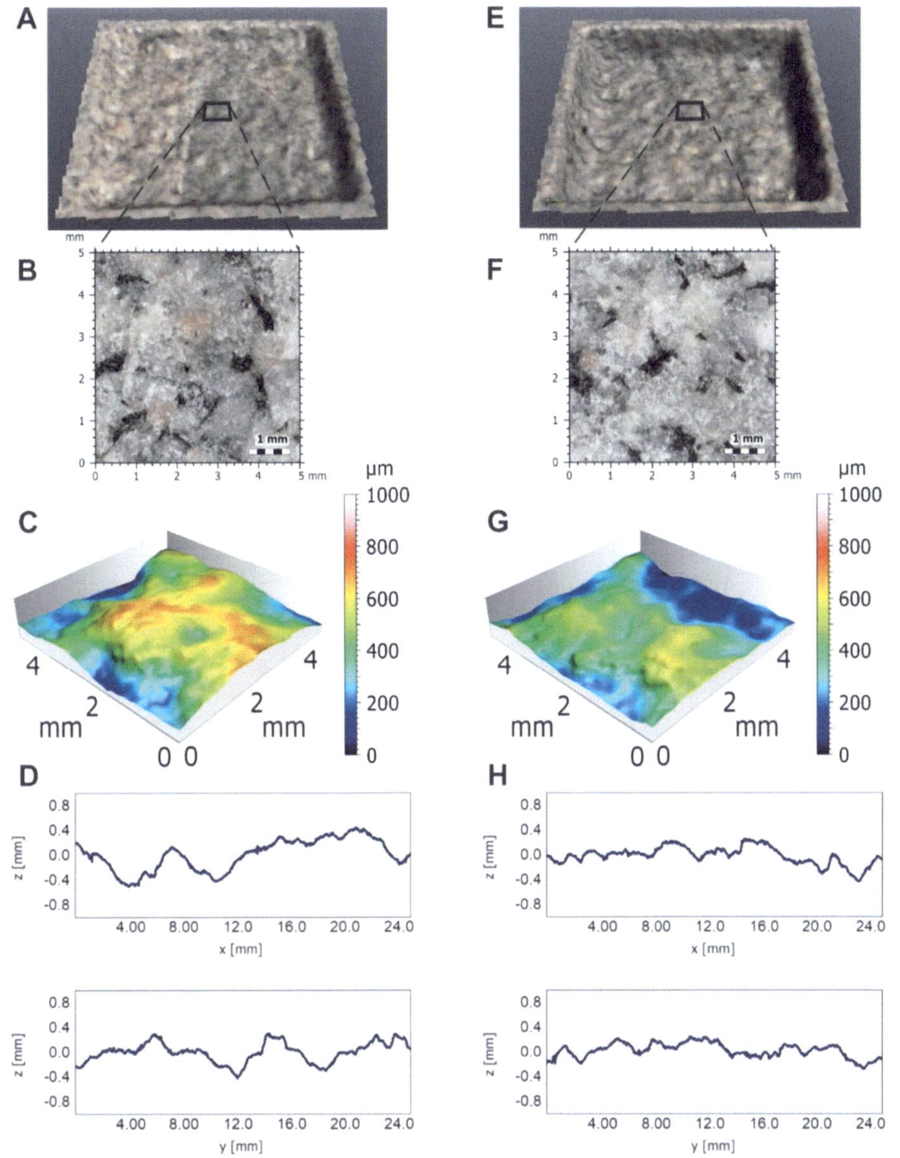

Fig. 14. Example of digital reconstructions of pockets eroded in Litice Granodiorite after one (left) and two (right) erosion cycles. (A, E – eroded pockets, B, F – details of obtained surfaces, C, G – height structure of details, D, H - profiles of eroded surfaces in two, mutually perpendicular directions).

layers caused uneven material removal when several erosion cycles were repeated at the same area [32]. The double volume removed after the second cycle suggests another interesting feature: while the flow impacted on a smooth surface of test samples during the first cycle, in the second cycle, the flow impacted on an irregular surface obtained

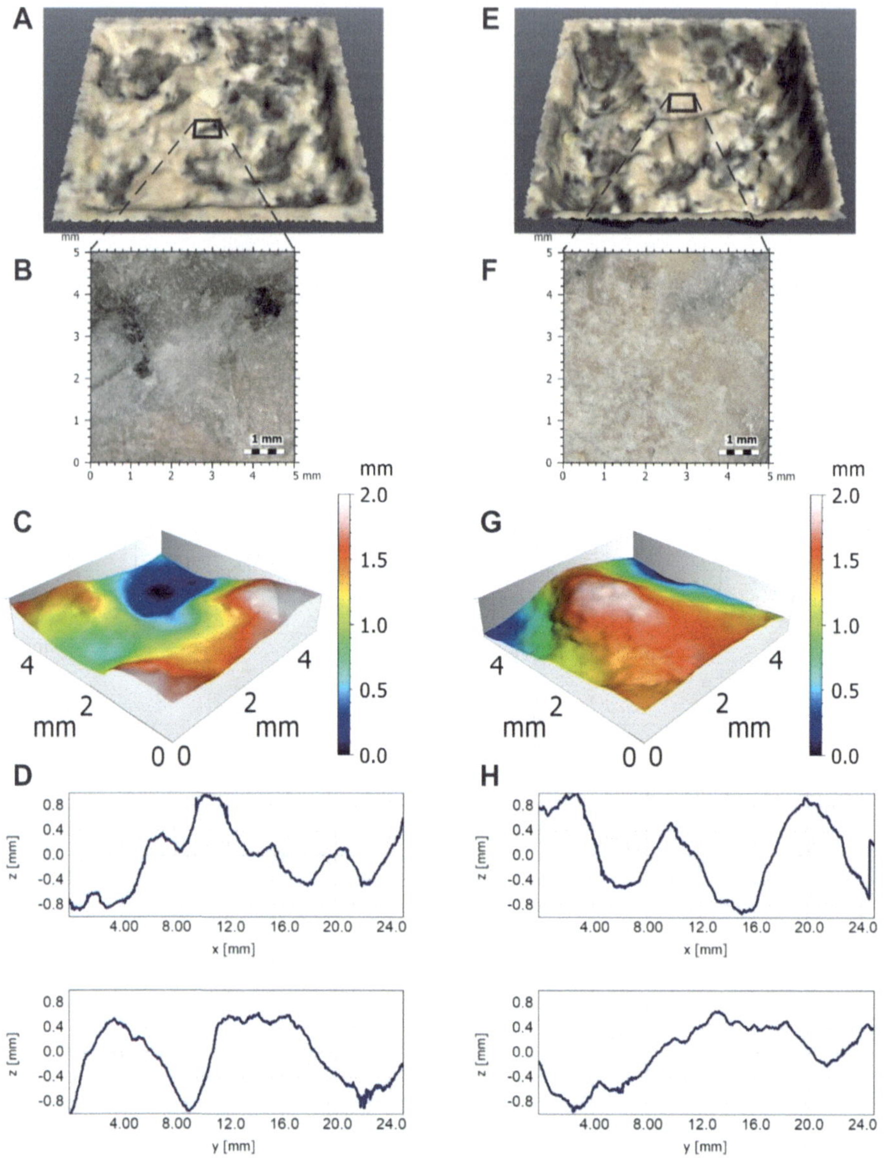

Fig. 15. Example of digital reconstructions of pockets eroded in Liberec Granite after one (left) and two (right) erosion cycles. (A, E – eroded pockets, B, F – details of obtained surfaces, C, G – height structure of details, D, H - profiles of eroded surfaces in two, mutually perpendicular directions).

after the first flow cycle. In case of the hydro-abrasive-cavitation erosion, the initial condition of the surface to be loaded (smooth vs. rough) thus does not play a significant role during the perpendicular impact of the flow.

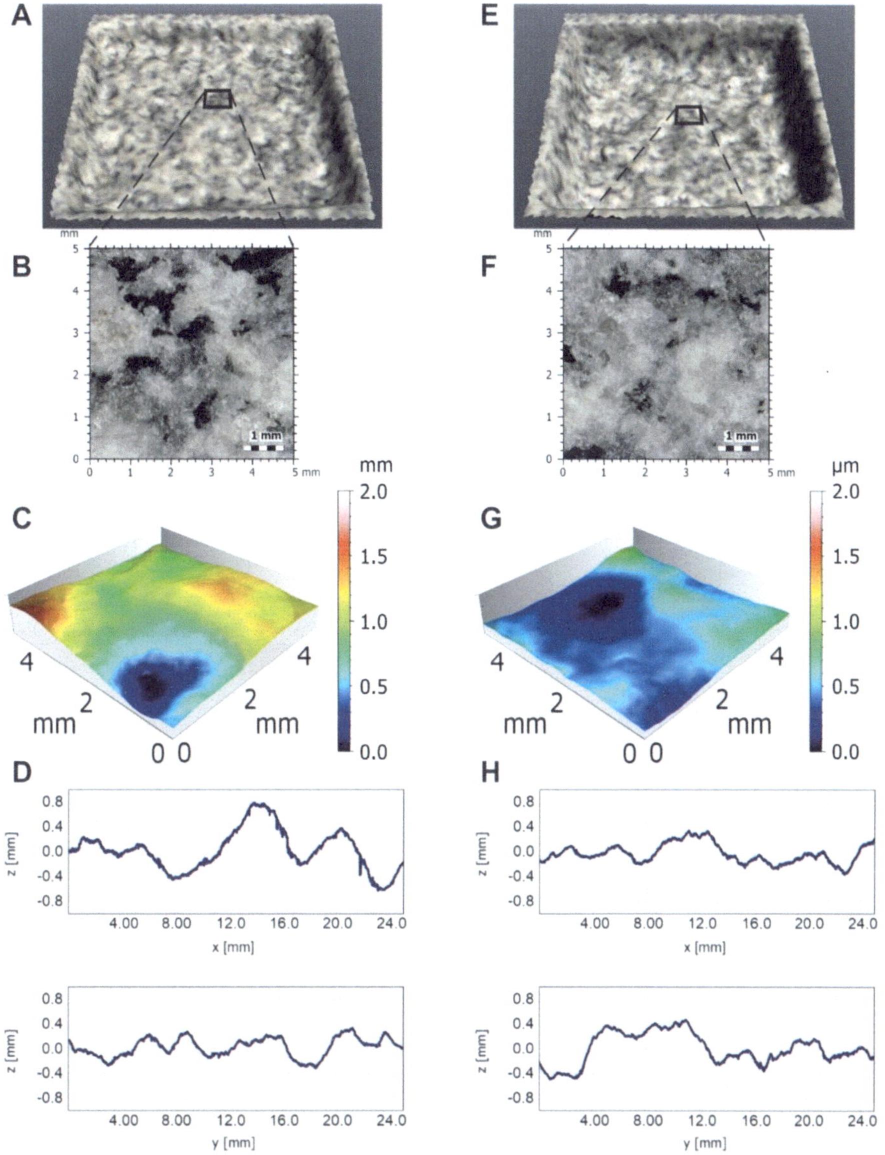

Fig. 16. Example of digital reconstructions of pockets eroded in Silesian Granodiorite after one (left) and two (right) erosion cycles. (A, E – eroded pockets, B, F – details of obtained surfaces, C, G – height structure of details, D, H - profiles of eroded surfaces in two, mutually perpendicular directions).

The grain size of rocks is not related to their resistance to the hydro-abrasive-cavitation erosion. The erosion resistance is rather affected by the resistance of individual

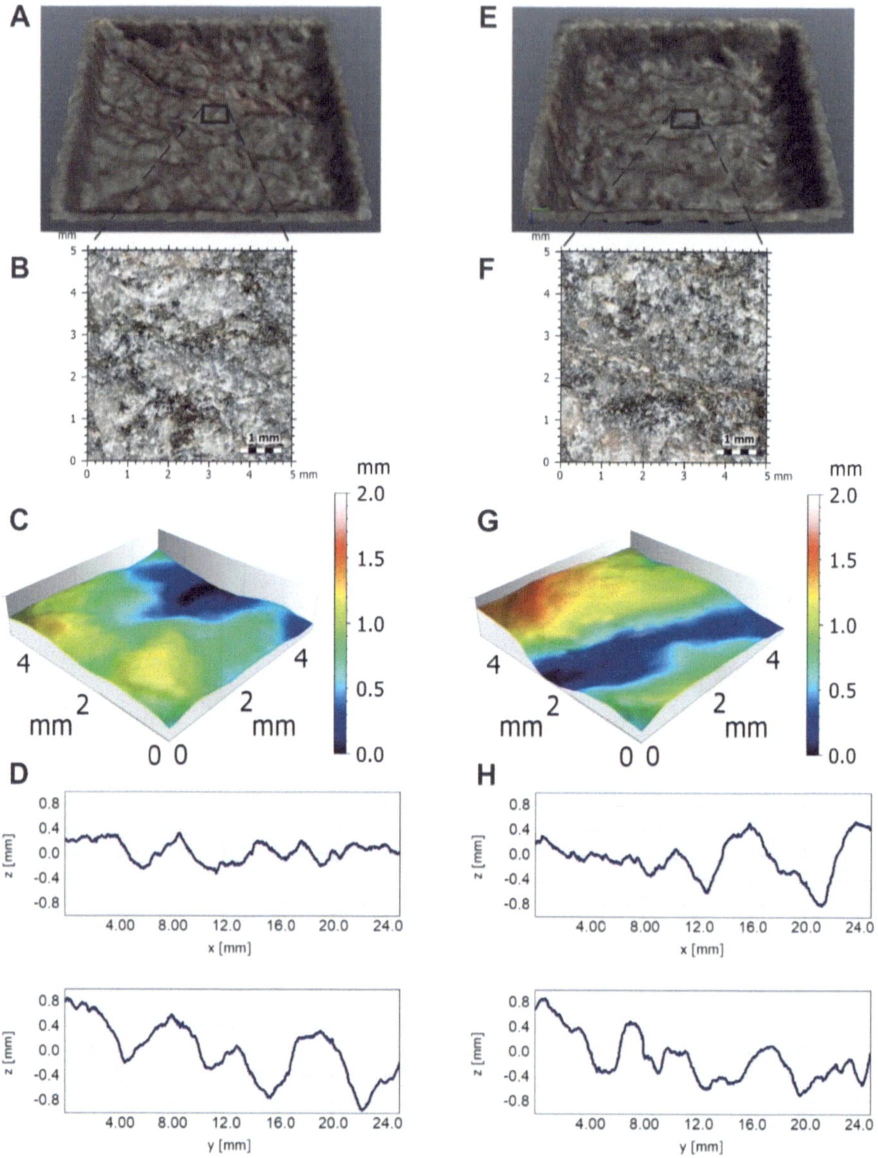

Fig. 17. Example of digital reconstructions of pockets eroded in Potštejn Gneiss after one (left) and two (right) erosion cycles. (A, E – eroded pockets, B, F – details of obtained surfaces, C, G – height structure of details, D, H - profiles of eroded surfaces in two, mutually perpendicular directions).

rock structural elements and the bounds between them (compare, e.g., Litice Granodiorite and Liberec Granite in Figs. 14 and 15). Similarly, no relation was found between

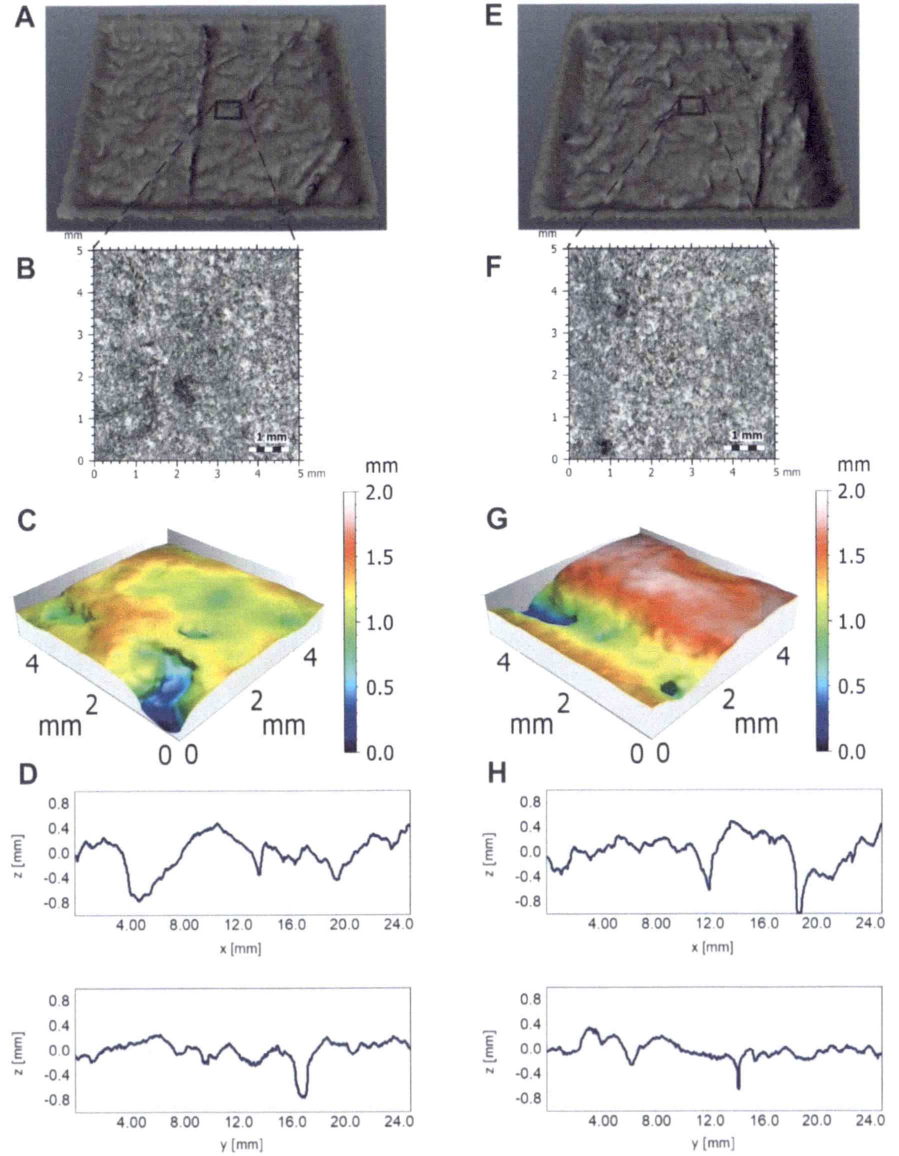

Fig. 18. Example of digital reconstructions of pockets eroded in Bílčice Basalt after one (left) and two (right) erosion cycles. (A, E – eroded pockets, B, F – details of obtained surfaces, C, G – height structure of details, D, H - profiles of eroded surfaces in two, mutually perpendicular directions).

petrographic type of rocks (igneous, metamorphic) and their resistance to the erosion caused by fast-flowing liquids with solid particles.

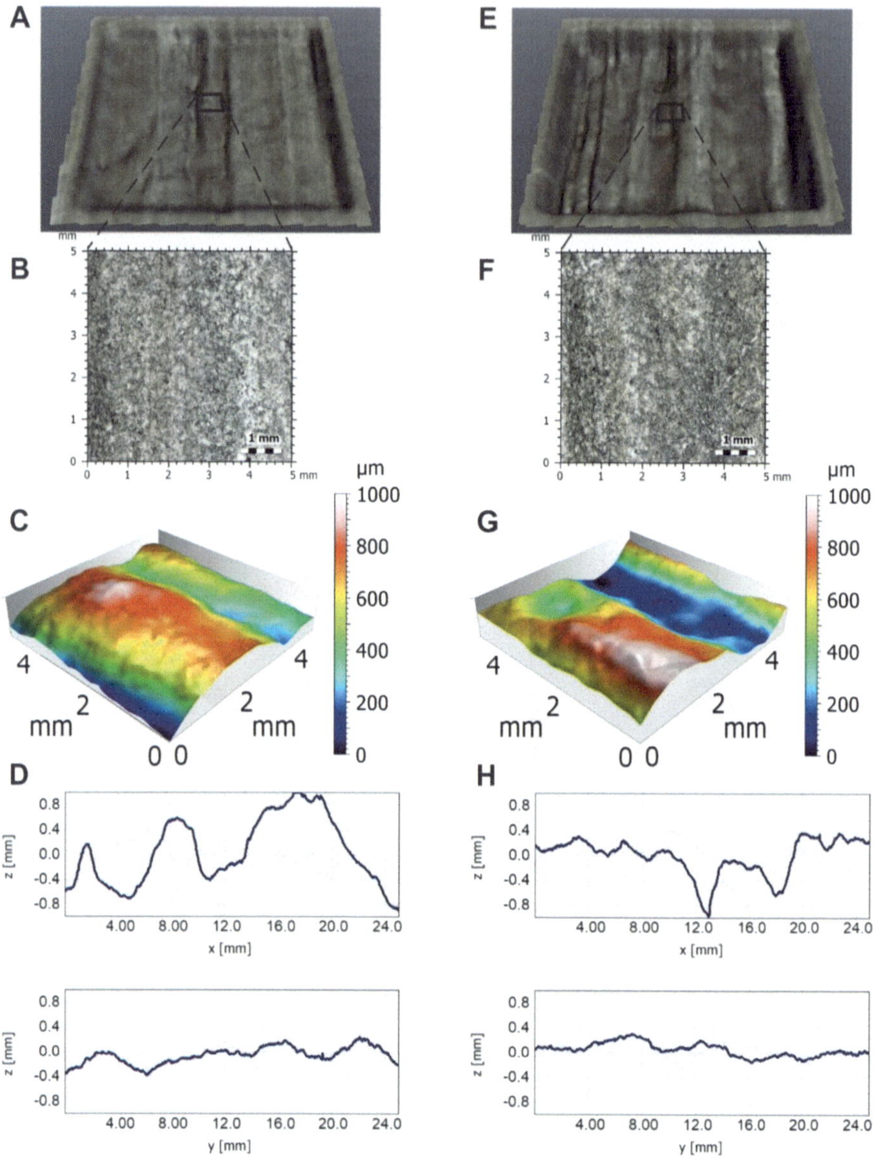

Fig. 19. Example of digital reconstructions of pockets eroded in Mistrovice Metabasite after one (left) and two (right) erosion cycles. (A, E – eroded pockets, B, F – details of obtained surfaces, C, G – height structure of details, D, H - profiles of eroded surfaces in two, mutually perpendicular directions).

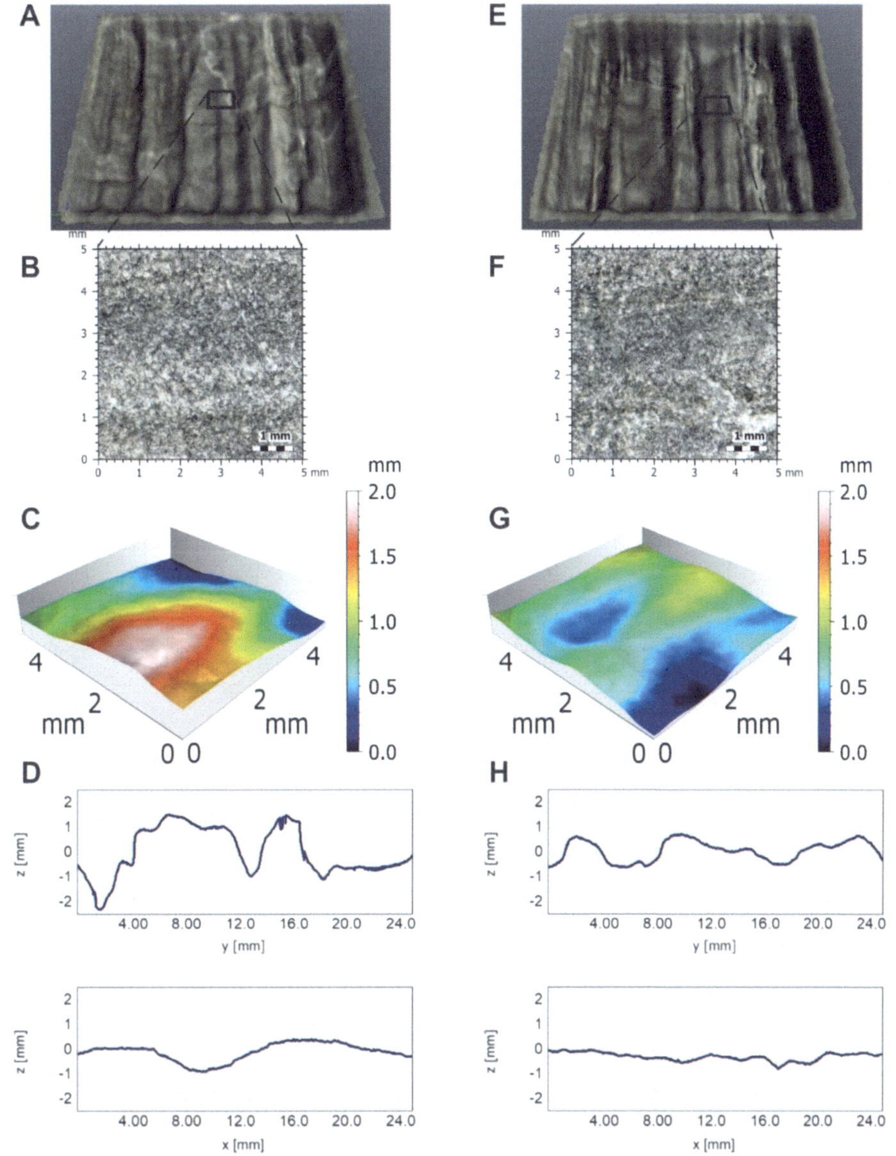

Fig. 20. Example of digital reconstructions of pockets eroded in Masty Metabasite after one (left) and two (right) erosion cycles. (A, E – eroded pockets, B, F – details of obtained surfaces, C, G – height structure of details, D, H - profiles of eroded surfaces in two, mutually perpendicular directions).

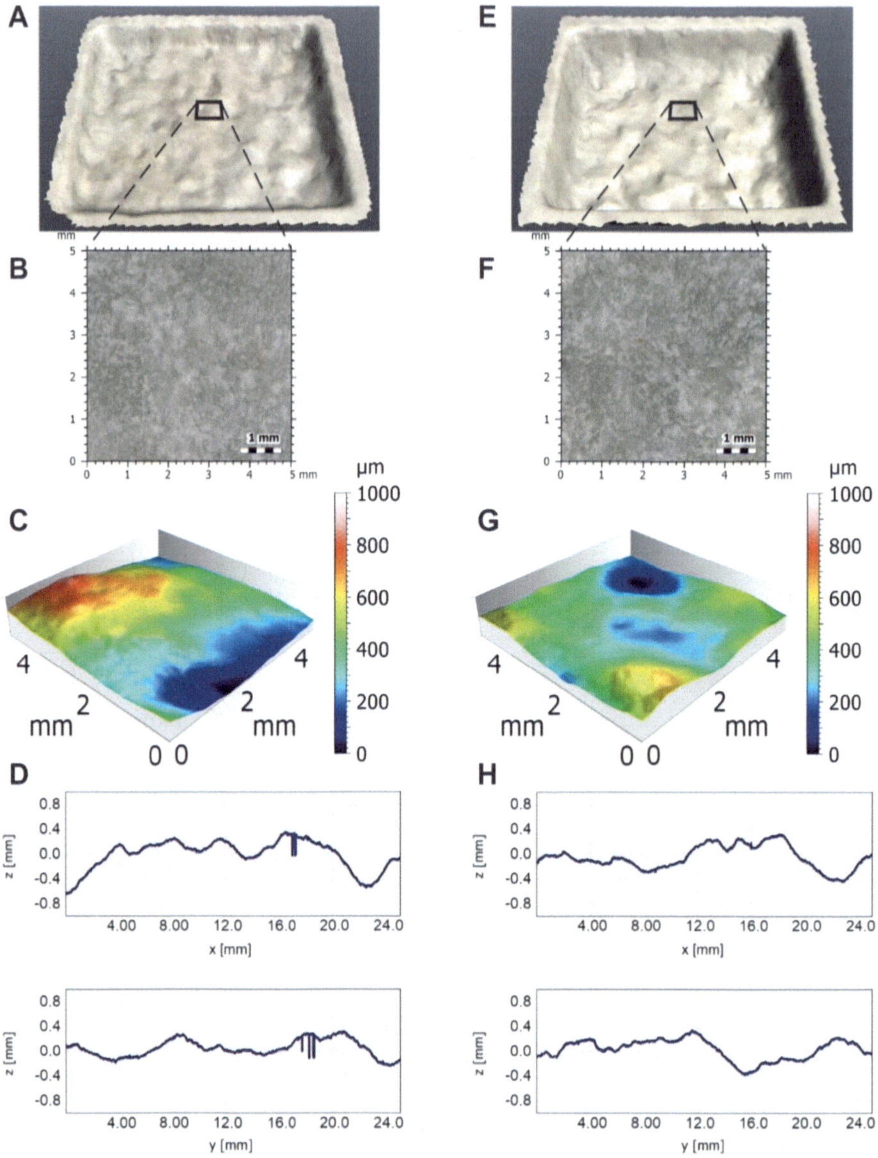

Fig. 21. Example of digital reconstructions of pockets eroded in Lipová Marble after one (left) and two (right) erosion cycles. (A, E – eroded pockets, B, F – details of obtained surfaces, C, G – height structure of details, D, H - profiles of eroded surfaces in two, mutually perpendicular directions).

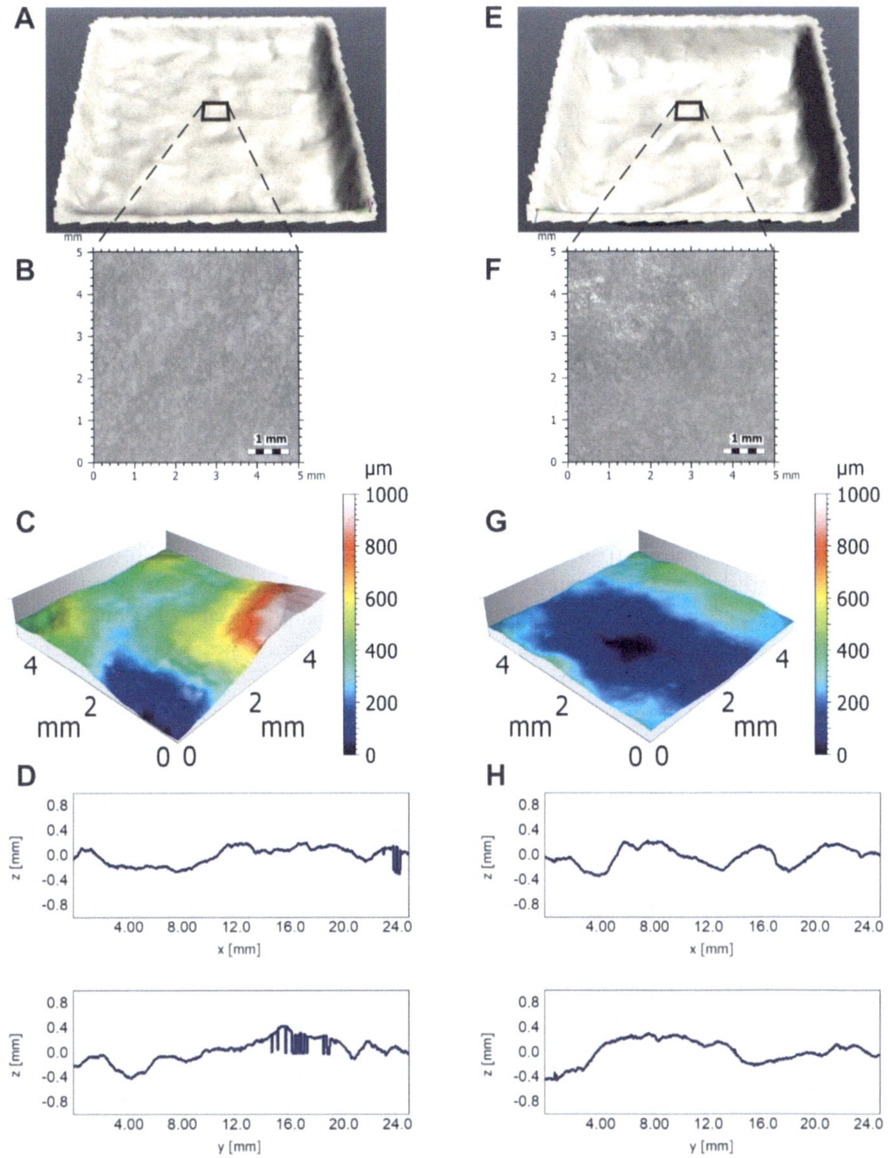

Fig. 22. Example of digital reconstructions of pockets eroded in Supíkovice Marble after one (left) and two (right) erosion cycles. (A, E – eroded pockets, B, F – details of obtained surfaces, C, G – height structure of details, D, H - profiles of eroded surfaces in two, mutually perpendicular directions).

5.2 Analysis of a Newly Created Surface After the Erosion Using the Hydro-Abrasive Flow

Pairs of erosional depressions (pockets) in all tested rocks created by the application of one or two erosion cycles show the same texture and identical features. Surface profiles created in erosion tests are clearly presented in Figs. 14 to 22. It is evident that the eroded surface is strongly dependent on the internal structure of a particular rock, its spatial anisotropy, and size and characteristics of individual grains which the rock consists of. While compact fine-grained rocks are eroded more or less uniformly throughout the sample volume and regardless of the flow direction (e.g., in Litice Granodiorite and Silesian Granodiorite – Figs. 14 and 16), in other rocks, significant surface irregularities can be observed depending on strength and erosion resistance of individual structural elements or on the overall anisotropy of a particular rock (e.g., in both metabasites tested – Figs. 19 and 20).

In all *granodiorites* and *granites* tested, outgrowths of orthoclase and white plagioclase form elevations (i.e. elevated areas more resistant to erosion). The outgrowths are conically eroded emphasizing the cleavage surfaces of feldspars. Depressions (cavities, i.e. areas less resistant to erosion) are formed by destroyed grey grains of quartz which are significantly eroded and fissured. Flakes of micas are destroyed in the direction of the basal cleavage and the extracted flakes are oriented parallel to the cleavage surface. The erosion affects more the quartz- and biotite-rich parts of the rock. The extent of erosion on crystals depends on the size of feldspar outgrowths.

Different texture of the eroded surface can be observed in *basalt*. Elevations consist of lighter accumulations of aggregates of fine lamellae of basal plagioclase with glassy phase, while the depressions consist of areas rich in glass and outgrowths of fissile olivine. Erosion mostly affects the rock parts rich in glassy phase. Deep erosion is also evidenced in parts with glassy intrusions in the massive rock. The erosion reveals millimetre-sized gas pores.

Marbles are rather homogenous and monomineralic rocks with regular eroded surface texture without directionality. Elevations are intersections of rhombohedron surfaces on calcite grains in direction of the crystallographic *c* axis of the crystal. Depressions are places of intersections of rhombohedron surfaces $\{10\bar{1}1\}$ of adjacent crystals. Erosion affects calcite on visible cleavage surfaces in form of fine erosion with chipped depressions along cleavage surfaces. At a macroscopic scale, compression twinning on the grains is evident.

In tested *gneiss,* elevations are formed by bands of light-coloured feldspar-rich melanosome with minor share of biotite and quartz. Top of the band is conically rounded. Depressions are formed by bands of quartz- and biotite-bearing leucosome where largely destructed grains of grey quartz get accumulated. They are significantly eroded and fissured. Biotite flakes are damaged in direction of the basal cleavage, as the hydro-abrasive flow acts perpendicular to the foliation surface. Erosion affects more the parts of rock with local accumulation of quartz and micas. The boundary between leucosome and melanosome bands is not sharp; the two bands interpenetrate each other and disappear.

Both tested *metabasites* show significant anisotropy. Here as well, the hydro-abrasive flow acted perpendicular to the surface of foliation. Elevations consist of thick lighter bands rich in feldspar and amphibole with a minor presence of epidote. Top of the band is conically rounded. In Masty Metabasite, the transitions between elevation and depression are gradual and rounded; whereas in Mistrovice Metabasite, significant difference between bands with elevation and the subsequent erosional depression is observed. These depressions form an erosion trough in the epidote-rich band with a minor presence of amphibole. Erosion affects more the rock parts with local accumulation of epidote. While the boundary between the bands in Masty Metabasite is not sharp and these two parts interpenetrate each other, the boundary between the bands in Mistrovice Metabasite is distinct, sometimes sharp.

The grain size of rocks has a significant effect on the overall profile of newly created surfaces. While the profiles of coarse-grained rocks are relatively irregular with higher roughness closely related to sizes of structural grains (e.g. Liberec Granite – Fig. 15), the profiles of fine-grained rocks are more regular, showing lower roughness (Litice Granodiorite – Fig. 14).

We were surprised by unusual erosion resistance of some structural grains occurring in the rock structure. The fragile but hard and non-cleavable quartz grains showed relatively low erosion resistance to fast-flowing liquids (e.g. Figures 14–16). However, quartz is considered, in general, to be very resistant to the abrasive erosion [44]. In contrast, feldspars (i.e. plagioclase and orthoclase), which are otherwise friable and cleavable materials [44], showed good erosion resistance (Figs. 14–17).

5.3 Comparison of Outputs of the AHACE Method with Standard Abrasion Tests and with Other Parameters

Based on the outputs of experiments carried so far [30–34], we supposed that the erosion resistance of rocks to fast-flowing liquids determined by the AHACE method would simply correlate with the strength or the abrasion resistance of rocks measured by standard methods. We also supposed that the evident relationship between erosion resistance and other parameters would be linear.

However, the efforts to simply correlate the erosion resistance of rocks to fast-flowing liquids with the rock strength characteristics or various abrasion resistances were unsuccessful. An example of linear correlation lines of the erosion resistance using the AHACE method related to selected strength characteristics is shown in Fig. 23. In Fig. 24, the linear correlation is related to some standard abrasion characteristics. Values of the R^2 coefficient of determination are very low, which proves no relationships. We also tried to derive other correlation dependencies (exponential, logarithmic, power, polynomial, etc.). Unfortunately, no correlations were found. The proposed AHACE method thus appears to be a unique way to determine relevant values of the resistance of materials to fast-flowing liquids, as it is based on similar principles and mechanisms occurring in real-life situations.

6 Conclusions

Resistance of nine types of rocks to fast-flowing liquids was investigated based on experiments with the accelerated hydro-abrasive erosion caused by high-velocity water flows with abrasive particles as erodents which moved along a predefined trajectory over a material to be eroded. The research outputs are summarised below point by point:

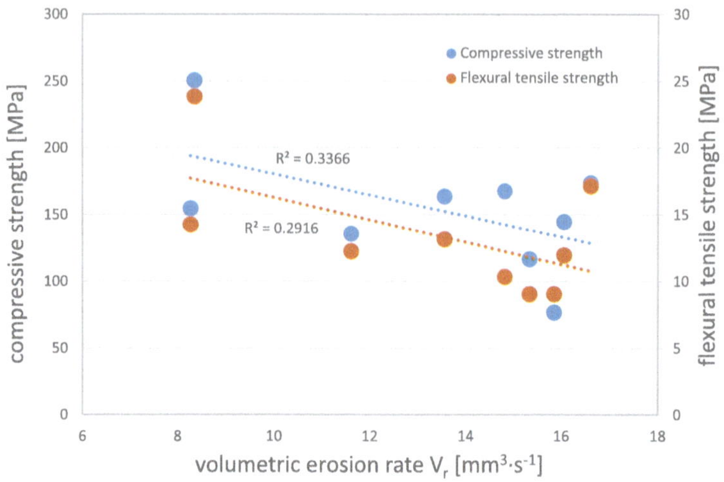

Fig. 23. Correlation between the erosion resistance of tested rocks using the AHACE method and the compressive and flexural tensile strengths (based on strength values from Table 1).

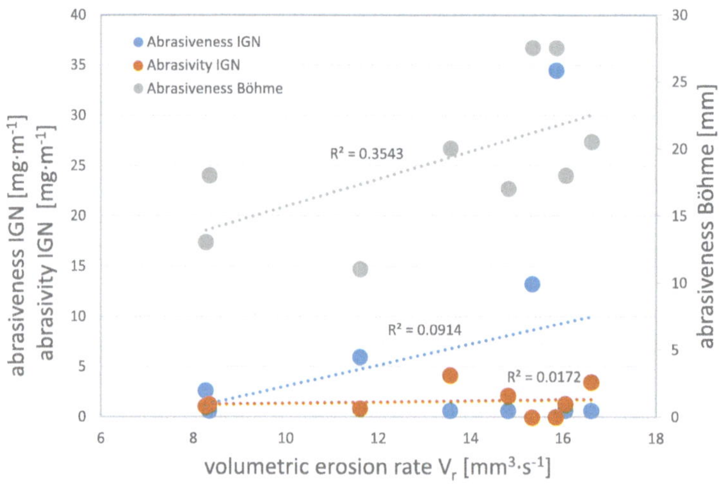

Fig. 24. Correlation between the erosion resistance of rocks using the AHACE method and the Böhme abrasiveness, IGN abrasiveness and IGN abrasivity (based on values of the rock abrasiveness and rock abrasivity presented in Table 1).

1. Tests of the erosion resistance of rocks have revealed that the basalt aggregate is the most erosion-resistant material (V_r = 8.4 mm^3·s^{-1}), followed by metabasites (V_r = 8.7 mm^3·s^{-1} or 11.8 mm^3·s^{-1}) and aggregates based on fine-grained Litice Granodiorite (V_r = 13.4 mm^3·s^{-1}), coarse-grained Liberec Granite (V_r = 14.9 mm^3·s^{-1}), Potštejn Gneiss (V_r = 15.4 mm^3·s^{-1}) and marbles (V_r = 15.5 mm^3·s^{-1} or 15.7 mm^3·s^{-1}, respectively). The fine-grained Silesian Granodiorite showed the worst, roughly half of the basalt resistance (V_r = 16.6 mm^3·s^{-1}).

2. Surprisingly, the fragile but hard and non-cleavable quartz grains occurring in the aggregate structure showed relatively low erosion resistance to fast-flowing liquids. In general, quartz is considered to be very resistant to the abrasive erosion. In contrast, feldspars, which are otherwise friable and cleavable materials, showed good erosion resistance.
3. No correlation was found between the size of mineral grains of rock and the resistance to the hydro-abrasive-cavitation erosion. The erosion resistance cannot be either correlated to the petrographic type of aggregate materials (igneous or metamorphic in our case).
4. Surfaces obtained after the application of the AHACE method are different for each type of rock. The final surface texture depends on the internal structure of a particular rock, erosion resistance of its structural elements and grain size.
5. It has been shown that the erosion resistance of rocks cannot be simply related to other parameters and properties of the material eroded (e.g. strength, abrasiveness and abrasivity).
6. Future research could benefit from comparing the AHACE method's results with those obtained through traditional erosion resistance testing methods and a more detailed statistical analysis (e.g. variability within rock samples or analysis between different rock types) could strengthen the conclusions drawn from this study.

Acknowledgement. This research was funded by Czech Science Foundation, grant number 23-07415S. The authors are very thankful for the support. The authors would also like to express their special thanks to Lucie Gurková for translation and language corrections. Lucy, you are the best!

References

1. American Concrete Institute, ACI 210R-93, reapproved 1998, Erosion of Concrete in Hydraulic Structures. In: ACI Committee 210 Report (1998)
2. Horszczaruk, E.K.: Hydro-abrasive erosion of high performance fiber-reinforced concrete. Wear **267**, 110–115 (2009)
3. Horszczaruk, E.K.: Abrasion resistance of high strength fibre-reinforced concrete. 6th RILEM Symposium on Fibre-Reinforced Concretes – BEFIB 2004, Varenna (2004)
4. Binici, H., Aksogan, O., Gorur, E.B., Kaplan, H., Bodur, M.N.: Hydro-abrasive erosion of concrete incorporating ground blast-furnace slag and ground basaltic pumice. Constr. Build. Mater. **23**, 804–811 (2009)
5. Jacobsen, S., Scherer, G.W., Schulson, E.M.: Concrete-ice abrasion mechanics. Cem. Concr. Res. **73**, 79–95 (2015)
6. Martinovic, S., Vlahovic, M., Boljanac, T., Dojcinovic, M., Volkov-Husovic, T.: Cavitation resistance of refractory concrete: Influence of sintering temperature. J. Eur. Ceram. Soc. **33**, 7–14 (2013)
7. Wang, X., Xie, T.Q.: Cavitation erosion behavior of hydraulic concrete under high-speed flow. Anti-Corrosion Methods and Materials **69**(1), 81–93 (2022)
8. Han, S.-C., Jo, Y., Yun, J.-I.: Chemical degradation of fly ash blended concrete with the seasonal variation of rainwater in a radioactive waste repository: A thermodynamic modeling approach. Cem. Concr. Res. **141**, 106326 (2021)

9. Liu, P., Chen, Y., Wang, W., Yu, Z.: Effect of physical and chemical sulfate attack on performance degradation of concrete under different conditions. Chemical Physics Letters **745** (2020)
10. Li, X., Lin, X., Lin, K., Ji, T.: Study on the degradation mechanism of sulphoaluminate cement sea sand concrete eroded by biological sulfuric acid. Constr. Build. Mater. **157**, 331–336 (2017)
11. ASTM C 1138-97, 2002. Standard test method for abrasion resistance of concrete (underwater method). In: Annual Book of ASTM Standards, vol. 04.02, ASTM International, West Conshohocken, PA (2002)
12. ASTM C944 / C944M-19. Standard Test Method for Abrasion Resistance of Concrete or Mortar Surfaces by the Rotating-Cutter Method, ASTM International, West Conshohocken, PA (2019)
13. ASTM C779 / C779M-19. Standard Test Method for Abrasion Resistance of Horizontal Concrete Surfaces, ASTM International, West Conshohocken, PA (2019)
14. Momber, A.W., Kovacevic, R.: Accelerated high speed water erosion test for concrete wear debris analysis. Tribol. Trans. **39**(4), 943–949 (1996)
15. Hu, X.G., Momber, A.W., Yin, Y.G.: Hydro-abrasive erosion of steel-fibre reinforced hydraulic concrete. Wear **253**, 848–854 (2002)
16. Hu, X.G., Momber, A.W., Yin, Y.: Erosive wear of hydraulic concrete with low steel fiber content. J. Hydraul. Eng. **132**, 1331–1340 (2006)
17. Momber, A.W.: Short-time cavitation erosion of concrete. Wear **241**, 47–52 (2000)
18. Momber, A.W.: Effects of erodent flow energy and local exposure time on the erosion of cement-based composites at high-speed hydro-abrasive flow. Wear **378–379**, 145–154 (2017)
19. Momber, A.W., Mohan, R.S., Kovacevic, R.: On-line analysis of hydro-abrasive erosion of pre-cracked materials by acoustic emission. Theoret. Appl. Fract. Mech. **31**, 1–17 (1999)
20. Momber, A.W.: Stress-strain relation for water-driven particle erosion of quasi-brittle materials. Theoret. Appl. Fract. Mech. **35**, 19–37 (2001)
21. Momber, A.W.: A probabilistic model for the erosion of cement-based composites due to very high-speed hydro-abrasive flow. Wear **368**, 39–44 (2016)
22. Hood, M., Nordlund, R., Thimons, E.: A Study of Rock Erosion using High-Pressure Water Jets. Int. J. Rock Mech. Min. Sci. & Geomech. Abstr. **27**, 77–86 (1990)
23. Dehkhoda, S., Hood, M.: An experimental study of surface and sub-surface damage in pulsed water-jet breakage of rocks. Int. J. Rock Mech. Min. Sci. **63**, 138–147 (2013)
24. Ozcelik, Y., Ciccu, R., Costa, G.: Comparison of the water jet and some traditional stone surface treatment methods in different lithotypes. Constr. Build. Mater. **25**, 678–687 (2011)
25. Ozcelik, Y., Tercana, A.E., Yilmazkayaa, E., Ciccu, R., Costa, G.: A study of nozzle angle in stone surface treatment with water jets. Constr. Build. Mater. **25**, 4271–7478 (2011)
26. Młynarczuk, M., Skiba, M., Sitek, L., Hlaváček, P., Kožušníková, A.: The research into the quality of rock surfaces obtained by abrasive water jet cutting. Arch. Min. Sci. **59**(4), 925–940 (2014)
27. Sitek, L., Foldyna, J., Martinec, P., Ščučka, J., Bodnárová, L., Hela, R.: Use of pulsating water jet technology for removal of concrete in repair of concrete structures. The Baltic Journal of Road and Bridge Engineering **6**(4), 235–242 (2011)
28. Hlaváč, L.M., Bodnárová, L., Janurová, E., Sitek, L.: Comparison of continuous and pulsing water jets for repair actions on road and bridge concrete. The Baltic Journal of Road and Bridge Engineering **7**(1), 53–59 (2012)
29. Bodnárová, L., Válek, J., Sitek, L., Foldyna, J.: Effect of high temperatures on cement composite materials in concrete structures. Acta Geodynamica et Geomaterialia **10**(2), 173–180 (2013)

30. Sitek, L., Hlaváček, P., Bodnárová, L., Hela, R., Martinec, P., Klichová, D., Foldyna, J., Foldyna, V.: Research activities in the field of concrete disintegration by high-speed water jets. Proceedings of the 12th pacific rim international conference on water jet technology (PRIC 2019), 8–9 Aug., Chiba (2019)
31. Sitek, L., Hlaváček, P., Bodnárová, L.: Use of high-speed water flows for accelerated mechanical modelling of erosive wear of concrete surfaces. MATEC Web of Conferences 244, 02007, In: Stančeková, D., Vaško, M., Rudawska, A., Čuboňová, N., Sapietová, A., Mrázik, J., Szabelski, J., Tlach, V. (eds.) Innovative Technologies in Engineering Production. Bojnice (2018)
32. Hlaváček, P., Sitek, L., Hela, R., Bodnárová, L.: Erosion test with high-speed water jet applied on surface of concrete treated with solution of modified lithium silicates. In: Hloch, S., Klichová, D., Krolczyk, G. M., Chattopadhyaya, S., Ruppenthalová, L. (eds.) Advances in Manufacturing Engineering and Materials. Proc. of the Int. Conf. on Manufacturing Engineering and Materials, LNME, pp. 135–143. Springer Nature Switzerland AG, Basel (2019)
33. Hlaváček, P., Sitek, L., Klichová, D., Bodnárová, L.: Effects of abrasives during accelerated simulation of mechanical corrosion of cement composites using abrasive water flow. In: Reiterman, P. (ed.) Acta Polytech. CTU Proc. 2019, vol. 22, pp. 31–37. Czech Technical University in Prague, Praha (2019)
34. Sitek, L., Hlaváček, P., Souček, K., Bodnárová, L., Foldyna, J., Zajícová, V., Berčáková, A., Foldyna, V.: X-Ray CT inspection of subsurface areas of concretes exposed to fast flowing liquids. In: Kotwica, K. (ed.) New trends in production engineering 2, pp. 450–459. Sciendo, Warsawa (2019)
35. Sitek, L., Bodnárová, L., Souček, K., Staš, L., Gurková, L.: Analysis of Inner Structure Changes of Concretes Exposed to High Temperatures using Micro X-ray Computed Tomography. Acta Geodynamica et Geomaterialia **12**(1), 79–89 (2015)
36. EN 1936:2006 Natural stone test methods - Determination of real density and apparent density, and of total and open porosity (2006)
37. EN 1926:2006 Natural stone test methods - Determination of uniaxial compressive strength (2006)
38. EN 12372:2006 Natural stone test methods - Determination of flexural strength under concentrated load (2006)
39. EN 13755:2008 Natural stone test methods - Determination of water absorption at atmospheric pressure (2008)
40. EN 14157:2017 - Natural stone test methods - Determination of the abrasion resistance (2017)
41. Mucha, K.: Application of Rock Abrasiveness and Rock Abrasivity Test Methods - A Review. Sustainability **15**(14), 11243 (2023)
42. ON 44 1121/81:1982 - Industry Standard - Solid rocks, Determination of abrasivity, Czechoslovak Federal Ministry of Fuels and Energy, Praha, VVUÚ Ostrava – Radvanice (1982)
43. Vašek, J.: Rock properties assessing when boom tunneling machines are to be used, Proceedings of the 14th International Symposium on Mine Planning and Equipment Selection, MPES 2005 and the 5th International Conference on Computer Applications in the Minerals Industries, CAMI 2005, pp. 1347–1373 (2005)
44. Szymański, A., Szymański, J.M.: Hardness Estimation of Minerals, Rocks and Ceramic Materials. Materials Science Monographs 49. Elsevier Science Publishers B.V., Amsterdam (1989)

Influence of Liquid Viscosity on the Origin and Propagation of Self-Excited Oscillations

Michal Zeleňák, Vladimír Foldyna(✉), Zdeněk Říha, Fernando Kevin Miranda, and Akash Nag

Fluid Jets Department, Institute of Geonics of the CAS, Ostrava, Czech Republic
vladimir.foldyna@ugn.cas.cz

Abstract. This paper deals with the study of the effect of fluid viscosity on the ability to generate and propagate self-excited oscillations in the inner space of a fluidic oscillator and at its outlet to the air environment. The generation and propagation of self-excited oscillations was monitored at several levels of inlet pressure values (5, 10, 15, 20 and 25 bar) and two flowing fluid consternations of the ethylene glycol (EG) (EG 100% and EG 90% with 10% of water). The fluidic oscillator was designed and manufactured with the possibility of connecting pressure sensors with a high sampling rate, allowing the measurement of the dynamic processes inside the oscillator. The experimental measuring circuit consisting of a middle-pressure pump, hose, and fluidic oscillator was equipped with flow and pressure diagnostics. The output profile of the self-excited fluid jet into the air environment was observed using a high-speed optical shadowgraphy technique. The dynamic viscosity values were measured in the temperature interval (18–46 °C) for selected concentrations of EG. The results of the observation of oscillations occurring inside the fluidic oscillator and at its output are discussed in the paper.

Keywords: Water jet · Self-Oscillated nozzle · Ethylene glycol

1 Introduction

Fluidic oscillators are devices where self-excited pressure and velocity oscillations occur in the flowing medium. This physical phenomenon causes a strong swirling motion of the fluid in the oscillating chamber. As a result, the jet of fluid exiting the oscillator breaks up into clusters and droplets. The separation mainly occurs as a result of the side-to-side oscillating motion of the exiting fluid jet. Fluidic oscillators are used for cleaning and surface treatment of various materials, measuring and fluid flow applications [1].

The fluidic oscillator consists of an input part of the oscillating chamber and an output part (i.e. output hole and stabilizer). The inlet part supplies pressurized liquid to the oscillating chamber. Self-excited pulsations of pressure and velocities occur in the oscillation chamber. The output hole with stabilizer then generates a side-to-side sweeping exiting jet of fluid. Several types of fluidic oscillators according to the shape of the internal space where the given fluid flows can be distinguished. The first type will

L. Sitek et al. (Eds.): Advances in Water Jetting II, LNME, pp. 258–271, 2025.
https://doi.org/10.1007/978-3-031-72778-8_19

be represented by fluidic oscillators with feedback channels, see Fig. 1. The oscillation chamber defined by the inlet and outlet openings also contains so-called feedback channels. The shape of the oscillation chamber causes the formation of flow instability, and the feedback channels ensure the regular course of the previously described instability [2]. Oscillators with feedback channels are very suitable for usage in the area of fluid flow measurement, where the value of the fluid flow is determined based on the evaluated frequency of the recorded pressure pulses [3].

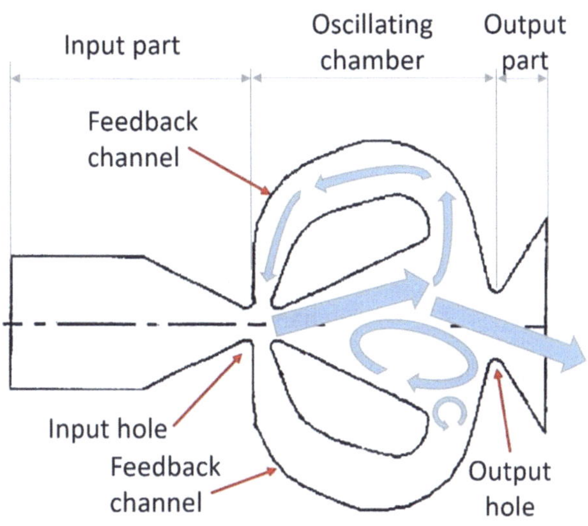

Fig. 1. Fluidic nozzle with feedback channels

The second type of fluidic oscillator (Fig. 2), on the other hand, is a device without any feedback. The oscillating chamber induces flow instability [4]. At the same time, the shape of the oscillation chamber ensures that the instability of the current field is repeated in time. At the output of the oscillator, we can see a jet of liquid oscillating from side to side. This type of fluidic oscillator has a very simple construction given the very simple shape of the oscillation chamber. The described type of fluidic oscillator is very suitable for manual spraying of liquids [5].

The third type of fluidic oscillator (Fig. 3) is a device with two or more input holes into the oscillating chamber [6]. Again, the oscillating chamber is here as a source of fluid flow instability. Instability of the flow field occurs due to the interaction of two or more fluid inlet streams and the shape of the oscillating chamber, see Fig. 3. At the output of the fluidic oscillator, a jet of fluid flows with a side to side sweeping motion. This type of fluidic oscillator also has feedback channels. The internal shape and overall construction can be very simple [7]. The mentioned device is very suitable for high-pressure applications, as the oscillating chamber will be over-pressured during operation [8]. This fact eliminates such phenomena as cavitation and evaporation, which could, very effectively, eliminate the self-excited oscillation in the oscillating chamber [9].

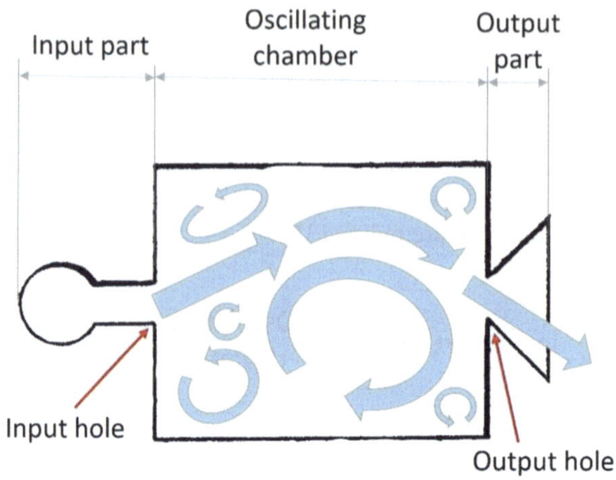

Fig. 2. Simple fluidic nozzle

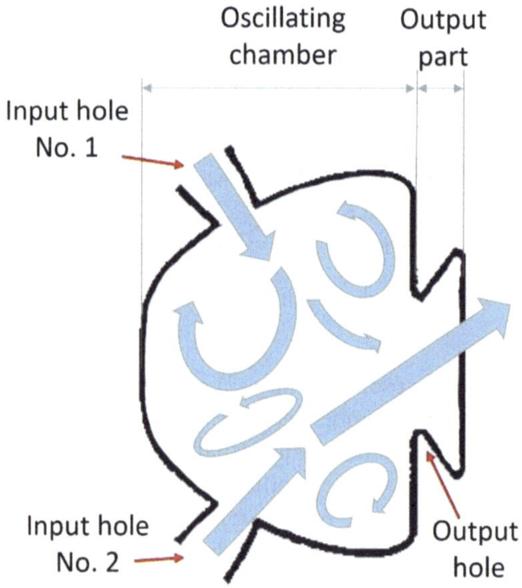

Fig. 3. Fluidic nozzle with two input holes

Viscosity is a physical quantity that indicates the ratio between the tangential stress and the change in velocity depending on the distance between adjacent layers of the flowing liquid. It is a quantity that characterizes the internal friction of a liquid and depends mainly on the attractive forces between particles. Liquids with greater attractive force between molecules have greater viscosity, which slows down the movement of the liquid or bodies in the liquid. Viscosity η is implicitly defined over the tangential stress τ by the formula (1):

$$\tau = \eta \frac{\Delta v}{\Delta x} \tag{1}$$

where Δv is the velocity difference and Δx is the layer spacing. Their shear expresses the velocity gradient (in 1D space). Viscosity defined in this way is called dynamic viscosity. The viscosity of liquids decreases with increasing temperature, at lower temperatures the molecules can group into larger clusters.

The fluidic oscillator with two inputs and one output was chosen to test the effect of viscosity on the propagation of self-excited oscillations. The selected type of oscillator enables achieving the overpressure of the oscillating chamber. During the overpressure, the oscillator maintains a stable output jet at relatively high values of oscillation frequencies. The overpressure of the oscillating chamber is significant when the oscillator is operated at higher pressures (5 MPa or more).

Ethylene glycol (EG) and a mixture of EG and water were chosen as test liquids, which allows changes in the dynamic viscosity over a wide range of values by adjusting the concentration mixture and temperature. Figure 4 shows the course of dynamic viscosity values in function of the temperature (interval from 0 °C to 95 °C) for different concentrations of the mixture of EG and water. The graphs show that a low temperature of around 5 °C and a high concentration of EG 100% increases the value of the dynamic viscosity up to 50 times compared to water.

To date, there has been very little investigation or recognition of the effect of self-excited liquid flow generation and propagation in various environments with respect to the change of viscosity properties. This knowledge gap can be formulated as follows:

- There is a lack of detailed studies describing the process of self-excited flow generation and propagation influenced by changing viscosity properties.
- The method using direct measurement of the self-excited vortices oscillation frequencies inside the nozzle has not yet been presented.
- The experimental assembly allowing direct monitoring of the fluid flow properties has not been presented.

This article investigates the change of self-excited oscillation under various liquid properties described by viscosity parameters. Methods of the fluid viscosity modification, measurement and analysis are presented in the paper. The goal is the verification of the stability of self-excited flow oscillations under various fluid conditions. The emergence and propagation of self-excited oscillations are observed for different levels of inlet pressure values (up to 25 bar) of the flowing liquid. The limitations describing attenuation and fading of self-excited flow propagation with respect to the measured fluid properties are discussed later in the paper. The dynamic viscosity value is assessed depending on the concentration of the water—EG mixture and its immediate temperature.

The presence of self-excited oscillations in the internal space of the oscillator is detected using high-frequency pressure sensors. The input energy of the flowing mixture is recorded by measuring the flow rate and pressure at the input to the fluidic oscillator. The shape of the output profile of the flowing capillary from the fluidic oscillator is measured using a high-speed camera. The experiment was carried out for five levels of inlet pressures (5, 10, 15, 20, and 25 bar) in combination pure 100% EG and 90% EG mixture. The temperature of the liquid was monitored during each test.

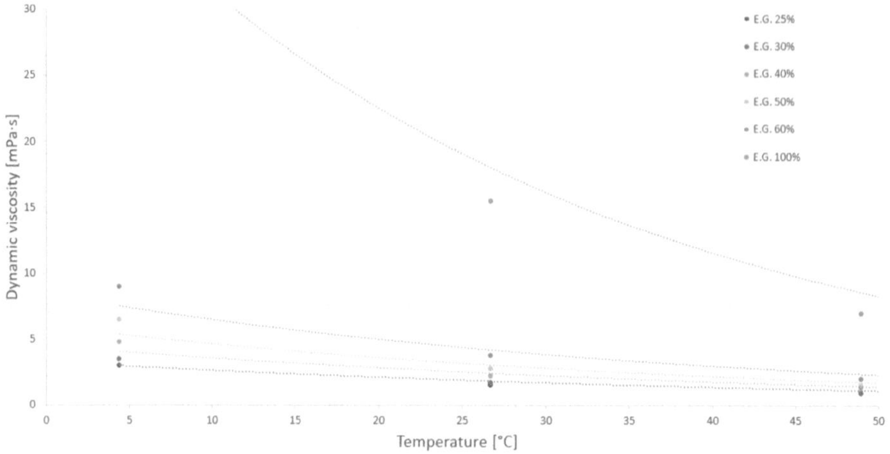

Fig. 4. Dependence of dynamic viscosity in the ratio of different concentrations of EG and water and temperature [10]

2 Materials and Methods

2.1 Implementation of a Fluidic Oscillator

For testing purposes, a fluidic oscillator was developed to measure the pressure of the oscillating chamber and it was printed with 3D printing technology, see Fig. 5. The fluidic oscillator contains an internal thread for the connection to the pressurized water line and to the inlet channel, which is further divided into two sub-channels. The mentioned sub-channels are then joined from the sides, into an oval-shaped oscillation chamber (width of one mouth 3.5 mm, height/depth of the chamber 2 mm). At the lower and upper part of the oscillation chamber, and orthogonal to it, there are openings (internal thread) that connect the chamber space with a high frequency pressure sensor. In the lower part of the chamber, there is an outlet opening (opening width 1.5 mm) through which the flowing liquid flows out. A high-speed output jet is then ejected, and it oscillates from side to side due to the self-excited swirling flow of the fluid in the oscillating chamber.

The designed part of the fluidic oscillator was printed by MultiJet printing technology on a ProJet MJP 2500 Plus printer. VisJet M2S-HT90 (pull-in strength limit 70–80 MPa) was chosen as the material. The fluidic nozzle made of the selected material should be functional up to a pressure of 10 MPa.

The design of the internal shape of the fluidic oscillator is based on CFD numerical calculations. The CAD model of the created oscillator is shown in Fig. 5.

The Höppler viscometer belongs to the category of ball viscometers, also called Stokes viscometers. A ball drops into a cylindrical container filled with the liquid to be measured. The time the ball falls along a certain path determines the dynamic viscosity. The measurement takes advantage of the fact that the velocity at which a ball falls in a given fluid depends on the viscosity of that fluid. The three forces acting on a body of mass m and radius r are the gravitational force G (2), the buoyancy force F_{vz} (3) and the environmental drag force F_o (4). For the magnitude of these forces, the dependence can

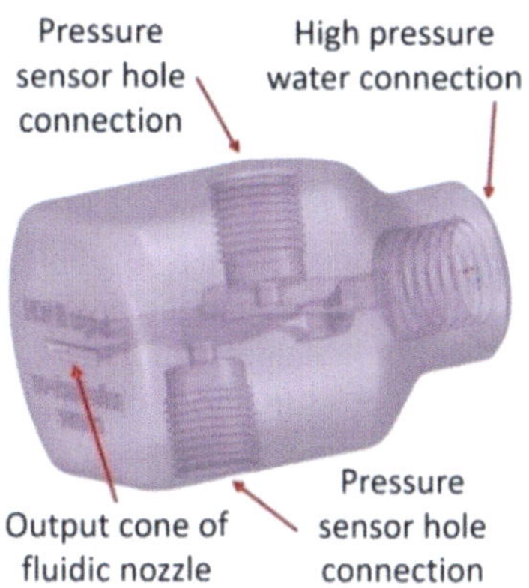

Fig. 5. CAD model of the tested fluidic nozzle

be expressed:

$$G = mg, \tag{2}$$

$$F_{vz} = \frac{4}{3}\pi r^3 \rho g, \tag{3}$$

$$F_o = 6\pi \eta r v \tag{4}$$

In the equations given, r is the density of the liquid, h is the dynamic viscosity of the fluid being measured, and v is the velocity of motion of the body in the fluid. Since the magnitude of the environmental drag force increases with velocity, the forces will (asymptotically) balance out and after a certain time, the solid will move in an essentially uniform rectilinear motion. The forces G, F_{vz}, and F_o are then in equilibrium and for their magnitudes.

$$G - F_{vz} = F_o \tag{5}$$

After substituting G, F_{vz}, F_o into this equation, we get the relation (6) for the viscosity

$$\eta = \frac{3m - 4\pi r^3 \rho}{18\pi r v} g = \frac{2r^2(\rho_{tel} - \rho)g}{9v} = \frac{2r^2(\rho_{tel} - \rho)g}{9l} t \tag{6}$$

where r_{tel} is a density of the sphere, l is the distance between the lines, t is the time of fall measured between the top and bottom mark. If all the quantities in Eq. (6) that are constant for a given viscometer and bead are included in the so-called bead constant K, Eq. (7) can be written as follow:

$$\eta = t(\rho_{tel} - \rho)K \tag{7}$$

2.2 Measurement of Hydraulic Parameters and Frequency Characteristics of Self-Excited Oscillations

The hydraulic measuring circuit for the generation of self-excited oscillations consisted of a Kärcher HD 7/18 pump. The fluidic oscillator with output orifice of 2 x1.5 mm (1.95 mm equivalent diameter) was mounted on a stiff holder. Online diagnostics were provided by a Kistler Kristal 0–100 MPa analogue pressure sensor and high-pressure turbine flow meter Küppers Electromechanics GmbH (1.2–10 l/min) including digital reading unit. The values of the inner pressure inside oscillators were monitored by a two Kistler 6211 pressure transducer (with a measured pressure range of 0–500 MPa) screwed directly into body fluid oscillator. The electrical signal generated by the piezoelectric sensor was amplified by a single channel charge amplifier. The signals from all sensors were then recorded and processed using NI USB 6251 BNC DAQ board transducers and NI LabView 2018 software [11]. A total of 100,000 samples were acquired at a sampling rate of 10,000 samples per second [12]. A schematic representation of the measurement circuit and hydraulic properties monitoring is shown in Fig. 6.

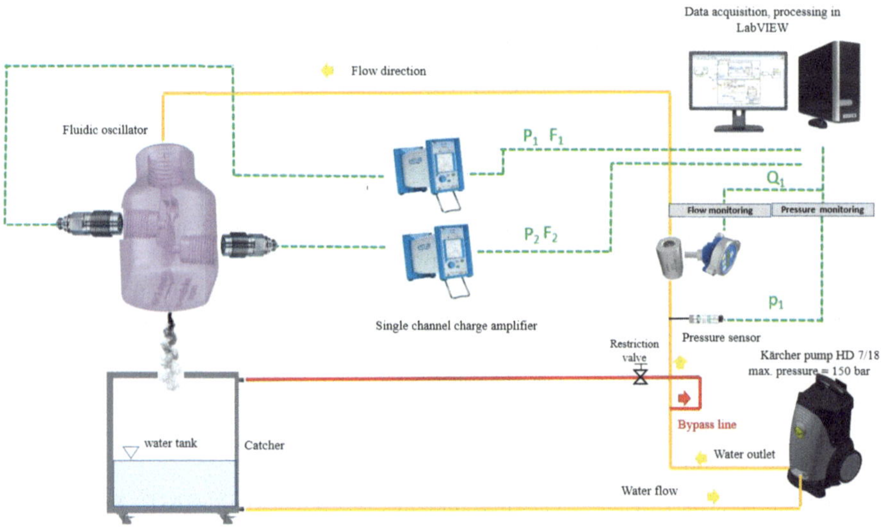

Fig. 6. Fluidic oscillator assembly for generating self-excited oscillations and their monitoring

Propagation of self-excited high-speed pulsations (at the output of the oscillator) was observed using high-speed shadowgraphs. The high-speed shadowgraphy line-up was composed of a Photron Phantom v2011 high-speed camera; a pulsed LED lights Veritas (120 W VERITAS) and a LaVision PTU X controller operating under DaVis 10 software as depicted in Fig. 7. The recording conditions of the high speed shadowgraphy are presented in Table 1.

Table 1. Table captions should be placed above the tables

Parameter	Value
area of interest (AIO)	144 × 512 pixels
camera dynamic range	12 bits
recording mode	Single frame
recording rate	98.23 kHz
camera exposure time	1 ms
LED illumination time	3 ms
LED maximal power	200 W strobed
luminous flux of one LED	20.000 lm strobed

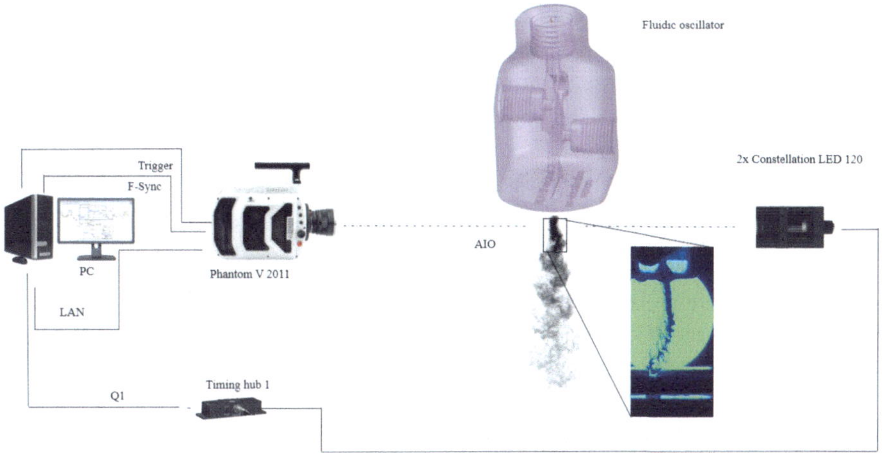

Fig. 7. The optical experimental setup for high-speed self-excited flow observation behind oscillator output

3 Analysis of Inner Pressure and Frequencies Inside the Fluidic Oscillator

The sequence in the time dependence of pressure inside the fluidic oscillator was recorded for operating pressures 5, 10, 15, 20 and 25 bars using EG 100% and EG 90% concentration fluids. An example of the baseline signal waveform recorded at the pressure of 27 bar is shown in Fig. 8 [13]. The signal obtained in this manner was then processed statistically as average values of all samples [14]. Subsequently, the time domain was converted to a frequency domain using an FFT (Fast Fourier Transform). Information about the static pressure amplitude and frequency were extracted from the frequency domain [15].

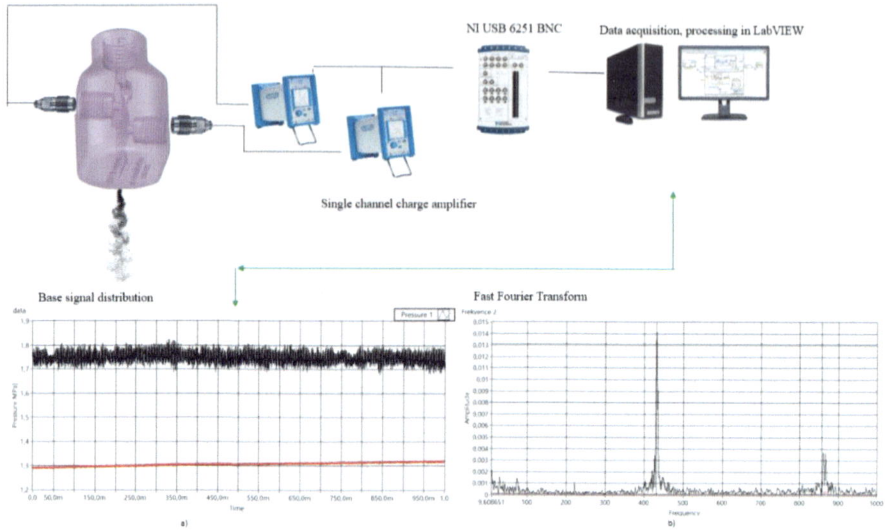

Fig. 8. Graphical illustration of the signals processing and frequency analysis

4 Results and Discussion

The results of dynamic viscosity measurements of EG 100% and EG 90% mixtures used for the study of the propagation of self-excited oscillations inside and at the output of the fluidic oscillator are presented graphically in Fig. 9. The graph shows the variation of dynamic viscosity for EG 100%. The individual points in the graph represent the average of three independent viscosity measurements with a minimum standard deviation of up to 5%. The dynamic viscosity in the temperature interval from 19 °C to 30 °C varies in the range of 16 to 21 mPa·s. A significant change in the dynamic viscosity is already evident for EG 90%, where the dynamic viscosity changes in the temperature interval from 19 °C to 30 °C in the interval from 10 to 14 mPa·s. An equidistant shift of the curves is evident from the graph.

Experiments performed to observe the effect of the liquid viscosity on the ability to generate self-excited oscillations were carried out over a temperature interval of 23–28 °C for EG 90% corresponding to a dynamic viscosity of 9–11 mPa·s and for EG 100% (within the same range of temperatures) corresponding to 15–19 mPa·s. Figure 10 shows a comparison of the frequencies generated inside the oscillator at the level pressures of 5, 10, 15, 20 and 25 bar for the two tested fluid mixtures EG 90% and EG 100%. From the comparison of the frequencies, it can be seen that a significantly lower frequency was achieved for the EG 100% mixture compared to the EG 90% mixture. However, the difference in frequency of the oscillator is within 5% and it can be concluded that there is no significant change.

The monitored flow rate of the fluid mixtures EG 100% and EG 90% through the oscillator for the selected pressure levels are shown in the graph in Fig. 11. The values of the hydraulic power in the oscillator are presented in the graph in Fig. 12. From these

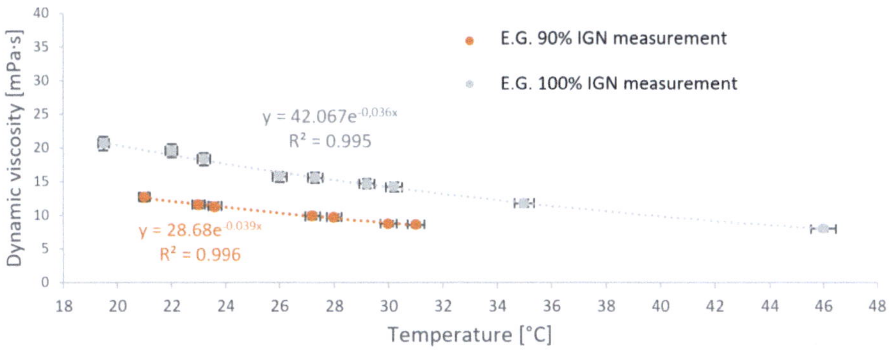

Fig. 9. Graphical dependence of dynamic viscosity measurements as a function of temperature for EG 90% and 100%

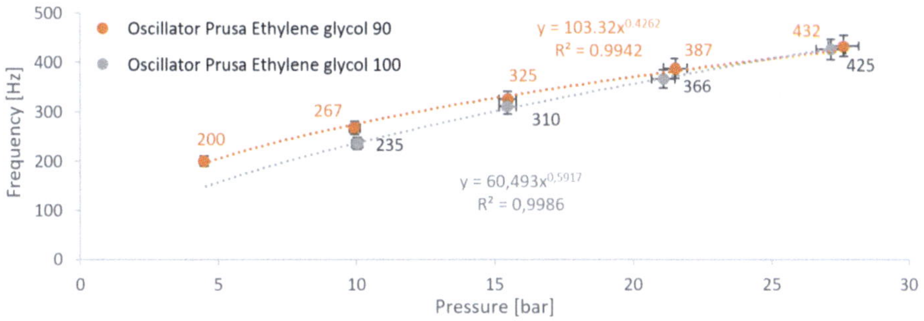

Fig. 10. Comparison of oscillation frequencies inside the oscillator for EG 90% and EG 100%

graphs, it can be seen that the more viscous the fluid, the lower the flow rate and the hydraulic power.

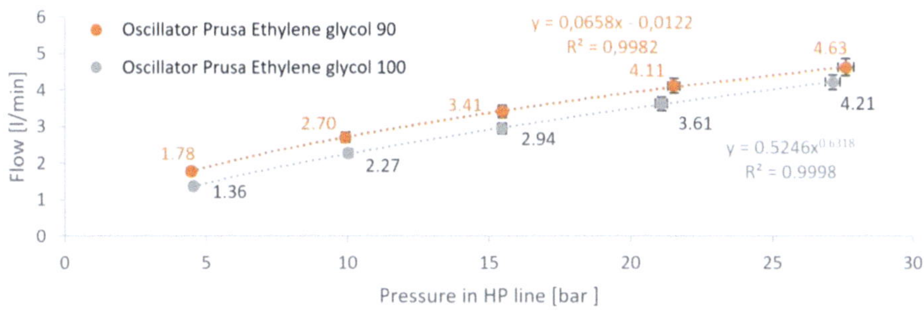

Fig. 11. Comparison of EG 90% and EG 100% mixture flow rate at the oscillator outlet

Figure 13 presents the measured values of pressure in the sensors located at the top and bottom of the oscillator. The individual points represent the average of the pressure calculated from 100 k of measured values. The plots show a linear distribution of pressure

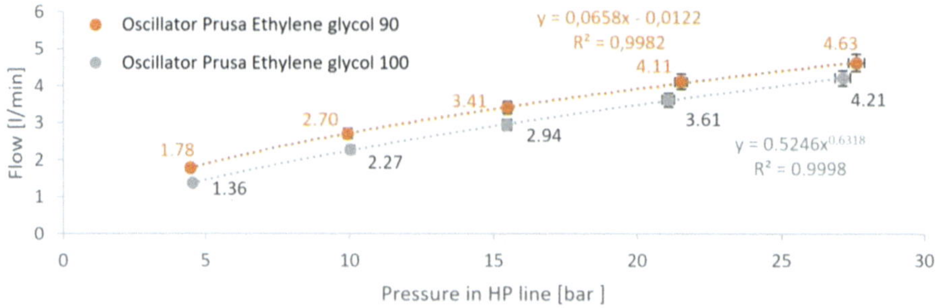

Fig. 12. Comparison of hydraulic performance at the oscillator outlet for EG 90% and EG 100%

inside the oscillator in both the upper and lower sensors. The linear pressure distribution is probably due to the small range of values (approx. 5–30 bar) of the measured pressure in the supply line.

Figure 14 shows the pressure amplitudes in the oscillation chamber in the lower sensor as a function of the liquid inlet pressure. The dependence is displayed for the two liquids used EG 90% and EG 100%. The frequency values corresponding to the pressure amplitude value is presented for the measured values. It is obvious that the change in frequency is relatively small compared to the viscosity. The pressure amplitude changes very significantly with change in viscosity. This is particularly true for higher values of liquid inlet pressure 10–25 bar.

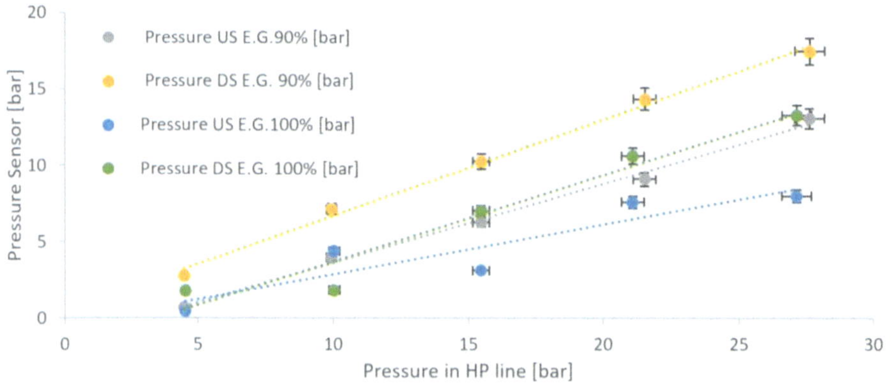

Fig. 13. Comparison of the internal pressure transformation in the upper sensor (US) and the lower sensor (DS) part of the oscillator for 90% and 100% ethylene glycol concentration

Experimental images of the flow visualization at the output of the fluidic oscillator are shown in Fig. 15. In the first row of Fig. 15a–f, the high-speed streams of the EG 90% concentration mixture are captured. Self-excited oscillations are formed for pressures equal or higher than 10 bar Fig. 15b. It can be assumed that the pulsation frequency at the output will be very close to the frequency measured inside the oscillator body. In the case of Fig. 15a, it is obvious that the input pressure does not make the jet to oscillate. The formation of small clusters along the jet axis is evident. The experimental images

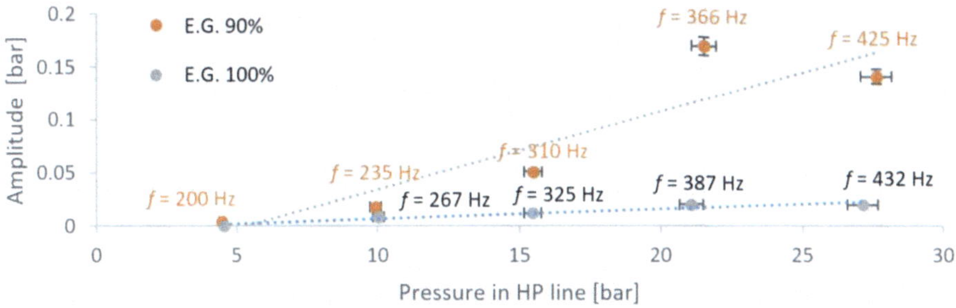

Fig. 14. Dependence of pressure amplitudes for EG 90% and EG 100% in the down sensor in relation to the input pressure of the flowing liquid, supplemented by the frequency values of self-excited oscillations

in Fig. 15b–f and the video sequences show that the jets present periodic oscillations at all tested input pressure values. In the second set of experimental images and video sequences taken in Fig. 15g–h for EG 100%, it is obvious that there is no formation of periodic oscillations at the output of the oscillator. This result does not correspond with the observation of the frequency within the oscillator acquired by the pressure sensors. Furthermore, in the image of Fig. 15g an almost laminar flow can be seen (the increment on the viscosity caused the oscillation activation pressure to be shifted to higher inlet pressures), while for the cases in the images of Fig. 15h–k a breaking of the jet structure can be observed due to the EG low surface tension and due to oscillations inside the oscillator. In all experimental images and video sequences of Fig. 15g–h a straight perpendicular flow was observed with no signs of oscillations of the exit jet profiles.

5 Conclusions

The generation and propagation of self-excited oscillations was monitored for different levels of inlet pressure (up to 25 bar) of the flowing fluid EG 90% and EG 100% concentration in the temperature interval 25–29 °C. Based on the experimental observations, the authors formulated the following conclusions:

- The formation of self-excited oscillations inside the fluid oscillator was observed for EG 90% concentration corresponds to a measured dynamic viscosity in the interval 9.2–10.8 mPa·s and EG100% in the interval of dynamic viscosity 14.8–25.14 mPa·s with pressure up to 25 bar.
- The generation of self-excited oscillations was observed at the output of the fluidic oscillator into the air environment from a pressure of 5 bar for an EG of 90% and a measured dynamic viscosity of 9.2–10.8 mPa·s. For EG 100%, no self-excited oscillations were observed at the output of the oscillator for the interval tested.
- In this contribution, the methodology and diagnostics of measurement and evaluation of properties of self-excited oscillations inside fluid oscillators of liquids was presented.

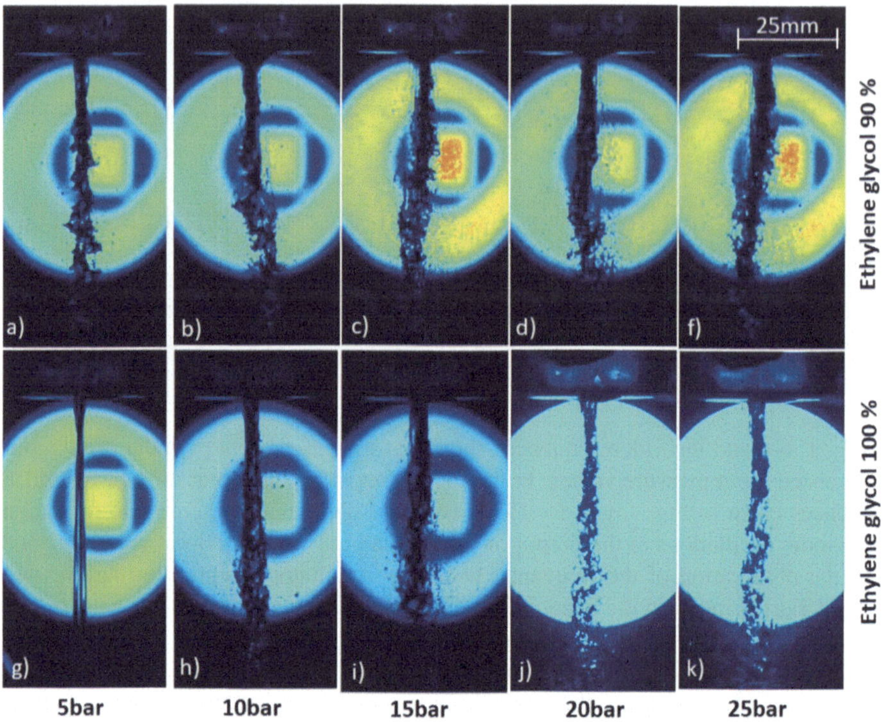

Fig. 15. Experimental images of flow visualization at the oscillator output for EG 90% a)-f) and EG 100% g)-h) at pressures of 5 bar (a,g), 10 bar (b,h), 15 bar (c,i), 20 bar (d,j) and 25 bar (f,h)

- The change in the dynamic viscosity of the fluid for EG 90% and EG 100% was indicated by a slight change in the main oscillation frequency of up to 5% but huge change in the value of the pressure amplitude up to 85%, see Fig. 10 and Fig. 14.

Acknowledgements. This work was supported by the Czech Science Foundation under grant project 23-05235S.

References

1. Liu, X., Liang, Z., Wen, G., et al.: Waterjet machining and research developments: a review. Inter-Natl. J. Adv. Manuf. Technol. **102**, 1257–1335 (2019)
2. Riha, Z., Rudolf, P., Pochyly, F., Haban, V.: Fluidicky ostrikovac, CZ patent No.: 286790 B6 (2000)
3. Adams, B.R.: Flowmeter, U.S. patent No.: 3640133 (1972)
4. Stoufer, R.D.: Low cost, low pressure fluidic oscillator, which is free of feedback, U.S. patent No.: 5213270 (1993)
5. Kondo, Y., Miura, K., Maeda, O., Kuroyanagi, M., Oshiro, T.: Fluid dispersal device U.S. Patent No.: 4721251 (1988)

6. Gopalan, S., Hartrant, E., Russell, G.: Cup-shaped fluidic curcuit with alignment tabs, nozzle assembly and methods, U.S. patent No.: 9089856 B2 (2015)
7. Gopalan, S.: Fluidic oscillator comprising three power inlet nozzles and an obstruction creating vortices. European patent No.: 1675686 B1 (2004)
8. Riha, Z., Zelenak, M., Kruml, T., Poloprudsky, J.: Comparison of the disintegration abilities of modulated and continuous water jets. Wear **478–479**, 203891 (2021)
9. Riha, Z., Zelenak, M., Nag, A., Poloprudsky, J., Kruml, T., Hloch, S.: A study of the erosion charac-teristics of an EN AE-6060 aluminium alloy processed using middle and high power continuous and modulated water jets. Wear **536–537**, 205154 (2024)
10. The Engineering ToolBox. Ethylene Glycol Heat-Transfer Fluid Properties (2003)
11. Zelenak, M., et al.: Theoreti-cal and practical research on micro hydrodynamic nozzles for modulated water jet generation as an innovation to replace continuous water jet in soft materials erosion. Measurement **221**, 113530 (2023)
12. Zelenak, M., Riha, Z., Jandacka, P.: Visualization and velocity analysis of a high-speed modulated water jet generated by a hydrodynamic nozzle. Measurement **159**, 107753 (2020)
13. Zelenak, M., et al.: Methods for the behaviour analysis of continuous flat water jet structures. Measurement **202**, 111886 (2022)
14. Leach, S.J., Walker, G.L.: The application of High Speed Liquid Jets to Cutting. Philosophical Transaction, 295–308, Royal Society London (1966)
15. Barker, C. R., Selberg, B. P.: Water Jet Nozzle Performance Tests. In: Proceedings of 4th Inter-national Symposium on Jet Cutting Technology: paper A1. Canterbury (1978)

Author Index

L. Sitek et al. (Eds.): Advances in Water Jetting II, LNME, pp. 273–274, 2025.
https://doi.org/10.1007/978-3-031-72778-8

The manufacturer's authorised representative in the EU is Springer Nature Customer Service Centre GmbH, Europaplatz 3, 69115 Heidelberg, Germany. If you have any concerns regarding our products, please contact ProductSafety@springernature.com

Printed and bound by CPI Group (UK) Ltd, Croydon, CR0 4YY
01/10/2026
02225851-0005